Advanced Data Analysis with SYSTAT

Advanced Data Analysis with SYSTAT

Kris N. Kirby, Ph.D.

Williams College

VNR VAN NOSTRAND REINHOLD
New York

Copyright © 1993 Multiscience Press, Inc.

All rights reserved. No part of this work covered by the copyright hereon may be reproduced or used in any form or by any means—graphic, electronic, or mechanical, including photocopying, recording, taping, or information storage and retrieval systems—without the written permission of the publisher.

Library of Congress Catalog Card Number 92-73266
ISBN 0-442-30860-4

I(T)P Van Nostrand Reinhold is a division of International Thomson Publishing. The ITP logo is a trademark under license.

Printed in the United States of America.

Designed and composed by Christopher Chabris & Associates.

Van Nostrand Reinhold
115 Fifth Avenue
New York, NY 10003

International Thomson Publishing
Berkshire House
168-173 High Holborn
London WC1V7AA, England

Thomas Nelson Australia
102 Dodds Street
South Melbourne 3205
Victoria, Australia

Nelson Canada
1120 Birchmount Road
Scarborough, Ontario
M1K 5G4 Canada

SYSTAT is a registered trademark of SYSTAT, Inc.

16 15 14 13 12 11 10 9 8 7 6 5 4 3 2 1

To my parents,
David and Betty Kirby

Contents

Preface xi

Introduction xv

1 Working with Files 1
   1.1 Creating and Using Files 1
   1.2 Importing Data from Text Files 8

2 Working with Data 15
   2.1 Data Manipulation Functions 15
   2.2 Math Functions 20

3 Describing Data in SYSTAT 55
   3.1 Summary Statistics 55
   3.2 Using Graphics to Describe Data 59

4 Meta-Analysis 101
   4.1 Converting Results to Common Metric 101
   4.2 Combining Studies 107
   4.3 Comparing Studies 113

5 Simple Linear Regression 117
   5.1 Preliminary Diagnostics: Look at the Raw Data 118
   5.2 Simple Regression 128
   5.3 Case Diagnostics 132
   5.4 Assumption Diagnostics 138
   5.5 Confidence Intervals 153
   5.6 The SHOES Data Set 160

6 Multiple Linear Regression 165

6.1 Preliminary Diagnostics: Look at the Raw Data 166
6.2 Multiple Regression 176
6.3 Collinearity Diagnostics 179
6.4 Case Diagnostics 188
6.5 Assumption Diagnostics 192
6.6 Confidence Intervals 204

7 Transformations 207

7.1 Transforming the Outcome Variable 208
7.2 Transforming the Predictor Variables 222

8 One-Way Analysis of Variance 235

8.1 The Example Data Set 235
8.2 Looking at the Data 237
8.3 One-Way ANOVA 238
8.4 Contrasts 242
8.5 Post-Hoc Tests 251
8.6 Checking Assumptions 258
8.7 Confidence Intervals 268
8.8 Graphing the Results 271
8.9 Test Scores Data Set 275

9 Fixed Factors Factorial ANOVA 279

9.1 The Example Data Set 279
9.2 Looking at the Data 282
9.3 Factorial ANOVA 283
9.4 Contrasts 286
9.5 Post-Hoc Tests 299
9.6 Checking Assumptions 301
9.7 Confidence Intervals 312
9.8 Graphing the Results 312
9.9 Unequal Cell Sizes 318

10 Nested Fixed Factors ANOVA 325

10.1 The Example Data Set 325
10.2 Looking at the Data 328
10.3 Nested ANOVA 330
10.4 Contrasts 337
10.5 Post-Hoc Tests 345
10.6 Checking Assumptions 348
10.7 Confidence Intervals 357
10.8 Graphing the Results 357

11 Random Factors ANOVA: Repeated Measures 365

11.1 The Example Data Set 366
11.2 Looking at the Data 369
11.3 Repeated Measures ANOVA 373
11.4 Contrasts 374
11.5 Post-Hoc Tests 385
11.6 Checking Assumptions 390
11.7 Graphing the Results 399
11.8 Unequal Cell Sizes 406

12 Nesting in Repeated Measures 409

12.1 Nesting Between Subjects 409
12.1 Nesting Within Subjects 422
12.3 Unequal Cell Sizes 432

13 Contingency Tables 435

13.1 The Data File 435
13.2 One-Way Tables 437
13.3 Two-Way Tables 441
13.4 Three-Way Tables 438
13.5 Graphing the Results 455

References 463

Index 465

Preface

This book is not a software manual, and it is not a statistics textbook. In teaching and using statistics I have found a need for a reference book that bridges the gap between the two, by providing the guidance of a statistics text without becoming mired in details and derivations, and interweaving thorough instructions on the use of statistical software without getting lost in an endless variety of procedures and options. This an attempt at such a book. I chose SYSTAT as the statistical software because I believe it has the best combination of power and user-friendliness of any program now available for personal computers. Anyone can start using it immediately, and few people will outgrow it, especially as SYSTAT itself keeps growing. I chose to focus on regression, ANOVA, and table analyses because these constitute the most basic and widely used statistical analyses. There simply wasn't room for much more in a single volume.

The SYSTAT manuals are some of the best software manuals that I have ever seen. They provide quite clear and exhaustive coverage of all of SYSTAT's features, and they contain very well-written tutorials on a number of statistical techniques. This book could not hope to cover all of the information contained in the SYSTAT manuals, and if it did, it would simply be redundant. However, out of necessity the SYSTAT manuals are organized around features of the software. For example, graphics and statistics procedures are divided into separate volumes, so to perform a complete regression analysis one might have to flip through a graphics chapter, the statistics chapter, the regression chapter, and the graphics chapter again. In contrast, because this is primarily an applied book, the chapters are organized around types of data sets and experimental designs rather than around specific statistical issues or software features. Each chapter provides the user with step-by-step instructions on how to carry out a complete analysis of a particular type of data set, from graphing the data, to summarizing the data, to performing statistical tests, to checking assumptions, to graphing the results. This is the way a data analyst's task is organized, and this book is aimed at making that task easier. So even though all of the SYSTAT procedures in this book are covered in the SYSTAT manuals, using this book will hopefully provide the reader with a more linear guide to their analysis, from start to finish, and help save the reader precious time and trouble. I try to give just enough SYSTAT instruction to do the analyses that the book covers, on the assumption that the reader will turn to the SYSTAT manuals when they want to learn more about the myriad of options available, or the vast array of additional analyses offered by SYSTAT.

There are a number of excellent textbooks on the market that cover statistical techniques in much more depth than they can be treated in this book. However, almost all textbooks are organized around specific procedures, and in this way have the same limitations for the user as do software manuals: it takes a lot of flipping around to find all of the things you want to do to a set of data. And from an application standpoint, statistics textbooks also have the drawback that they contain much more detail about statistical procedures than one needs for everyday use. I am not suggesting that it is O.K. not to know what is going on when one instructs the computer to crank out a result. On the contrary, as a statistics teacher I have taken great pains to ensure that people are uncomfortable with such blind analyses. I am merely suggesting that you can get that kind of information from other sources, and once you understand it, you needn't spend time reading through derivations and justifications every time you analyze data. I try to give just enough statistical background to remind the reader why they are doing what they are doing. When you need to consult a derivation or justify a procedure, the statistics text will be there.

In laying out procedures for analyzing data sets "from start to finish," in a few places I have taken the liberty of expressing some rather strong suggestions about how data *should* be analyzed. In some cases these "suggestions" are simply built into the structure of the chapters. For example, I think data analysts should be in the habit of always looking carefully at their raw data before they carry out any significance testing, and that they should always check the assumptions underlying their analyses. Therefore, from Chapters 5 through 13, each chapter is organized in this sequential way. I hope that this structure will ultimately encourage readers to proceed in this way. In other instances my opinions and suggestions are stated quite explicitly. For example, I see little to recommend significance tests of effects with more than one degree of freedom, and I say so quite often, instead emphasizing the use of contrasts. Unfortunately, to attempt to fully justify and defend such suggestions would turn this book too much into a statistics text and detract from its primary objective. Fortunately, readers who hold contrary views on how to go about a particular analysis will still benefit from most of the material in a given chapter, even while deviating from my suggestions as they see fit.

Acknowledgments

To the extent that the material in this book is helpful and correct, credit goes to my teachers at Harvard University: William K. Estes, Robert Rosenthal, Donald Rubin, Hal Stern, and Douwe Yntema. I am grateful for their years of generous and patient instruction, and I cannot imagine that anyone has been more fortunate in having access to such a wonderful group of teachers.

A number of people have been instrumental is helping this book along. Alan Rose at Multiscience Press was a perfect publisher: supportive, frantic, and patient all at once. Leland Wilkinson at SYSTAT was unfailing in his support, and along with Eve Goldman generously kept me up-to-date in SYSTAT software. Special thanks also go to those who kindly reviewed and commented on assorted proposals and chapters with little compensation: David C. Howell, Hal Stern, Robert Rosenthal, Wilma Koutstaal, and Monica Harris.

Finally, some special notoriety must go to Christopher Chabris, who made the book look great, and without whom the whole thing would never have happened. For these reasons, he should be held accountable for any and all errors contained herein.

Williamstown, MA
May 1993

Introduction

This book was designed to assist both students and advanced data analysts. The chapters are organized around types of data sets, and Chapters 5 to 13 lay out a series of procedures for analyzing those data sets from graphing the raw data to graphing the results, and everything in between. Students can benefit from this structure, because it helps give them a feel for "the big picture," as well as steady guidance in how to proceed from one step to the next. Advanced users can benefit from the efficiency provided by this structure, hopefully saving them a lot of time and effort.

It is difficult to satisfy all users, and there are a couple of trade-offs worth noting. First, the book may have too much explanation of procedures for some advanced users who just want to know *how* to do what they want to do as efficiently as possible, and not enough explanation for some students who are trying to learn *what* to do. I have tried to strike a balance, assuming that even advanced users can sometimes profit from brief reminders about what a procedure does and what the reasons are for doing it, and that students can turn to a textbook for more detailed information and thorough explanations.

Second, the book contains a significant amount of redundancy between chapters, so that each chapter can stand on its own as much as possible. Some discussions are repeated nearly verbatim in a number of places throughout the book. This will greatly annoy any advanced users who read the book all the way through (like my reviewers), but I assume that, in practice, few advanced users will ever read more than one chapter at a time. It seemed to me that what an advanced user would want most out of a book like this is to have to interact with it to the least extent possible. This means minimizing page-turning between chapters and looking things up in the index and table of contents. For example, someone who is sitting down to analyze a two-way ANOVA wants to read about calculating an effect size estimate in the appropriate place in the two-way ANOVA chapter, and not have to go looking for it elsewhere in the book. And although students are more likely to read straight through successive chapters, they often can profit from the repetition. Therefore, wherever feasible, I have tried to repeat relevant discussions and include all necessary commands in each chapter. This, of course requires some compromise: I could have made each section of the book completely self-contained, but this would make the book absurdly long. I hope that I have achieved a happy medium.

How to Use this Book

Using this book should be straightforward. Chapters 4 through 13 are organized around analyzing specific example data sets. So if you want to analyze a real data set, you should determine which chapter most closely represents the type of data set that you want to analyze, and jump right to that chapter. These chapters are intended to be as self-contained as possible, but they do contain some cross-references to material in the first three chapters, which is used throughout the book and is too voluminous to repeat in each chapter. Chapter 1 brings together basic procedures for getting data into SYSTAT and working with files. Chapter 2 contains procedures for manipulating and transforming data, and concludes with a section on how to find exact probabilities for various test statistics, and vice versa. Chapter 3 shows how to compute basic summary statistics and how to graph data. So even though the commands for the procedures shown in these first three chapters are usually shown in their appropriate places when they are used in later chapters, for details, menu equivalents, and options you may have to refer back to these chapters.

Throughout the book, every step of an analysis is given in its SYSTAT command form. With extremely few exceptions (which are noted in the text) the same commands work in the Macintosh version, the DOS version, and the Windows version of SYSTAT. In this book SYSTAT commands look like this:

>THIS IS A SYSTAT COMMAND

Accompanying these commands throughout the book are their menu equivalents, presented separately for the Macintosh and for DOS (Windows users should see the next section). The menu commands are shown in a bracket in the margin of the page, with a heavy black arrowhead along the side that points to the beginning of the corresponding set of commands. In places where commands are illustrated that are virtually identical to a command that was just presented, for example, differing by only a single variable or option, the commands are repeated but their menu equivalents are not.

Even die-hard menu users should consider paying attention to the commands and trying to learn and understand them. Very quickly you will learn to automatically translate commands into their menu equivalents, but more importantly, you will find that a command or two will often do the work of a host of menu selections. Menus are invaluable when you can't remember a command, but when you can remember the command, it will usually be faster to use it.

Microsoft Windows

The Windows version of SYSTAT was released just after this book was completed, and it was too late at that point to cover it in the text. Fortunately, for two reasons this is not a serious handicap for Windows version users. First, the commands in the Windows version are the same as those in the DOS version, so the command lines shown throughout this book can simply be used in the Windows version as well. Just select **Command Prompt** from the **Window** menu and enter commands after the prompt (>). If your first response to this suggestion is "If I wanted to type commands, I wouldn't have bought the Windows version, now would I?", do not despair. Your second lucky break is that the Windows menus and dialog boxes are nearly identical in content to their Macintosh counterparts. So to find the appropriate Windows menus instructions in this book, use the ones shown under the Macintosh headings.

The differences in most cases are very slight, and you will soon learn to translate from Macintosh to Windows menus. For example, in section 9.3, you are shown how to conduct a two-way ANOVA using the following seven Macintosh menu commands:

① Choose the menu item **Stats/MGLH/Fully Factorial (M)ANOVA**.
② Select SCORE from the **Dependent Variable(s)** list.
③ Select GENDER and GROUP from the **Factor(s)** list.
④ Click on **More...**
⑤ Turn on **Means & Std. Errors**, **Extended Output**, and **Save Residuals**.
⑥ Click **OK**.
⑦ Type "twoway.res" into the **Save residual as** box, and click **OK**.

In Windows the first three commands are identical except for trivial labeling differences:

① Choose the menu item **Stats/MGLH/Fully Factorial (M)ANOVA**.
② Select SCORE from the **Dependent** variable list.
③ Select GENDER and GROUP from the **Factors** list.

With the fourth and fifth steps, things start to change, because in the Windows version there is no "More..." button. But in fact you don't need it, because all this button does in the Macintosh version is display the options listed in step 5, which are normally hidden from view in the Macintosh version. However, in the Windows version two of these options, **Means and Std. Errors** and **Save File** (the same thing as **Save Residuals** in the Macintosh version), are already shown, so you simply click on them. Up to this point, the two versions are only trivially different.

This leaves as the only substantial difference the problem that the **Extended Output** option is not offered in the MANOVA window in Windows, and to turn this option on you must either select **Extended (long)** under "Results to print" in the **Data/Formats...** dialog box, or type "print = long <return>" in the command window. Unfortunately, you needed to do this prior to opening the MANOVA window. So either hit **Cancel** and start over, or finish using the default output with the menu commands

④ ⑤ Turn on **Means and Std. Errors** and **Save File**.
⑥ Click **OK**.
⑦ Type "twoway.res" into the **Save a File** box, and click **Save**.

Now if you want to repeat the analysis to show the extended results as suggested in the Macintosh menus, do the following in Windows:

① Choose the menu item **Data/Formats...**, click on **Extended (long)**, and hit **OK**.
② Press **Ctrl-R** (or select the menu item **Data/Redo last**) to repeat the analysis.

As in the Macintosh and DOS versions, you may choose to save the residuals file in single or double precision, and you will be prompted to choose from a variety of residuals in the "Residual information to Save" box, to which you would simply use the default by clicking OK in this example.

To summarize, most of the differences between the Windows and Macintosh menus in the examples used in this book are simply trivial differences in the way items are labeled in the menus and dialog boxes, such as "Save File" (Windows) versus

"Save Residuals" (Macintosh). The only procedural difference in the above example was that the extended output option had to be selected before opening the MANOVA window in the Windows version, because the option to turn it on is not included in that window as it is in the Macintosh version. You will discover these small differences as you use the book, but you will quickly learn to recognize them and carry out the rest of the analysis without difficulty. And if you get stuck, you can always use SYSTAT's command interface.

1
Working with Files

1.1 Creating and Using Files

1.1.1 Using Menus and Commands

SYSTAT provides at least two ways of doing just about everything: either by typing commands into the *command window*, or by choosing menu items and working with dialog boxes. On the Macintosh, both methods are available simultaneously, and you can use whichever is easier for any given procedure. To see the command window on the Mac, choose the menu item **Window/Command**. I normally resize this window and place it near the bottom of the screen.

In DOS you can easily toggle back and forth between menu control and the command window by typing "menus" into the command window to turn on the menus, or toggling the menu item **Utilities/Commands/Menu On** to turn off the menus.

Menus are invaluable for learning to use a program, but experienced users find that commands can often save a lot of time. In SYSTAT, some simple commands can do the work of many menu item selections, so it may be worth your while to gradually learn some of the commands (of course, if you ever forget one, it will still be there in the menus). Fortunately, SYSTAT makes learning commands very easy. Both Mac and DOS versions allow you to see the commands that are being generated when you use menus. On the Macintosh, you simply need to have the command window open, and then most of the menu commands that you use will be displayed automatically in their command form in this window. In DOS, you choose the menu item **Utilities / Commands / Review On** and SYSTAT will show you each command before it is performed. In this mode, choosing **Go!** from a menu generates the command(s) for the desired procedure, and then you hit <F10> to execute those commands.

In this book, whenever practicable, I give the commands for every procedure in command window form (which is identical for DOS and Macintosh), and also show how to do the same thing with DOS menus and Mac menus. By giving both menu instructions and commands I hope to satisfy users who prefer one or the other, and to aid menu users in gradually learning the commands. When a number of variations on a command are shown in close proximity, I sometimes will not repeat the menu description for each variable, but a relevant description can always be found nearby.

In the remainder of this chapter I present some basic information on entering and importing data into SYSTAT. One can find much of this information in the *Getting Started* and *Data* SYSTAT manuals. However, because every SYSTAT user must input data, a brief summary is in order, along with a few notes and tips. The rest of the chapter contains three parts: entering data, importing data from text files, and importing data from other programs.

1.1.2 Entering Data

1.1.2.1 Entering Data in the Data Editor Window

SYSTAT can open and work with data files with or without showing the file to the user. But to enter, view, or alter data it is a good idea to open the data file using the *data editor window*. This window displays the data in a row × column format that allows you to move easily through the data file. Open a new data file in the data editor window simply by typing into the command window

MACINTOSH

① Choose the menu item **File/New**.

② If you do not see the new data editor window, choose the menu item **Window/Editor**.

MS-DOS

① Choose the menu item **Data / Edit**.

② If you already had a file in use, type <esc> and then "new" to create a new one.

▶ `>EDIT`

This command opens a new worksheet in which the rows correspond to cases (or "records" or "lines") and the columns correspond to variables. (In DOS, if you already have a file in use, type "edit," <esc>, and then "new.") The same thing can be accomplished with the menu commands shown at left.

The data editor worksheet looks very much like a spreadsheet and, in fact, SYSTAT can perform many spreadsheetlike functions. However, the cells of this table may contain only character or numeric information: They cannot contain formulas as spreadsheets can. Before entering data into a column you must first label the column, indicating whether the column will contain a character variable or a numeric variable: Once this is set you cannot change the type of a column. These issues are discussed further in section 1.1.2.4. To enter column labels or data into the data editor window, select the desired cell, type the entry, and then push **<return>**, **<tab>**, or **<enter>**. The next cell in the data set is then highlighted and ready to receive typing (see section 1.1.2.3 for controlling the order of data entry).

1.1.2.2 Entering Data in the Command Window

You may enter variable labels and data directly into the command window, without ever looking at the actual data file. In the command window you use the input command followed by the variable labels, and then each of the data cases on successive lines. For example, the sequence of commands

```
>SAVE FILENAME
>INPUT SCORE1 SCORE2 NAME$
>RUN
>05  21  GREG
>12  12  DAVE
>02  25  BETTY
>~
```

creates a new data file with three variables, two numeric and one character (see section 1.1.2.4), with three cases. The tilde (~) signals the end of the data to SYSTAT.

1.1.2.3 Controlling Data Entry and Data Display

In the data editor window, you can choose the direction of data entry, vertical or horizontal, and the number of decimal points displayed. I also recommend that you tell SYSTAT to display values near zero in exponential notation. This can help you keep track of which zeros in the file are real zeros (for example, when zeros are used in dummy variables) and which numbers are just very close to zero. The command for controlling these settings is:

```
>FORMAT= N /UNDERFLOW
```

where N is the number of decimals you wish to display for values in nonscientific notation, and the *underflow* option tells SYSTAT to show values near zero in exponential notation.

MACINTOSH

① Choose the menu item **Edit/Preferences...**

② Enter the desired number of decimal places into the **Decimal places** box.

③ Click on *display numbers near zero in exponential notation*.

④ Click on the desired order of data entry: *Next case* causes the cell highlight to move vertically down a column when you press <enter>, <tab>, or <return>, whereas *Next variable* causes the cell highlight to move horizontally across each row when you press <enter>, <tab>, or <return>.

⑤ Click **OK**.

MS-DOS

① Choose the menu item **Utilities/Output/Format/Places**.

② Select the desired number of decimal places and hit <ret>.

③ Choose the menu item **Utilities/Output/Format/Underflow** and hit <ret>.

④ Use the arrow keys to move horizontally or vertically through the data editor worksheet.

1.1.2.4 Columns

Each column in a SYSTAT data file contains a separate variable, although you may use the same label for more than one column. You enter variable labels, just as you would enter data, into the blank cells at the top of each column (the unnumbered top row). SYSTAT will not let you enter data into a column before you give that column a label. Variable labels in SYSTAT must begin with a letter, and may contain any combination of letters, numbers, and the underscore (_). Letters may be typed in either upper- or lowercase, but will only appear in uppercase. SYSTAT files may contain up to 99 columns on the Macintosh and 256 columns in DOS.

Variables in SYSTAT can be either *character* or *numeric*, both of which are described below. When you establish a column in a data file as character or numeric, you can never change its column type. Instead, you may create a new, correctly labelled column in the blank column at the end of the data set. On the Macintosh, the incorrectly labelled column may be deleted by selecting the entire column and pressing <**delete**>. In DOS you are stuck with the unwanted columns. Unfortunately, SYSTAT allows you to delete rows and columns, but never to insert them except at the end of the data set. If you wish to change the order of your columns, perhaps the best way is to save the file and then read the file back into a new SYSTAT file with the variables reordered. For example, if you have a file with the variables in the order A, B, and C, you can read the file into SYSTAT by typing into the command window

```
>USE OLDFILENAME (C B A)
>SAVE NEWFILENAME
>RUN
```

This reads the old file and saves the data into a new file containing the variables C, B, and A, in that order. You can also use this method to read in only the columns that you want to keep, which is a useful way to get rid of unwanted columns in DOS.

1.1.2.4.1 Character Variables

*Character variable labels must end with a "$."* Any variable label for a column that is going to contain data using any characters other than numbers, including blank spaces, must have a dollar sign at the end. You may use as many as eight characters in a character variable name, not including the "$" character.

Character cell entries may contain up to 12 characters, including letters, numerals, spaces, punctuation marks, and other symbols. If you wish to begin a character cell entry with a numeral or a period you must precede the entry with a double or single quote. The quotes will disappear when you hit return and do not count towards the length of the entry. If a single or double quote is a part of the entry itself, the entire cell must be enclosed in double or single quotes, respectively. For example, if you wanted to enter the text *Joe's data* into a cell in a character column, you must enclose it in double quotes: "Joe's data." If you wanted to enter *respd "yes"* into a cell, you would have to enclose it all in single quotes: 'respd "yes"'. Although the outer quotes do not count toward the 12-character limit, any quotes inside them do. Missing or empty character cells must contain at least one blank space; to enter a blank space you must enclose it in single or double quotes: " ".

1.1.2.4.2 Numeric Variables

Numeric variable labels must not *end with a "$."* Any variable that is going to contain numbers that are to be treated as numbers, i.e., have calculations performed with them, must not end with a dollar sign. These variable labels may contain up to eight characters and may also be subscripted by placing subscripts in parentheses at the end of the variable label. For example, if you wanted three numeric variables that all began with the label "TIME," you could name these variables TIME(1), TIME(2), and TIME(3). The parentheses and subscripts do not count towards the eight-character maximum for the label length.

Numeric values may be up to eleven digits long, ten if a decimal point is used. Missing or empty cells in numeric columns must contain a single period ('.'). Using scientific notation you may enter values ranging from plus or minus $0.000000001 \times 10^{-35}$ to $999,999 \times 10^{35}$. In SYSTAT's notation, you may enter data in scientific notation simply by following the initial digits by the letter 'e' (upper- or lowercase) and the appropriate power of ten. You cannot enter powers of ten higher than 35, although SYSTAT can display powers up to 41. Up to six significant digits can be shown in scientific notation. SYSTAT shows very small numbers that exceed the number of digits displayed as zero unless you use the underflow option (section 1.1.2.3).

1.1.2.5 Rows

Each row in a SYSTAT data file contains a separate case (sometimes called "lines" or "records"). Rows may contain one or more numeric or character data entries, or character entries describing data entries. Like columns, rows may be deleted by selecting the case number at the beginning of the row and pressing <delete>. However, new rows cannot be inserted between existing rows. You may add new rows only into the blank row at the end of the data set. Your data set may contain as many rows as you have disk space to hold them.

HINT: VARIABLE LABELS

Column types cannot be changed once entered. To get your columns labelled correctly at the outset, it is a good idea to write out the column labels in the correct order on a sheet of paper. Remember that once they are entered into SYSTAT you cannot change their position except by re-entering the column at the end of the data set.

1.1.3 Saving Files

1.1.3.1 Writing a Data File

After you have the data entered the way that you want it, you should write the data to a SYSTAT file. In DOS and older Mac versions of SYSTAT this was done using the *write* command

▶ >WRITE FILENAME

In Mac version 5.2, this command has been replaced with the *esave* command

>ESAVE FILENAME

Both commands are hot commands that write the data to the file immediately. If a file of the name you specify already exists, SYSTAT will ask you whether you want to write over it.

Normally you would want to save files as SYSTAT files so that they can be read quickly by SYSTAT, even though they cannot be opened by other programs. If you want to save a file in ASCII text, use the *put* command:

▶ >PUT FILENAME

With the *esave* (*write*) and *put* commands you can choose to store the data in *single precision* or *double precision*. Single precision storage takes up only about half the disk space of double precision, and stores numbers up to approximately seven significant digits, which is plenty for most purposes. It is a good idea to get in the habit of storing data in single precision: It can save a lot of disk space in the long run.

MACINTOSH

① Choose the menu item **File/Save** (or **File/Save as...** when the file is new).

② Type the filename into the dialog box if necessary and click **OK**.

MS-DOS

① Type "save *filename*" when you are in the data editor.

② When you are not in the data editor: Choose the menu item **Files/Save** and the name of the file when you are saving to an existing file, or choose the menu item **Files/Save**, hit <esc>, and type the new filename when you are saving a new file.

MACINTOSH

① Choose the menu item **File/Save as...**

② Type the new filename into the dialog box.

③ Select the **Text** button and click **OK**.

MS-DOS

① Choose the menu item **Files/ Imp/Export / Export/Type/ASCII**.

② Choose the menu item **Files/ Imp/Export / Export/File/** and type the new text filename.

③ Choose the menu item **Files/ Imp/Export / Export/Go!**

HINT: "ESAVE" VERSUS "SAVE"

The commands *write* (*esave* in the new Mac versions) and *save* both save data to a file, but they differ in that *write* is a hot command that saves the current data editor window, whereas *save* is a cold command that is not executed until the next hot command, such as *sort*. For example, if you changed a cell in the data editor window and wished to save the change, type

```
>WRITE (OR ESAVE) FILENAME
```

If you wish to sort a data set, you must enter the *systat* module, and then use the *save* command followed by a *sort* command:

```
>SYSTAT
>SAVE FILENAME
>SORT VARIABLE1 VARIABLE2 ...
```

1.1.3.2 Saving Results of Procedures

Some procedures generate output that can be directly saved into a file when the procedure is executed. The *save* command is a cold command that tells SYSTAT where to save the results of the next procedure. Sometimes the next command will simply be *run*. For example, in section 1.1.2.4 we saw how to open a file and save it into a new file with the order of the columns changed:

```
>USE OLDFILENAME (C B A)
>SAVE NEWFILENAME
>RUN
```

The *save* command is not executed until the *run* command is typed.

HINT: SAVING DATA ENTRY IN PROGRESS (MACINTOSH ONLY)

When entering data it is very important to save the file every few lines so that the entries will not be lost if the computer loses power or otherwise exits the program. To save time you can use the command-key equivalent of *esave*: On the Mac just press the key with the apple/cloverleaf and the 's' key, that is, **<command>-s**.

As with *write* and *esave*, with the *save* command you can choose to store the data in *single precision* or *double precision* by typing "single" or "double" after a slash at the end of the command line.

1.1.4 Opening SYSTAT Files

SYSTAT files can be opened in two ways: one that allows you to see the data and one that does not. The first method opens the data file in the data editor window. In the command window type

▶ >EDIT *FILENAME*

The second method allows you to *use* the data file for statistical analyses, but does not open the data editor window. Simply type

▶ >USE *FILENAME*

into the command window. Because large data sets can take a while to open with the editor and can sometimes get in the way, the 'use' option can be useful. However, most of the time you will probably want to see the data set, if for no other reason than to be reassured that it is all there. On a Macintosh, if you simply *use* the file with the command window open, SYSTAT will report the type of file that was opened (for example 'RECT' for rectangular) and will list all of the variable labels. If you *use* a file with the data editor window open, SYSTAT will show the data in the window just as if you had typed the *edit* command.

1.2 Importing Data from Text Files

SYSTAT can read ASCII text files as well as files from some other database programs (see section 1.2.2 below). If your data reside in a file created by another program that SYSTAT will not read, such as a word-processing program, try to save the data into an ASCII text file using that program, and then read it into SYSTAT using the procedure described below.

1.2.1 Opening Text Files

SYSTAT can read ASCII text files produced by itself or by other programs. SYSTAT will prompt you to select a file and to choose the file type. You must tell SYSTAT how the values and cases in the data set are separated. The characters used to separate cells and cases are called *delimiters*. A text file must

---

MACINTOSH

① Choose the menu item **File/Open...**
② Locate and select the desired file.
③ Click "on" the **Edit** option, if it is not already on.
④ Double-click on the filename or click **OK**.

MS-DOS

① Choose the menu item **Files/Use** and select the name of the file to open, or hit <esc> and type the name of the file.
② Choose the menu item **Data/Edit**!

MACINTOSH

① Choose the menu item **File/Open...**
② Locate and select the desired file.
③ Click on **Use**.

MS-DOS

① Choose the menu item **Files/Use**, and select the name of the file to open, or hit <esc> and type the name of the file.

have two types of delimiters: delimiters that separate cells within a case, and delimiters that separate cases. When SYSTAT saves text files it uses commas to separate cells within each case and returns to separate cases. Other programs sometimes use spaces, or tabs, or "whitespace" (any number of contiguous spaces and/or tabs), as delimiters. If you are not sure how the file you wish to open is delimited, you can open it with a text editor and examine it or simply try importing it into SYSTAT in a few different ways. Note that if the file uses spaces as delimiters you cannot have two spaces together or SYSTAT will think that a blank cell has been encountered. In such a case you have to open the file with a text editor to remove the additional blank spaces before importing the file into SYSTAT.

WARNINGS: TEXT FILES

- Make sure that the last row of the text file ends with a return. If it does not, SYSTAT will not read that line.

- If you put variable labels on the first line of a text file to be imported into SYSTAT, make sure that the variable labels do not contain spaces or illegal characters.

- If your text file contains all character data, and the first row does not contain variable labels, append a single numeric entry to the end of the first row to force SYSTAT to read that row as data. The resulting numeric column can be deleted in SYSTAT's editor.

- Do not allow negative numeric values to appear in the first row of a text file because SYSTAT will read the minus sign as a character entry and create a column of character type. To avoid this, either move a different case to the first line of the text file before importing, or delete the minus sign before importing and re-enter it in the newly created SYSTAT file.

1.2.1.1 Setting Delimiters on the Macintosh

When SYSTAT imports a text file, it does not import into an open data editor window; rather, it imports the text file into a new SYSTAT file. The imported data can be saved as a SYSTAT file, and can be opened as was discussed in the previous section. If the data does not appear correct, for example if there are blank cells where there should not be, open the original text file using a text editor and see if you can find the problem. On the Mac, try reimporting the text file using a different

> delimiter setup. Delimiters can be set using the menu selections shown at left on the Mac.

MACINTOSH
① When you ask SYSTAT to import a file, it will prompt you with the delimiter dialog box.
② Select the delimiter you want and click OK.

1.2.1.1.1 Spaces

When the text file uses spaces as delimiters, if you have multiple spaces in a row, SYSTAT will assume that there are missing data points between those spaces. If you want to include spaces in a character variable while using spaces as delimiters, you must enclose the character entry in quotes. Note that these must be regular straight-up-and-down quotes (" or ')—ASCII numbers 034 for double quotes or 039 for single quotes, respectively. You cannot use "smart quotes" ("" or '') in the text file.

If you use spaces as delimiters, then quotes are unnecessary. Leading and following spaces in a numeric entry are ignored. But note that leading and following spaces in a character entry are treated as part of the character entry itself. For example, suppose you have a character variable "gender" with two categories: "male" and "female." If your text file inadvertently has a leading space before one of the "male" entries, SYSTAT will read the file and think that you have three genders: male, female, and <space>male. You must be very careful when importing text files that the files do not contain extra spaces or other unwanted characters.

1.2.1.1.2 Tabs

Tabs can also be used as delimiters. As with any other delimiter, however, if you have two tabs in a row, SYSTAT will interpret the place they come together as a missing cell. If you are using tabs as delimiters, you should check the text file to make sure that there are no unwanted spaces following those tabs because the spaces may be read as part of character entries (see previous section). When you are not using tabs as delimiters, they are ignored by both character and numeric columns.

1.2.1.1.3 Commas and Other Delimiters

SYSTAT allows you to use commas or other characters that you specify as delimiters. These function similarly to spaces and tabs: Any that you want to use as data entries must be enclosed in quotes (see section 1.2.1.1a), and any two in a row indicate a missing entry in between.

1.2.1.1.4 Whitespace

Unlike in some other programs, *whitespace*, defined as any succession of spaces and/or tabs, is not recognized as a delimiter per se in SYSTAT. The spaces or tabs that make up the whitespace, however, may be treated as delimiters. For example, the whitespace item space-space-tab-tab-space would be read by SYSTAT as three delimiters with two missing cell entries if spaces were selected as delimiter, or two character spaces, one missing entry, and a character space if tabs were selected as delimiters.

1.2.1.1.5 Recommendations

If you have purely numeric data, any of the delimiters are as good as the next. However, tabs may make the text file most readable. But when you have character data it may be best to use a character delimiter such as commas. This will allow you to (a) use spaces in your character cell entries if you wish without quotes (but make sure the file has no unwanted spaces), and (b) use tabs to make the text file more readable. The tabs will be ignored when SYSTAT imports the file.

1.2.1.2 Text Files with Variable Labels

When SYSTAT reads a text file with no variable labels specified, it treats the first row of the file as variable labels, so long as that row contains no *illegal* variable labels. SYSTAT uses the entries in the second row to determine each column's type. SYSTAT will determine the variable types automatically, or you can put dollar signs into the text file to force columns to be character variables. You cannot force a variable to be numeric: Its type will be determined by the type of entry in the first or second row. If this entry is character and you left the dollar sign off the variable label, SYSTAT will put a dollar sign on the label for you and treat the entire variable as character. If a character occurs in any numeric variable after the second row, the entry is treated as missing. An exception to this rule is the minus sign, which is treated as numeric when it occurs before numbers in an entry. Note that if a period, which SYSTAT usually interprets as a missing numeric entry, occurs in the second row of the text file, SYSTAT will treat it as a *character* and initialize the entire column as a character column. This is unfortunate because it means that you must not have any missing numeric data on the first line of data in order for SYSTAT to read your file properly. You can trick SYSTAT, however, by creating a bogus first row of data that contains character and numeric entries in the correct columns. This will force SYSTAT to properly initialize all of your variables.

   If the text file contains variable labels, but one or more of these labels is illegal, SYSTAT assumes that the entries in the first row are not labels, but rather the first entries in character or numeric columns. For example, suppose you have a text file in which the first row contains the variable labels "char," "num," and "1." Because the label "1" is illegal, SYSTAT assumes that all three of these items are data entries in the first case of the data set. SYSTAT will initialize these variables as two character and one numeric variable, respectively, regardless of what follows in the other cases (see the next section).

If all of the variables in a text file are character, SYSTAT will think that the first line of the data set contains variable labels, whether it actually does or not. To avoid this, simply append an illegal label, such as a numeric entry, at the end of the first row of the text file. This will make SYSTAT read the first row as data, and the unwanted numeric column can then be deleted when the newly created SYSTAT file is opened with the editor.

1.2.1.3 Text Files without Variable Labels

If variable labels are not included in the first row of a text file, or if an illegal variable label such as a numeric data entry occurs in the first row, SYSTAT will read the entire first line as data entries in the first case of the data file. It uses the entries themselves to initialize the column types. It will then give each column a generic name corresponding to that column's type; for example, the first column would be 'COL1' or 'COL1$'. You will have to relabel all of the variables in the file, which is annoying if you have a large number of columns.

As noted above, if your text file has a missing numeric entry in the first row, SYSTAT will initialize the entire column as a character variable. Because real data sometimes has no cases with no missing numeric entries, one cannot always begin a text data file with a real case. Instead, you can trick SYSTAT by creating a first row of bogus data that contains character and numeric entries in the correct columns. This will force SYSTAT to properly initialize all of your variables. A lone space is also treated as a missing character entry and causes the same difficulty. Interestingly, a space in the first row followed by a numeric entry is ignored, and the column is initialized as a numeric column.

1.2.1.4 The Entries

After the columns are initialized SYSTAT treats the rest of the entries according to those initializations. Any non-numeric characters except leading and trailing spaces will produce a missing entry when they occur in a numeric column. An exception to this is the letter 'e', which may be read as indicating scientific notation when it occurs in a numeric column. Spaces that occur in a character column are treated as part of the character entry, unless spaces are used as delimiters. Numbers that occur in character columns are treated as character numerals: They cannot be used to perform mathematical operations.

1.2.2 Importing Data from Other Programs

SYSTAT can also open files generated by a number of other programs, such as Microsoft Excel, Lotus, Symphony, and dBase, as well as other file formats such as DIF, map, and portable files. Each of these file formats may present difficulties for SYSTAT, and often will require a bit of trial and error until you figure out how to get SYSTAT to read the file. For example, to read an Excel file, the file must have variable labels on the first row of the file. SYSTAT then determines the column type automatically according to entries in the second row of the file. Most of the rules for text files (section 1.2.1) apply: For example, if the Excel file has blank cells in a numeric column, these cells must contain periods ".". The other programs themselves may have their own idiosyncratic rule: For example, if an Excel file contains empty cells in a character column, they must contain spaces enclosed in quotes, " ". This is because Excel will not retain spaces by themselves in its cells.

Importing files from other programs has caused me the most frustration of any of SYSTAT's procedures. This may be due to my using versions of the other programs that are out of sync with SYSTAT. Usually, I end up cutting and pasting the values from the other program into SYSTAT. For large data files I sometimes save them as text files and then read them into SYSTAT, which seems to work more reliably.

2
Working with Data

2.1 Data Manipulation Functions

SYSTAT's data editor is equipped with a large array of data manipulation functions. I will not discuss all of them, but I will illustrate some of the most useful functions. The rest should be fairly straightforward. The data set in Table 2.1 will be used to illustrate SYSTAT's data manipulation functions.

2.1.1 Sorting Data

From an already existing data file SYSTAT allows you to create a new data file in which the cases are arranged in ascending (or descending) numeric or alphabetic order of the entries within a particular column or columns. If you select more than one column, SYSTAT initially sorts the cases according to the first column selected and then does a nested sort of the entries in the second column *within* each group in the first column. The sorting function is of some use in itself, but more importantly,

Table 2.1 An example data set used to illustrate SYSTAT's data manipulation functions.

A	B$	C
3	D	25
1	C	21
4	A	32
3	B	15

some of SYSTAT's other procedures require that you sort the data before using them.

To illustrate the sorting procedure, create a new data file containing the data in Table 2.1 (see section 1.1 for more information on creating data files). The cases in this file are not in any particular order. Suppose that you wanted to create a new file that is sorted by the second column, column B. To do this, you will have to create a new file into which to save this newly sorted data. The sort procedure requires two commands

>SAVE *FILENAME.SRT*
>SORT B$

The *save* command gives the name of the new file (I will always use the 'srt' suffix to designate sorted files), the *sort* command lists the name(s) of the column(s) by which the file is to be sorted and executes the procedure. You must type the save command before the sort command.

Unfortunately, you cannot sort a data file within an open window, but must save the sorted data into a new file, which must then be opened in the usual way. This constraint appears to be inherited from the days in which computers did not have enough RAM to sort large files in memory. If you are using an older version of SYSTAT on the Mac, SYSTAT will prompt you to choose whether to save the file in single or double precision, after it prompts you for a filename (the default name simply appends the suffix SORTED to the end on the original filename).

When the sort is successful, SYSTAT reports that the sorted file has been saved. This file can then be opened as usual (see section 1.1.4). Table 2.2 shows the data from Table 2.1 sorted by the second column in ascending order.

Nested sorting is easily accomplished by specifying each of the sort variables in order after the sort command. For example, to sort the data in Table 2.1 by the first column, and

---

MACINTOSH

① Choose the menu item **Data/Sort...**
② Select the variable(s) on which the file is to be sorted, in this case *'B$'*.
③ Type the name of the file to be created in the dialog box, in this case *filename.srt*.
④ Select *single* or *double* precision and click **OK**.

MS-DOS

① Choose the menu item **Data/Sort/Variables** and select the variable *'B$'*.
② Choose the menu item **Data/Sort/Save**, hit <ret>, and type the name of the new file.
③ Select *single* or *double* precision and hit <esc>.
④ Choose the menu item **Data/Sort/Go!**

Table 2.2 Example data set from Table 2.1 sorted in ascending order by the second column.

A	B$	C
4	A	32
3	B	15
1	C	21
3	D	25

then sort the data within each category in the first column by the second column, type

```
>SAVE NEWNAME.SRT
>SORT A B$
```

Notice that the filename I used in the *save* command is different from the one used previously. SYSTAT will not let you sort into a file with the same name as the open file. The resulting data file is shown in Table 2.3.

Table 2.3 Example data set from Table 2.1 sorted in ascending order by the first and second columns.

A	B$	C
1	C	21
3	B	15
3	D	25
4	A	32

2.1.2 Replacing Data with Ranks

SYSTAT also allows you to create a new file in which the entries in one column have been replaced by their ranks in ascending order. For example, if the data in Table 2.1 were ranked according to the third column, the '15' in the second row, being the smallest entry, would be replaced by a '1', and the '32' in the last row would be replaced by a '4'. The sequence of commands for creating a column of ranks is:

```
>SAVE FILENAME.RNK
>RANK C
```

Here, the *rank* command replaces the entries in column C by the *case number* that they would have if the data were sorted on that column, except that ties are replaced by the average of the case numbers that the tied entries would have. I will always use the suffix 'rnk' to designate files with ranked variables.

Ranking data can be particularly useful when the spacing of the original data entries is not important, or when the distribution of the values in a column poses problems for further analysis, for example, when usual assumptions about the distribution of the data cannot be made. Ranked data can sometimes look more normally distributed than the original data, and some data analysis techniques, for example, the

MACINTOSH
① Choose the menu item **Data/Sort...**
② Select the sorting variables in order, in this case *'A'* and then *'B$'*.
③ Type the name of the file to be created in the dialog box, in this case *filename.srt*.
④ Select *single* or *double* precision and click **OK**.

MS-DOS
① Choose the menu item **Data/Sort/Variables** and select the variable *'A'*.
② Choose the menu item **Data/Sort/Variables** and select the variable *'B$'*.
③ Choose the menu item **Data/Sort/Save**, hit <ret>, and type the name of the new file.
④ Select *single* or *double* precision and hit <esc>.
⑤ Choose the menu item **Data/Sort/Go!**

MACINTOSH
① Choose the menu item **Data/Rank...**
② Select the variable you wish to convert to ranks, in this case *'C'*.
③ Type the name of the file to be created in the dialog box, in this case *filename.rnk*.
④ Select *single* or *double* precision and click **OK**.

MS-DOS
① Choose the menu item **Data/Rank/Variables** and select the variable *'C'*.
② Choose the menu item **Data/Rank/Save**, hit <ret>, and type the name of the new file.
③ Select *single* or *double* precision and hit <esc>.
④ Choose the menu item **Data/Rank/Go!**

Mann-Whitney and Kruskal-Wallis tests, work directly with ranks.

2.1.3 Standardizing Data

Another very useful SYSTAT function converts the raw cell entries in a column into *standard scores*. (These are sometimes referred to as "z-scores," but standard scores are not actually distributed as Z, i.e., *normally*, unless the original data was normally distributed in the first place. Standardization is a linear transformation, and therefore does not change the shape of a distribution.) These scores are computed by subtracting the mean of the column from each cell entry and then dividing the results by the sample standard deviation (based on n, not $n-1$). Thus, the new column of standard scores has a mean of 0 and a standard deviation of 1, meaning that the cell entries have been converted into standard deviation units. For example, a standard score of 2.0 means that the original data point is two sample standard deviations above the mean of its column.

As with the two preceding functions, SYSTAT requires you to put the standardized data into a new file. The standard scores must be saved into a new file that can then be opened in the usual manner. The commands for standardizing the third column, C, in Table 2.3 are

```
>SAVE FILENAME.STD
>STAND C
```

I will always use the suffix 'std' to designate files with standardized data. The resulting data set is shown in Table 2.4. The values of C from Table 2.3 have been replaced by their standard scores.

2.1.4 Transposing Rows and Columns

For some procedures you may wish to convert the variables (columns) in your data set into cases (rows) and the cases into variables. SYSTAT's *transpose* command allows this to be done for files that either (1) contain only numeric data or (2) that contain character data in the first column only. If the first column contains character data, you must rename the first column "LABEL$" before you transpose the file, and this column will be used as the column headings in the new file. You cannot transpose a file that has character data in any but the first column.

As an example, suppose that for display purposes we wished to transpose the rows and columns in Table 2.1, that is, make

---

MACINTOSH

① Choose the menu item **Data/Standardize...**

② Select the variable you wish to convert to standard scores, in this case *'C'*.

③ Type the name of the file to be created in the dialog box, in this case *filename..std*.

④ Select *single* or *double* precision and click **OK**.

MS-DOS

① Choose the menu item **Data/Standardize/Variables** and select the variable *'C'*.

② Choose the menu item **Data/Standardize/Save**, hit <ret>, and type the name of the new file.

③ Select *single* or *double* precision and hit <esc>.

④ Choose the menu item **Data/Standardize/Go!**

Table 2.4 Example data set from Table 2.3 with the data in column C standardized.

A	B$	C
1	C	–0.31532
3	B	–1.15618
3	D	0.24525
4	A	1.22625

each column a row and each row a column. Because the second column of Table 2.1 contains character entries, you must move that column to the first column of the file and give it the name "LABEL$" rather than 'B$'. Unfortunately, you cannot insert columns to the left of existing columns. One of the easiest ways to move columns is to read the data file into a new file with the columns in a different order (section 1.1.2.4). Table 2.5 shows the data in Table 2.1 with the second column moved to the first column and named "LABEL$." Now the file can be transposed and saved into a new file. The commands are

>SAVE *FILENAME.TSP*
>TRANSPOSE

I always use the suffix 'tsp' to designate files containing transposed data. The transposed data set from Table 2.5 is shown in Table 2.6. Notice that the original column headings, LABEL$, A, and C, are no longer retained in the file.

Table 2.5 The data set in Table 2.1 with column B$ moved to the first column and relabelled "LABEL$."

LABEL$	A	C
D	3	25
C	1	21
A	4	32
B	3	15

MACINTOSH

① Choose the menu item **Data/Transpose...**
② Type the name of the file to be created in the dialog box, in this case *filename.tsp*.
③ Select *single* or *double* precision and click **OK**.

MS-DOS

① Choose the menu item **Data/Transpose/Save**, hit <ret>, and type the name of the new file, in this case *filename.tsp*.
② Select *single* or *double* precision and hit <esc>.
③ Choose the menu item **Data/Transpose/Go!**

Table 2.6 The data file in Table 2.5 transposed.

D	C	A	B
3	1	4	3
25	21	32	15

2.2 Math Functions

SYSTAT contains a large number of mathematical functions that can be used to transform or generate numbers in data files. You can either transform data within a column, or create columns containing the new data. Rather than duplicate the description of these functions that can be found in the *Getting Started* volume of the SYSTAT manuals, the sections below will briefly outline how SYSTAT transforms columns of data, then show how you can use SYSTAT's math functions to find *p*-values from significance tests and to find the critical values on distributions corresponding to *p*-values. These procedures employ many of the features of SYSTAT's mathematical transformation capabilities, and more can be found in Chapter 4 on meta-analysis. After looking through these examples you should be well equipped to perform other kinds of transformations on a set of data.

Note that, in DOS, you must use the command window for math transformations, so no menu descriptions are given for DOS in this section.

2.2.1 Transforming Columns

2.2.1.1 Basic Operators

To begin, you should become familiar with SYSTAT's basic arithmetic operators. They are, in order of evaluation precedence:

```
^       exponentiation (raises number to a power)
-       unary minus (makes number negative)
*, /    multiplication, division
+, -    addition, subtraction
```

Along with these operators, parentheses can be used to group parts of expressions and to control the order of evaluation.

The command that SYSTAT uses for all transformations is the *let* command. This command is followed by the name of a column that you want to set equal to something, an equal sign, and then the expression that you want to set the column equal to. This expression may contain numbers, column names, operators, function names, and parentheses. Generically this command has the form

```
>LET SOME COLUMN = SOME EXPRESSION
```

For example, suppose you wanted to demean (subtract the mean of) the third column in Table 2.1. The mean of this column is 23.25. You have two options—you can replace the values in the third column with their demeaned values, or you can create a new column containing the demeaned values. To replace a column, you set it equal to some function of itself. Therefore, to replace the third column with its own demeaned values, type

```
>LET C = C - 23.25
```

The resulting data set is shown in Table 2.7. The original values in column C have been replaced with their demeaned values. If you want to retrieve the original values, you must either close the file without saving the changes, or remean the data by adding 23.25 back to it. In general, it's pretty risky to replace columns with transformed values unless you are sure you will no longer need the original values.

> **MACINTOSH**
> ① Choose the menu item **Data/Math...**
> ② Select the variable that you want to replace with transformed values so that it shows in the "let variable" box on the left, in this case the variable '*C*'.
> ③ In the "=variable or expression" box on the right, enter the transformation expression, in this case "*C* – *23.25*".
> ④ Click **OK**.

HINT: SELECTING VERSUS TYPING FUNCTIONS

When you are using menus, SYSTAT conveniently gives you a list of all of its transformation functions, and, if you select a function from this list, it will enter the function name followed by parentheses into the expression box. This only saves typing, at most, five characters. Often it takes more effort to select the function and ensure that it was inserted properly than it would have just to type the function into the box. I often use the function list to remind myself of the name of the function and then proceed to type it into the box.

The safer way to transform data is to create a new column to receive the transformed values. You can first create the column and then perform the transformation, or you can just enter a new column name on the left side of the equation that SYSTAT doesn't recognize, and it will create a new column with that name automatically. For example, if you type

```
>LET DEMEAN = C - 23.25
```

> **MACINTOSH**
> ① Choose the menu item **Data/Math...**
> ② Type the new column name that you wish to contain the transformed values so that it shows in the "let variable" box on the left, in this case the name '*DEMEAN*'.
> ③ In the "=variable or expression" box on the right, enter the transformation expression, in this case "*C* – *23.25*".
> ④ Click **OK**.

Table 2.7 Example data set from Table 2.1 with the third column demeaned.

A	B$	C
3	D	1.75
1	C	–2.25
4	A	8.75
3	B	–8.25

SYSTAT will create a new column called "DEMEAN" that contains the newly transformed values. Such a file is shown in Table 2.8.

Table 2.8 Example data set from Table 2.1 with a new column containing the demeaned values from the third column.

A	B$	C	DEMEAN
3	D	25	1.75
1	C	21	−2.25
4	A	32	8.75
3	B	15	−8.25

2.2.1.2 Built-In Functions

In addition to the basic arithmetic operators, SYSTAT has a large array of other built-in functions, including logarithms, trigonometric functions, absolute values, and so on. Most of these functions take a single argument in parentheses, but these arguments may be complex expressions.

For example, suppose you wanted to take the natural log of the first column in Table 2.8. First, you must know that the natural log function in SYSTAT is LOG(). (This is one time when menus can be a lot of help, because they list all of SYSTAT's math functions.) Then type

▶ >LET LOG_A = LOG(A)

where the name *log_A* on the left is the name of the new column that will contain the transformed values, and *A* is the column of values to transform. The results are shown in Table 2.9.

MACINTOSH

① Choose the menu item Data/Math...
② Type the new column name in the "let variable" box on the left, in this case the column name '*LOG_A*'.
③ In the "=variable or expression" box on the right, enter the transformation expression, in this case '*LOG(A)*'.
④ Click **OK**.

Table 2.9 Example data set from Table 2.8 with a new column containing the log of the first column.

A	B$	C	DEMEAN	LOG_A
3	D	25	1.75	1.09861
1	C	21	−2.2	0.000005
4	A	32	8.75	1.38629
3	B	15	−8.2	1.098615

HINT: LONG EXPRESSIONS IN THE DIALOG BOX (MACINTOSH ONLY)

When you attempt to type long expressions into the "=variable or expression" box (within the **Data/Math...** dialog box) you will soon notice that the expression overflows the edge of the box. Don't worry, the expression is all there. In order to see or edit the part of the expression that has disappeared, hold the mouse button down anywhere in the box and then drag the mouse in the direction of the hidden text. This will scroll the contents of the box and allow you to check and edit what you have typed. When you are satisfied that the expression is correct, click OK.

As a final example, suppose you knew in advance that you wanted to take the logs of the absolute values of the demeaned values shown in the fourth column of Table 2.9. You could compute them in steps, first demeaning the values, then taking the absolute values, and then taking the logs. Or you could save two steps by performing all of these transformations with the same command. To do this, simply nest the absolute value and subtraction operations within the parentheses of the log function

```
>LET LOG_DEM = LOG( ABS(C - 23.25) )
```

This expression is evaluated from the inside out: first 23.25 is subtracted from every row of column C, then the absolute values of this column are computed, and, finally, the logs of the absolute values are taken. The results appear in Table 2.10.

MACINTOSH

① Choose the menu item **Data/Math...**
② Type the new column name in the "let variable" box on the left, in this case the column name '*LOG_DEM*'.
③ In the "=variable or expression" box on the right, enter the transformation expression, in this case '*LOG(ABS (C – 23.25))*'.
④ Click **OK**.

Table 2.10 Example data set from Table 2.9 with a new column containing the log of the absolute values of the demeaned third column.

A	B$	C	DEMEAN	LOG_A	LOG_DEM
3	D	25	1.75	1.09861	3.06805
1	C	21	–2.2	0.000005	3.23868
4	A	32	8.75	1.38629	2.67415
3	B	15	–8.2	1.098615	3.44999

2.2.2 Conditional Transformations

SYSTAT allows you to make your transformations conditional on values in columns. You still use the *let* command exactly as described in the previous section, but you precede it with an "if" and some condition that must be satisfied for the *let* command to be executed. Thus, the generic form of a conditional transformation is

```
>IF SOME CONDITION THEN LET SOME COLUMN = SOME
EXPRESSION
```

The condition may contain numbers, column names, expressions, the logical operators *and* and *or*, and parentheses for controlling the order of evaluation of the logical operators. The elements of the condition can be related using the following relational operators:

=	equal to
<> or ><	not equal to
<	less than
>	greater than
<= or =<	less than or equal to
>= or =>	greater than or equal to

To illustrate with the example data set in Table 2.1, suppose you wanted to find the log of the third column, but only when the row has a '3' in column A. The following command creates a new column D that contains this transformed value:

```
>IF A=3 THEN LET D = LOG(C)
```

The fourth column of Table 2.11 shows this newly created column: Note that the rows that failed to meet the condition are given missing cells in column D.

You can set multiple conditions using the *and* operator. For example, if you wanted the logs of column C for rows with values of A equal to 3 and B$ equal to "D," you could use the following command:

```
>IF A=3 AND B$="D" THEN LET E = LOG(C)
```

Note that the condition on the column B$ must be placed in quotes because it is a character variable (see section 1.1.2.4.1). The result is shown in column E of Table 2.11. Only the first row met both conditions.

You may specify alternative conditions using the *or* operator. For example, if you wanted the log of column C for rows that had *either* column A equal to 3 *or* column B$ equal to "C," you could use the command:

```
>IF A=3 OR B$="C" THEN LET F = LOG(C)
```

The result is shown in column F of Table 2.11.

Conditions can become quite complex when you use mul-

MACINTOSH

① Choose the menu item **Data/Recode...**

② Select a variable name in the "If Test variable" box, in this case "**A**."

③ Select a relational operator, in this case "**=**."

④ Enter the condition into the "Variable or expression" box, in this case type "**3**."

⑤ Select or type a (new) column name in the "let variable" box, in this case the column name '**D**'.

⑥ In the "=variable or expression" box, enter the transformation expression, in this case '**LOG(C)**'.

⑦ Click **OK**.

MACINTOSH

① Choose the menu item **Data/Recode...**

② Click on **Complex recode...**

③ Enter the first condition into the "If test expression" box, in this case "**A=3**."

④ Click on **And**.

⑤ Enter the second condition into the "If test expression" box, in this case " **B$="D"** ."

⑥ Select or type a (new) column name in the "then let variable" box, in this case the column name '**E**'.

⑦ In the "=variable or expression" box, enter the transformation expression, in this case '**LOG(C)**'.

⑧ Click **OK**.

tiple *and*'s or *or*'s. You should think through these conditions carefully and use parentheses liberally to clarify and control the order of evaluation. For example, the command

```
>IF A=3 OR (A=1 AND B$="C") THEN LET G = LOG(C)
```

evaluates as true under different conditions than the command

```
>IF (A=3 OR A=1) AND B$="C" THEN LET H = LOG(C)
```

although all that has been changed are the locations of the parentheses. The outputs from these two commands are shown in columns G and H, respectively, of Table 2.11. None of the rows satisfied the condition in the second command, so column H was set entirely to missing cells.

Table 2.11 The example data set with new columns generated by conditional transformations.

A	B$	C	D	E	F	G	H	I
3	D	25	3.219	3.219	3.219	3.219	.	3.219
1	C	21	.	.	3.045	.	.	3.045
4	A	32	.	.	.	.	.	.
3	B	15	2.708	.	2.708	2.708	.	2.708

The expressions within the condition can be as complicated as you like. For example, the command

```
>IF SQR(A*10) >= LOG(C+10) THEN LET I = LOG(C)
```

is a perfectly valid command (SQR is SYSTAT's square-root function), and the result is shown in column I of Table 2.11.

Finally, keep in mind that the column of *case numbers* can be used as a variable. The command

```
>IF CASE>1 AND CASE<4 THEN LET J = LOG(C)
```

would compute the logs of column C in Table 2.11 for the second and third rows only and put them in a new column labelled J (not shown).

2.2.3 Finding Probabilities and Critical Values

Some of the most useful mathematical functions in SYSTAT are those functions that allow you to compute *p*-values from points on distributions, such as F and χ^2, and related functions that allow you to compute critical values on distributions corresponding to *p*-values. By using these functions you will never have to look up significance levels in tables ever again.

And these functions do tables one better by reporting the exact *p*-values and critical values, rather than just giving cutoffs.

These functions are essentially transformation functions like any others in SYSTAT. They have three-letter names and take arguments in parentheses. I will present five of the most commonly used distribution functions below, the normal, *F*, *t*, χ^2, and the binomial. To make reference easier, these sections have been made somewhat redundant so that each section stands on its own.

2.2.3.1 The Normal Distribution Functions

This section describes two of SYSTAT's normal distribution functions: ZCF(), for *Z cumulative function*, which converts points on the normal distribution into probabilities, and ZIF(), for *Z inverse function*, which converts probabilities into points on the normal distribution. ZRN, which generates random numbers from a normal distribution (see Section 5.6), and ZDF(), which can be used for plotting normal curves (see Figure 2.1), are not discussed in this section. In this section, I use 'Z' to denote points on a normal distribution with $\sigma = 1$: This *Z* is not to be confused with standard scores, or "z-scores" (see section 2.1.3).

2.2.3.1.1 Finding Probabilities from Z

One-tailed (directional) probabilities. ZCF() is a *cumulative* function, which gives the area under the normal curve from minus infinity *up to* the value placed within the parentheses of ZCF(); that is, ZCF() takes one argument, any positive or negative number, and calculates the proportion of a normal distribution that would fall below that value, assuming a standard deviation of 1.0 and mean of 0. To illustrate, Figure 2.1 shows a normal distribution with $\sigma = 1$. If 2.0 is given as an argument to ZCF(), this function will compute the proportion of the distribution to the *left* of +2.0 on the *X*-axis in Figure 2.1. In fact, this corresponds to .977 of the distribution. The area of the curve to the right of +2.0 in Figure 2.1, therefore, corresponds to 1 – .977 = .023 of the distribution. This number is the "*p*-value" for a *Z* of 2.0: There is only a .023 chance of finding at random a value of *Z* greater than, or equal to, 2.0 in a normal distribution with $\sigma = 1$.

In SYSTAT, you can perform *p*-value computations in the data editor window. First, you should create a new column in your data set, or open a temporary data file to contain the *p*-values. (If using a new file, you must give at least one column

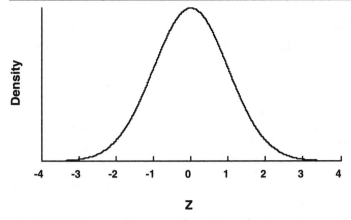

Figure 2.1 A normal curve with σ = 1 and mean zero.

a label, say 'P', and fill at least one cell with a missing value or a number.) Then you can use the ZCF() function to fill this column with the desired *p*-value(s). Suppose, for example, that you want to find the one-tailed *p*-value corresponding to $Z = 2.0$: Simply type into the command window

```
>LET P = 1-ZCF(2.0)
```

The cell or cells in column P will be filled with the number 0.023 (assuming you are using three decimal places), the desired one-tailed *p*-value.

If you want to find more than one *p*-value, it is easy to compute them all at once. Simply create a column containing the *Z*'s for which you want the probabilities, and then you can compute the entire column of *p*-values with one command. Table 2.12 shows a column of *Z*'s, called 'MY_ZS', that has been transformed into corresponding one-tailed *p*-values with the command

```
>LET P = 1-ZCF(MY_ZS)
```

◀ **MACINTOSH**
① Choose the menu item **Data/Math...**
② Type the new column name in the "let variable" box on the left, in this case the column name '**P**'.
③ In the "=variable or expression" box on the right, enter the transformation expression, in this case '*1 – ZCF(2.0)*'.
④ Click **OK**.

Table 2.12 The second column contains the one-tailed probabilities of a *Z* greater than, or equal to, the value given in the first column.

MY_ZS	P
0.241	0.405
1.960	0.025
1.388	0.083
2.414	0.008

If you want the one-tailed probability of finding a value of Z in the *lower* (negative) tail of the distribution, just leave off the "1–" on the right side of the equation in the foregoing command. For example, the command

```
>LET P = ZCF(-1.64)
```

gives the one-tailed probability of finding a Z less than (more negative than) –1.64. The resulting *p*-value is equal to 0.05.

Two-tailed (nondirectional) probabilities. A *p*-value in Table 2.12 corresponds to the probability of finding a Z of a certain value, or larger, in the upper tail of the normal distribution. Often, however, you will be interested in finding the probability of finding a Z of a given *magnitude* or larger, positive or negative, due to chance. This means that you are interested in adding the areas from both tails of the normal distribution corresponding to values more extreme than the Z of interest. To do this, rather than computing both tails and adding them, you can simply multiply the one-tailed *p*-values by two. For example, to transform the Z's in the first column of Table 2.12 into their corresponding two-tailed probabilities, type

```
>LET P_TT = 2*(1-ZCF(MY_ZS))
```

where P_TT is just my mnemonic label for "probability two-tailed," and MY_ZS is the label of the column of Z's. The resulting data table is shown in Table 2.13. The second Z in Table 2.13, 1.960, corresponds to the traditional .05 level of significance, two-tailed.

2.2.3.1.2 Finding Critical Values of Z from α

One-tailed (directional) critical values. Sometimes you may want to find the critical value of Z corresponding to a given probability. This is frequently done in meta-analysis, for example (Chapter 4). This is easy in SYSTAT using ZIF(), the *Z inverse function*. This function takes a proportion (probability) as an argument, and calculates the value of Z corresponding to

Table 2.13 The second column contains the two-tailed probabilities of a Z with magnitude greater than, or equal to, the value given in the first column.

MY_ZS	P_TT
0.241	0.810
1.960	0.050
1.388	0.165
2.414	0.016

the point on the normal distribution that would cut off that proportion of the distribution in the lower tail; that is, ZIF() gives the value of Z to the left of which (in Figure 2.1) falls the given proportion of the distribution. For example, a value of Z equal to -1.96 cuts off the bottom 2.5% of the lower tail, whereas a value of Z equal to $+1.96$ cuts off the bottom 97.5% of the "lower tail."

Therefore, if you want the value of Z that cuts off the *top* 5% of the *upper* tail, you must ask SYSTAT to give the Z that cuts off the lower 95% of the distribution; that is, the argument to ZIF() should be "$1 - p$," where p is the desired probability. You can either do the subtraction yourself and put .95 in the parentheses of the ZIF() function or ask SYSTAT to do it for you in the command

```
>LET Z = ZIF(1-.05)
```

SYSTAT will put the value 1.645 into the column labelled Z. Using the "$1 - p$" method is especially useful if you want to convert a column of p's into their corresponding positive Z's. Table 2.14 shows the results of typing

```
>LET Z_CRIT = ZIF(1-P)
```

where Z_CRIT is my label for the new column of critical Z values, and 'P' is the column of one-tailed probabilities. If you compare Tables 2.12 and 2.14, you will see that I just converted the p's back into the Z's that I started with.

To find a critical *lower-tail* value of Z that cuts off a given proportion of the lower tail of the normal distribution, simply give the probability as an argument to ZIF(). For example, the Z corresponding to the lower 5% of the distribution can be found by typing

```
>LET Z = ZIF(.05)
```

With this command, SYSTAT will put the value -1.645 into the column labelled Z.

Table 2.14 The second column contains the upper-tailed values of Z corresponding to the one-tailed probabilities given in the first column.

P	Z_CRIT
0.405	0.241
0.025	1.960
0.083	1.388
0.008	2.414

Two-tailed (nondirectional) critical values. Finding values of Z corresponding to two-tailed probabilities is identical to doing so for one-tailed probabilities, but you must remember to divide the probability by two before entering it into the ZIF() function. For example, if you want to find the critical magnitude of Z that cuts off the extreme 5% of the distribution *regardless of sign*, you must find the magnitude of the values that cut off the upper 2.5% and lower 2.5% of the distribution. The following command divides the *p*-value by two within the argument to ZIF():

```
>LET Z_CRIT = ZIF(1-P/2)
```

Again, Z_CRIT is my label for the new column of critical Z magnitudes, and 'P' now is a column of *two-tailed* probabilities. Table 2.15 shows these two columns, using the probabilities from Table 2.14. The values of the Z's in the column Z_CRIT are the absolute values of the Z's, positive or negative, that cut off the extreme 2.5% of their respective tails.

Table 2.15 The second column contains the magnitudes of Z corresponding to the two-tailed probabilities given in the first column.

P	Z_CRIT
0.405	0.833
0.025	2.241
0.083	1.734
0.008	2.652

2.2.3.2 The *t* Distribution Functions

This section describes two of SYSTAT's *t* distribution functions: TCF(), for *t cumulative function*, which converts points on the *t* distribution into probabilities, and TIF(), for *t inverse function*, which converts probabilities into points on the *t* distribution. Each of these functions is discussed below. TRN(), which generates random numbers from a *t* distribution, and TDF(), which can be used for plotting *t* distributions (see Figure 2.2), are not discussed in this section. In this section, I will use '$t_{(df)}$' to denote points on a *t* distribution with degrees of freedom equal to the *df* in the subscript.

2.2.3.2.1 Finding Probabilities from *t*

One-tailed (directional) probabilities. TCF() is a *cumulative* function, which gives the area under the *t* curve from minus infinity *up to* the value placed within the parentheses of TCF(); that is, TCF() takes any positive or negative number, along with its degrees of freedom, and calculates the proportion of that *t* distribution that would fall below that value. To illustrate, Figure 2.2 shows a *t* distribution with $df = 12$. If "2.0,12" is given as an argument to TCF(), this function will compute the proportion of the distribution to the *left* of +2.0 on the X-axis in Figure 2.2. In fact, this corresponds to .966 of the distribution. The area of the curve to the right of +2.0 in Figure 2.2, therefore, corresponds to $1 - .966 = .034$ of the distribution. This number is the "*p*-value" for a $t_{(12)}$ of 2.0: There is only a .034 chance of finding at random a value of $t_{(12)}$ greater than, or equal to, 2.0.

In SYSTAT, you can perform *p*-value computations in the data editor window. First, you should create a new column in your data set, or open a temporary data file to contain the *p*-values. (If using a new file, you must give at least one column a label, say 'P', and fill at least one cell with a missing value or a number.) Then you can use the TCF() function to fill this column with the desired *p*-value(s). Suppose, for example, you want to find the one-tailed *p*-value corresponding to $t_{(12)} = 2.0$: Simply type into the command window

```
>LET P = 1-TCF(2.0,12)
```

The cell or cells in column P will be filled with the number 0.034 (assuming you are using three decimal places), the desired one-tailed *p*-value.

If you want to find more than one *p*-value, it is easy to compute them all at once. Simply create a column containing

MACINTOSH

① Choose the menu item **Data/Math...**

② Type the new column name in the "let variable" box on the left, in this case the column name **'P'**.

③ In the "=variable or expression" box on the right, enter the transformation expression, in this case **'1 – TCF(2.0,12)'**.

④ Click **OK**.

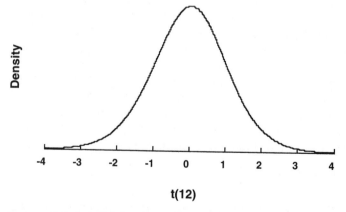

Figure 2.2 A *t* distribution with 12 *df*.

the *t*'s of which you want the probabilities, and a second column containing their degrees of freedom. You can then compute the entire column of *p*-values with one command. Table 2.16 shows a column of *t*'s labelled T, and their degrees of freedom in a column labelled DF. The third column shows the corresponding upper-tailed *p*-values, which were calculated with the command

```
>LET P = 1-TCF(T,DF)
```

Table 2.16 The third column contains the upper-tailed probabilities of a *t* greater than, or equal to, the value given in the first column.

T	DF	P
0.241	12	0.407
1.960	8	0.043
1.388	36	0.087
2.414	14	0.015

If you want the one-tailed probability of finding a value of *t* in the *lower* (negative) tail of the distribution, just leave off the "1–" on the right side of the equation in the foregoing command. For example, the command

```
>LET P = TCF(-1.78,12)
```

gives the one-tailed probability of finding a *t* less than (more negative then) –1.78. The resulting *p*-value is equal to 0.05.

Two-tailed (nondirectional) probabilities. A *p*-value in Table 2.16 corresponds to the probability of finding a *t* of a certain value, or larger, in the upper tail of the *t* distribution. Often, however, you will be interested in finding the probability of finding a *t* of a given *magnitude* or larger, regardless of sign, due to chance. This means that you are interested in adding the areas from both tails of the *t* distribution corresponding to values more extreme than the *t* of interest. To do this, rather than computing both tails and adding them, you can simply multiply the one-tailed *p*-values by two. For example, to transform the *t*'s in the first column of Table 2.16 into their corresponding two-tailed probabilities, type

```
>LET P_TT = 2*(1-TCF(T,DF))
```

where P_TT is just my mnemonic label for "probability two-tailed," and T and DF are the labels for the columns of *t*'s and degrees of freedom, respectively. The resulting data table is shown in Table 2.17. The t in the second row of Table 2.17, 2.306, corresponds to the traditional .05 level of significance, two-tailed.

Table 2.17 The third column contains the two-tailed probabilities of a t with magnitude greater than, or equal to, the value given in the first column.

T	DF	P_TT
0.241	12	0.814
2.306	8	0.050
1.388	36	0.174
2.414	14	0.030

2.2.3.2.2 Finding Critical Values of t from α

One-tailed (directional) critical values. Sometimes you may want to find the critical value of t corresponding to a given probability. This is easy in SYSTAT using TIF(), the *t inverse function*. This function takes a proportion (probability) and degrees of freedom as arguments, and calculates the value on the $t_{(df)}$ distribution that would cut off that proportion of the distribution in the lower tail. That is, TIF() gives the value of $t_{(df)}$ below which falls the given proportion of the distribution. For example, a value of $t_{(12)}$ equal to –2.306 cuts off the bottom 2.5% of the lower tail, whereas a value of $t_{(12)}$ equal to +2.306 cuts off the bottom 97.5% of the "lower tail."

Therefore, if you want the value of $t_{(df)}$ that cuts off the *top* 5% of the *upper* tail, you must ask SYSTAT to give the $t_{(df)}$ that cuts off the lower 95% of the distribution. That is, the argument to TIF() should be "1 – p," where p is the desired probability. You can either do the subtraction yourself and put .95 in the parentheses of the TIF() function, or ask SYSTAT to do it for you in the command

```
>LET T = TIF(1-.05,DF)
```

where DF is either the value of, or the column containing, the corresponding degrees of freedom. For $df = 12$, SYSTAT will put the value 1.782 into the column labelled T. Using the "1 – p" method is especially useful if you want to convert a column of p's into their corresponding positive t's. Table 2.18 shows the results of typing

```
>LET T_CRIT = TIF(1 P,DF)
```

where T_CRIT is my label for the new column of critical t values, P is the column of one-tailed probabilities, and DF is the column of degrees of freedom. If you compare Tables 2.16 and 2.18, you will see that I just converted the p's back into the t's that I started with.

Table 2.18 The third column contains the upper-tailed values of t corresponding to the one-tailed probabilities given in the first column.

P	DF	T_CRIT
0.407	12	0.241
0.043	8	1.960
0.087	36	1.388
0.015	14	2.414

To find a critical *lower-tail* value of t that cuts off a given proportion of the lower tail of the t distribution, simply give the probability as an argument to TIF(). For example, the $t_{(12)}$ corresponding to the lower 5% of the distribution can be found by typing

```
>LET T = TIF(.05,12)
```

With this command, SYSTAT will put the value -1.782 into the column labelled T.

Two-tailed (nondirectional) critical values. Finding values of t corresponding to two-tailed probabilities is identical to doing so for one-tailed probabilities, but you must remember to divide the probability by two before entering it into the TIF() function. For example, if you want to find the critical magnitude of t that cuts of the extreme 5% of the distribution *regardless of sign*, you want the magnitude of the values that cut off the upper 2.5% and lower 2.5% of the distribution. The following command divides the p-value by two within the argument to TIF():

```
>LET T_CRIT = TIF(1-P/2,DF)
```

Again, T_CRIT is my label for the new column of critical t magnitudes, P now is a column of *two*-tailed probabilities, and DF is the column of degrees of freedom. Table 2.19 shows these three columns, using the probabilities from Table 2.18. The values of the t's in the column T_CRIT are the absolute values of the t's, positive and negative, that cut off the extreme 2.5% of their respective tails.

Table 2.19 The third column contains the magnitudes of t corresponding to the two-tailed probabilities given in the first column.

P	DF	T_CRIT
0.407	12	0.859
0.043	8	2.403
0.087	36	1.759
0.015	14	2.771

2.2.3.3 The F Distribution Functions

This section describes two of SYSTAT's F distribution functions: FCF(), for *F cumulative function*, which converts points on the F distribution into probabilities, and FIF(), for *F inverse function*, which converts probabilities into points on the F distribution. Each of these functions is discussed below. FRN(), which generates random numbers from an F distribution, and FDF(), which can be used for plotting F distributions (see Figure 2.3), are not discussed in this section. In this section, I will use '$F_{(n,d)}$' to denote points on an F distribution with n numerator degrees of freedom and d denominator degrees of freedom.

2.2.3.3.1 Finding Probabilities from F

Nondirectional ("two-tailed") probabilities. FCF() is a *cumulative* function, which gives the area under the F curve from zero *up to* the value placed within the parentheses of FCF(). That is, FCF() takes any positive number, along with its numerator and denominator degrees of freedom, and calculates the proportion of that F distribution that would fall below that value. To illustrate, Figure 2.3 shows an F distribution with 3 df in the numerator and 12 df in the denominator. If "4.0,3,12" is given as an argument to FCF(), this function will compute the proportion of the distribution to the *left* of 4.0 on the X-axis in Figure 2.3. In fact, this corresponds to .965 of the distribution. The area of the curve to the right of 4.0 in Figure 2.3, therefore, corresponds to $1 - .965 = .035$ of the distribution. This number is the "p-value" for an $F_{(3,12)}$ of 4.0: There is only a .035 chance of finding at random a value of $F_{(3,12)}$ greater than, or equal to, 4.0.

In SYSTAT, you can perform *p*-value computations in the data editor window. First, you should create a new column in your data set or open a temporary data file to contain the *p*-values. (If using a new file, you must give at least one column

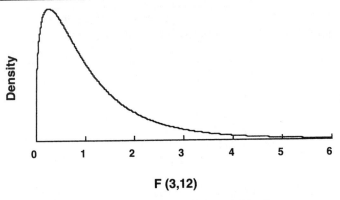

Figure 2.3 An *F* distribution with 3 and 12 *df*.

a label, say 'P', and fill at least one cell with a missing value or a number.) Then you can use the FCF() function to fill this column with the desired *p*-value(s). Suppose, for example, you want to find the *p*-value corresponding to $F_{(3,12)} = 4.0$: Simply type into the command window

```
>LET P = 1-FCF(4.0,3,12)
```

The cell or cells in column P will be filled with the number 0.035 (assuming you are using three decimal places), the desired *p*-value.

If you want to find more than one *p*-value, it is easy to compute them all at once. Simply create a column containing the *F*'s for which you want the probabilities, a second column containing their numerator *df*, and a third column containing their denominator *df*. You can then compute the entire column of *p*-values with one command. Table 2.20 shows a column of *F*'s labelled 'F', and their degrees of freedom in columns labelled NUM_DF and DENOM_DF. The fourth column shows the corresponding *p*-values, which were calculated with the command

```
>LET P = 1-FCF(F,NUM_DF,DENOM_DF)
```

Although we are usually only interested in the upper tail of the *F*, where values may be considered too large to be due to chance, *F*'s also have a lower "tail" or "end" (see Figure 2.3). If you want the probability of finding a value of *F* in the lower end of the distribution, just leave off the "1–" on the right side of the equation in the foregoing command. For example, the command

```
>LET P = FCF(.044,3,12)
```

MACINTOSH

① Choose the menu item **Data/Math...**

② Type the new column name in the "let variable" box on the left, in this case the column name *'P'*.

③ In the "=variable or expression" box on the right, enter the transformation expression, in this case *'1 – FCF(4.0,3,12)'*.

④ Click **OK**.

Table 2.20 The fourth column contains the probabilities of an F greater than, or equal to, the value given in the first column.

F	NUM_DF	DENOM_DF	P
0.058	1	12	0.814
3.490	3	12	0.050
2.333	4	45	0.070
1.657	2	30	0.216

gives the probability of finding an $F_{(3,12)}$ less than or equal to 0.044. The resulting *p*-value is equal to 0.013. This end of the distribution is rarely of interest, but values very near zero may be considered too small to be due to chance given the variance in the data. This has occasionally been used to help detect cooked data.

Directional ("one-tailed") probabilities for $F_{(1,d)}$. Importantly, there is a sense in which an F distribution with 1 *df* in the numerator has two-tails *within* its upper tail. An $F_{(1,d)}$ is just the square of a $t_{(d)}$. Squaring the $t_{(d)}$ distribution (see Figure 2.2), essentially "folds over" the negative tail onto the positive tail, and stretches them both out. Thus, an upper-tailed test on an $F_{(1,d)}$ distribution is equivalent to a two-tailed test on a *t* distribution with *d* degrees of freedom.

To illustrate, the data in Table 2.21 shows the results of a hypothetical study in which subjects' heart rates were monitored in two medical treatment conditions. The difference in mean heart rate between subjects in the two treatments is 14.7. You can compute an $F_{(1,14)}$ on the means in Table 2.21 to determine whether this difference is significantly different. The result is $F_{(1,14)} = 3.986$. As discussed in the previous section, you can find the probability of a value this large or larger due to chance with the command

>LET P = 1-FCF(3.986,1,14)

Table 2.21 Results from a hypothetical study in which subjects' heart rates were monitored in two medical treatment conditions.

	n	Mean heart rate	Variance
Treatment A	8	84.5	232.6
Treatment B	8	69.8	201.1

giving $p = .066$, which is not quite statistically significant at the traditional level. This p-value is the probability of finding *any difference* of the size in Table 2.21 or larger due to chance, regardless of the direction of the difference—regardless of which treatment it favors. Suppose, instead, that your theory made a strong prediction that the subjects in treatment B would have lower heart rates than the subjects in treatment A. In this case, you would classify differences in which the mean heart rate in treatment A was lower, along with nonsignificant differences, as failing to support your theory. Because only half of the differences, those favoring treatment B, would support your theory, the odds are only half as great that you will find support for your theory due to chance. Thus, the p-value given above is twice what it ought to be. To find the directional, "one-tailed," probability from the $F_{(1,14)}$ distribution, divide the p-value in half using the command

```
>LET P = (1-FCF(3.986,1,14))/2
```

The resulting $p = .033$, which is significant at the traditional .05 level. This is equivalent to taking the square root of 3.986 and finding the upper-tail probability on a $t_{(14)}$ distribution (see section 2.2.3.2.1), which can be done using the following command

```
>LET P = 1-TCF( SQR(3.986),14 )
```

This command computes the square root within the TCF() function using SYSTAT's SQR() function. Again, the resulting $p = .033$. When you are computing one-tailed tests from an $F_{(1,d)}$, it is usually best if you convert into $t_{(d)}$ and use a minus sign to indicate results in a direction opposite to your prediction or to other results. Simply reporting a "one-tailed test" from an $F_{(1,d)}$ will confuse many of your readers.

On one hand, there is a lot of potential to abuse one-tailed tests, which is why some folks cringe at their use. It is very tempting to look at your own data and say "Oh, of course! I could have predicted that!" and then proceed to divide all of your p-values by two. Keep in mind that fooling yourself about the existence of an effect in your data can waste your own time, as well as other people's. On the other hand, one-tailed tests have the advantage of greater power than two-tailed tests, so when a one-tailed test is justified, it would be a mistake not to use it. Fooling yourself about the nonexistence of an effect in your data can waste your own time, as well as other people's, and also cause you to miss potentially important results.

2.2.3.3.2 Finding Critical Values of F from α

Nondirectional ("two-tailed") critical values. Sometimes you may want to find the critical value of F corresponding to a given probability. This is easy in SYSTAT using FIF(), the *F inverse function*. This function takes a proportion (probability), a numerator *df*, and a denominator *df* as arguments and calculates the value on the $F_{(n,d)}$ distribution that would cut off that proportion of the distribution from zero up to that value; that is, FIF() gives the value of $F_{(n,d)}$ below which falls the given proportion of the distribution. For example, a value of $F_{(3,12)}$ equal to 3.490 cuts off the lower 95% of the distribution.

Therefore, if you want the value of $F_{(n,d)}$ that cuts off the *top* 5% of the distribution, you must ask SYSTAT to give the $F_{(n,d)}$ that cuts off the *lower* 95% of the distribution. The argument to FIF() should be "$1 - p$," where *p* is the desired probability. You can either do the subtraction yourself and put .95 in the parentheses of the FIF() function, or ask SYSTAT to do it for you in the command

```
>LET F = FIF(1-.05,NUM_DF, DENOM_DF )
```

where *num_df* is the corresponding numerator degrees of freedom, *denom_df* is the corresponding denominator degrees of freedom, and F is the column that will receive the $F_{(n,d)}$ value. For *df* = 3 and 12, SYSTAT will put the value 3.490 into the column labelled *F*. Using the "$1-p$" method is especially useful if you want to convert a column of *p*'s into their corresponding positive $F_{(n,d)}$'s. Table 2.22 shows the results of typing

```
>LET F_CRIT = FIF(1-P,NUM_DF, DENOM_DF)
```

where F_CRIT is my label for the new column of critical $F_{(n,d)}$ values, P is the column of probabilities, NUM_DF is the column of numerator degrees of freedom, and DENOM_DF is the column of denominator degrees of freedom. If you compare Tables 2.20 and 2.22, you will see that I just converted the *p*'s back into the *F*'s that I started with.

Table 2.22 The fourth column contains the upper-tailed values *F* corresponding to the probabilities given iin the first column.

P	NUM_DF	DENOM_DF	F_CRIT
0.814	1	12	0.058
0.050	3	12	3.490
0.070	4	45	2.333
0.216	2	30	1.657

To find a critical "lower-tail" value of $F_{(n,d)}$ that cuts off a given proportion of the lower end of the $F_{(n,d)}$ distribution, simply give the probability as an argument to FIF(). For example, the $F_{(3,12)}$ corresponding to the lower 5% of the distribution can be found by typing

```
>LET F_CRIT = FIF(.05,3,12)
```

With this command, SYSTAT will put the value 0.114 into the column labelled F_CRIT.

Directional ("one-tailed") critical values for $F_{(1,d)}$. Finding values of $F_{(1,d)}$ in the upper end of the distribution corresponding to one-tailed probabilities is identical to doing so for two-tailed probabilities, but you must remember to multiply the desired probability by two before entering it into the FIF() function. For example, if you want to find the value of $F_{(1,14)}$ corresponding to the 5% of the outcomes that result from a directional prediction, you must ask for the value that cuts off the upper 10% of the entire distribution. The following command multiplies the *p*-value by two within the argument to FIF():

```
>LET F_CRIT = FIF(1-P*2,1,14)
```

Again, F_CRIT is my label for the new column of critical $F_{(1,14)}$ magnitudes, and P now is a column of *one*-tailed probabilities. The resulting $F_{(1,14)} = 3.102$, which is the value of $F_{(1,14)}$ that cuts off the extreme 10% of the upper tails. Because only half of the values in this region correspond to the directional prediction, this value of $F_{(1,14)}$ corresponds to the one-tailed *p*-value of .05. See section 2.2.3.3.1 for further discussion of "one-tailed" *F* tests.

2.2.3.4 The χ^2 Distribution Functions

This section describes two of SYSTAT's χ^2 distribution functions: XCF(), for χ^2 *cumulative function*, which converts points on the χ^2 distribution into probabilities, and XIF(), for χ^2 *inverse function*, which converts probabilities into points on the χ^2 distribution. Each of these functions is discussed below. XRN(), which generates random numbers from a χ^2 distribution, and XDF(), which can be used for plotting χ^2 distributions (see Figure 2.4), are not discussed in this section. In this section, I will use '$\chi^2_{(df)}$' to denote points on a χ^2 distribution with degrees of freedom equal to the *df* in the subscript.

2.2.3.4.1 Finding probabilities from χ^2

Nondirectional ("two-tailed") probabilities. XCF() is a *cumulative* function, which gives the area under the χ^2 curve from zero *up to* the value placed within the parentheses of XCF(); that is, XCF() takes any positive number, along with its degrees of freedom, and calculates the proportion of that χ^2 distribution that would fall below that value. To illustrate, Figure 2.4 shows a χ^2 distribution with $df = 3$. If "7.0,3" is given as an argument to XCF(), this function will compute the proportion of the distribution to the *left* of 7.0 on the X-axis in Figure 2.4. In fact, this corresponds to .928 of the distribution. The area of the curve to the right of 7.0 in Figure 2.4, therefore, corresponds to $1 - .928 = .072$ of the distribution. This number is the "*p*-value" for a $\chi^2_{(3)}$ of 7.0: There is only a .072 chance of finding at random a value of $\chi^2_{(3)}$ greater than, or equal to, 7.0.

In SYSTAT, you can perform *p*-value computations in the data editor window. First, you should create a new column in your data set, or open a temporary data file to contain the *p*-values. (If using a new file, you must give at least one column a label, say 'P', and fill at least one cell with a missing value or a number.) Then you can use the XCF() function to fill this column with the desired *p*-value(s). Suppose, for example, you want to find the *p*-value corresponding to $\chi^2_{(3)} = 7.0$: Simply type into the command window

```
>LET P = 1-XCF(7.0,3)
```

The cell or cells in column P will be filled with the number 0.072 (assuming you are using three decimal places), the desired *p*-value.

MACINTOSH

① Choose the menu item **Data/Math...**

② Type the new column name in the "let variable" box on the left, in this case the column name *'P'*.

③ In the "=variable or expression" box on the right, enter the transformation expression, in this case *'1 – XCF(7.0,3)'*.

④ Click **OK**.

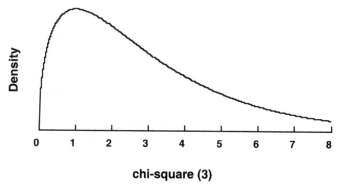

Figure 2.4 A χ^2 distribution with 3 *df*.

If you want to find more than one p-value, it is easy to compute them all at once. Simply create a column containing the χ^2's for which you want the probabilities and a second column containing their degrees of freedom. You can then compute the entire column of p-values with one command. Table 2.23 shows a column of χ^2's labelled X2 and their degrees of freedom in a column labelled DF. The third column shows the corresponding upper-tailed p-values, which were calculated with the command

```
>LET P = 1-XCF(X2,DF)
```

Table 2.23 The third column contains the upper-tailed probabilities of a χ^2 greater than, or equal to, the value given in the first column.

X2	DF	P
4.136	4.000	0.388
7.815	3.000	0.050
8.111	9.000	0.523
2.856	1.000	0.091

Although we are usually only interested in the upper tail of the χ^2, where values may be considered too large to be due to chance, χ^2's also have a lower "tail" or "end" (see Figure 2.4). If you want the probability of finding a value of χ^2 in the lower end of the distribution, just leave off the "1–" on the right side of the equation in the foregoing command. For example, the command

```
>LET P = XCF(.15,3)
```

gives the probability of finding a $\chi^2_{(3)}$ less than or equal to 0.15. The resulting p-value is equal to 0.015. This end of the distribution is rarely of interest, but values very near zero may be considered too small to be due to chance given the variance in the data. This has occasionally been used to help detect cooked data.

Directional ("one-tailed") probabilities for $\chi^2_{(1)}$. Importantly, there is a sense in which the $\chi^2_{(1)}$ distribution has two tails *within* its upper tail. A $\chi^2_{(1)}$ is just the square of a normal distribution. Squaring the normal distribution (see Figure 2.1) essentially "folds over" the negative tail onto the positive tail and stretches them both out. Thus, an upper-tailed test on a

$\chi^2_{(1)}$ distribution is equivalent to a two-tailed test on a normal distribution.

Table 2.24 Results from a hypothetical survey in which 20 Democrats and 16 Republicans were asked a simple yes or no question.

	Yes	No	Total
Democrats	16	4	20
Republicans	8	8	16
Total	24	12	36

To illustrate, the data in Table 2.24 shows the results of a hypothetical survey in which 20 Democrats and 16 Republicans were asked a simple yes or no question. 80% of the democrats responded yes, whereas only 50% of the Republicans responded yes. You can compute a $\chi^2_{(1)}$ on the table of counts in Table 2.24 to determine whether these proportions are significantly different. The result is $\chi^2_{(1)} = 3.60$. As discussed in the previous section, you can find the probability of a value this large or larger due to chance with the command

```
>LET P = 1-XCF(3.6,1)
```

giving $p = .058$, which is not quite statistically significant at the traditional level. This is the probability of finding any *difference* of the size in Table 2.24 or larger due to chance, regardless of the direction of the difference, that is, regardless of which party it favors. Suppose, instead, that your theory made a strong prediction that proportionally more Democrats would respond yes in this survey. In this case, you would classify differences favoring the Republicans, along with nonsignificant differences, as failing to support your theory. Because only half of the possible differences, those favoring Democrats, would support your theory, the odds are only half as great that you will find support for your theory due to chance. Thus, the p-value given above is twice what it ought to be. To find the directional, "one-tailed," probability from the $\chi^2_{(1)}$ distribution, divide the p-value in half using the command

```
>LET P = (1-XCF(3.6,1))/2
```

The resulting $p = .029$, which is significant at the traditional .05 level. This is equivalent to taking the square root of 3.6 and finding the upper-tail probability on a normal distribution (see section 2.2.3.1.1), which can be done using the following command

```
>LET P = 1-ZCF( SQR(3.6) )
```

This command computes the square root within the ZCF() function using SYSTAT's SQR() function. Again, the resulting $p = .029$. When you are computing one-tailed tests from a $\chi^2_{(1)}$, it is usually best if you convert into Z's and use the minus sign to indicate results in a direction opposite to your prediction or to other results. Simply reporting a "one-tailed test" from a $\chi^2_{(df)}$ will confuse many of your readers.

To repeat what I have stated previously on one-tailed tests, there is a lot of potential to abuse one-tailed tests, which is why some folks cringe at their use. It is very tempting to look at your own data and say "Oh, of course! I could have predicted that!" and then proceed to divide all of your p-values by two. Keep in mind that fooling yourself about the existence of an effect in your data can waste your own time, as well as other people's. However, one-tailed tests have the advantage of greater power than two-tailed tests, so when a one-tailed test is justified, it would be a mistake not to use it. Fooling yourself about the nonexistence of an effect in your data can waste your own time, as well as other people's.

2.2.3.4.1 Finding Critical Values of χ^2 from α

Nondirectional ("two-tailed") critical values. Sometimes you may want to find the critical value of χ^2 corresponding to a given probability. This is easy in SYSTAT using XIF(), the χ^2 *inverse function*. This function takes a proportion (probability) and degrees of freedom as arguments, and calculates the value on the $\chi^2_{(df)}$ distribution that would cut off that proportion of the distribution from zero up to that value. That is, XIF() gives the value of $\chi^2_{(df)}$ below which falls the given proportion of the distribution. For example, a value of $\chi^2_{(3)}$ equal to 7.815 cuts off the lower 95% of the distribution.

Therefore, if you want the value of $\chi^2_{(df)}$ that cuts off the *top* 5% of the distribution, you must ask SYSTAT to give the $\chi^2_{(df)}$ that cuts off the lower 95% of the distribution; that is, the argument to XIF() should be "$1 - p$," where p is the desired probability. You can either do the subtraction yourself and put .95 in the parentheses of the XIF() function, or ask SYSTAT to do it for you in the command

```
>LET X2 = XIF(1-.05,DF )
```

where *df* is either the value of, or the column containing, the corresponding degrees of freedom, and X2 is the column that will receive the $\chi^2_{(df)}$ value. For $df = 3$, SYSTAT will put the value 7.815 into the column labelled X2. Using the "1 – p" method is especially useful if you want to convert a column of *p*'s into their corresponding positive $\chi^2_{(df)}$'s. Table 2.25 shows the results of typing

```
>LET X2_CRIT = XIF(1-P,DF)
```

where X2_CRIT is my label for the new column of critical $\chi^2_{(df)}$ values, P is the column of probabilities, and DF is the column of degrees of freedom. If you compare Tables 2.23 and 2.25, you will see that I just converted the *p*'s back into the $\chi^2_{(df)}$'s that I started with.

Table 2.25 The third column contains the upper-tailed values of $\chi^2_{(df)}$ corresponding to the probabilities given in the first column.

P	DF	X_CRIT
0.388	4.000	4.136
0.050	3.000	7.815
0.523	9.000	8.111
0.091	1.000	2.856

To find a critical "lower-tail" value of $\chi^2_{(df)}$ that cuts off a given proportion of the lower end of the $\chi^2_{(df)}$ distribution, simply give the probability as an argument to XIF(). For example, the $\chi^2_{(3)}$ corresponding to the lower 5% of the distribution can be found by typing

```
>LET X2 = XIF(.05,3)
```

With this command, SYSTAT will put the value 0.352 into the column labelled X2.

Directional ("one-tailed") critical values for $\chi^2_{(1)}$. Finding values of $\chi^2_{(1)}$ in the upper end of the distribution corresponding to one-tailed probabilities is identical to doing so for two-tailed probabilities, but you must remember to multiply the desired probability by two before entering it into the XIF() function. For example, if you want to find the value of $\chi^2_{(df)}$ corresponding to the 5% of the outcomes that result from a directional prediction, you must ask for the value that cuts off the upper

10% of the entire distribution. The following command multiplies the *p*-value by two within the argument to XIF():

```
>LET X2_CRIT = XIF(1-P*2,1)
```

Again, X2_CRIT is my label for the new column of critical $\chi^2_{(1)}$ magnitudes, and P now is a column of *one-tailed* probabilities. The resulting $\chi^2_{(1)} = 6.251$, which is the value of $\chi^2_{(1)}$ that cuts off the extreme 10% of the upper tails. Because only half of the values in this region correspond to the directional prediction, this value of $\chi^2_{(1)}$ corresponds to the one-tailed *p*-value of .05. See section 2.2.3.4.1 for further discussion of "one-tailed" $\chi^2_{(1)}$ tests.

2.2.3.5 The Binomial Distribution Functions

This section describes two of SYSTAT's binomial distribution functions: NCF(), for *biNomial Cumulative Function*, which converts points on the binomial distribution into probabilities, and NIF(), for *biNomial Inverse Function*, which converts probabilities into points on the binomial distribution. NRN(), which generates random numbers from a binomial distribution, and NDF(), which can be used for plotting binomial distributions (see Figure 2.5), are not discussed in this section. Binomial events can have either of two mutually exclusive outcomes, like heads or tails in a coin toss, and I will refer to these generically as "successes" and "failures," respectively. (Gregory D. Kirby, my brother, once dropped a quarter that came to rest on its edge on a flat countertop. So much for binary outcomes.) In this section, I will use '$N_{(n, p)}$' to denote points on a binomial distribution N with n trials and a probability of success on any trial equal to p. For example, 20 coin tosses would give the binomial distribution $N_{(20, 0.5)}$.

2.2.3.5.1 Finding Probabilities of Binomial Outcomes

One-tailed (directional) probabilities. In SYSTAT, you can perform *p*-value computations in the data editor window. First, you should create a new column in your data set, or open a temporary data file to contain the *p*-values. (If using a new file, you must give at least one column a label, say 'P', and fill at least one cell with a missing value or a number.) Then you can use the NCF() function to fill this column with the desired *p*-value(s).

NCF() takes three arguments: the number of successes, the number of trials, and the probability of a chance success, and calculates the proportion of that binomial distribution less

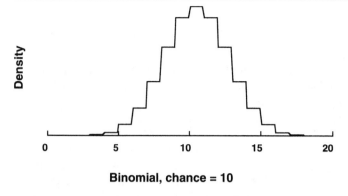

Figure 2.5 A binomial distribution with 20 trials and the probability of chance successes equal to 0.5. (The vertical lines look slanted, but really go "straight up.")

than or equal to the given number of success; that is, NCF() is a *cumulative* function that gives the sum of the probabilities for each number of successes in *n* trials, from zero *up to and including* the number of successes given in the first argument to NCF(). To illustrate, Figure 2.5 shows a binomial distribution with 20 trials and a probability of chance success equal to 0.5. If "14, 20, 0.5" is given as an argument to NCF(), the function will add up the probabilities of each number of successes from 0 through 14. This is equivalent to computing the proportion of the distribution to the *left* of 15 on the X-axis in Figure 2.5. This proportion is .979 of the distribution, which is also the probability of obtaining *14 or fewer* successes due to chance in this distribution. The probability of obtaining *more than (but not including) 14* successes equals 1 − .979 = .021.

Usually, we are more interested in the probability of obtaining *x or more* successes in *n* trials. NCF() does not give us this probability directly, but we can calculate it by making use of the fact that the probability of *x* or more successes in *n* trials is equal to the probability of *n − x or fewer failures* in *n* trials. Thus, we essentially redefine the distribution in terms of failures and have SYSTAT calculate the probability of finding *n − x or fewer* failures in *n* trials, with the probability of failure equal to 1 − *p*, where *p* is still the probability of chance successes. Generically, the command looks like this:

```
>LET P_VALUE = NCF( N-X, N, 1-P)
```

where P_VALUE is the label of the column that will contain the *p*-value(s). For example, the probability of 14 or more successes (6 or fewer failures) in 20 trials with the probability of a chance success equal to 0.5, can be computed by

---

MACINTOSH

① Choose the menu item **Data/Math...**

② Type the new column name in the "let variable" box on the left, for example, the column name *'P_VALUE'*.

③ In the "=variable or expression" box on the right, enter the transformation expression, for example, *'NCF(N−X, N, 1−P)'*.

④ Click OK.

```
>LET P_VALUE = NCF( 20-14, 20, 1-0.5)
```

which will fill the column P_VALUE with the probability 0.058.

If you want to compute more than one *p*-value, it is easy to compute them all at once. Simply create a column containing the numbers of trials, a column containing the numbers of success, and a column containing the probabilities of chance successes. You can then compute the entire column of *p*-values with one command. Table 2.26 shows a column of numbers of trials labelled N, a column of successes labelled X, and a column of probabilities of chance successes labelled P_SUCCES. The fourth column shows the corresponding probabilities of obtaining *X* or more successes, which were calculated with the command

```
>LET P_VALUE = NCF( N-X, N, 1-P_SUCCES )
```

In this table, all of the *n*'s equal five and all of the probabilities of successes equal approximately 1/6. Therefore, the values in column P_VALUE correspond to the probabilities of rolling five dice and obtaining, say, *x* or more 1's. For example, the probability of rolling five dice and obtaining three or more 1's rounds to .0355. It would only happen about 1 in 28 times.

Table 2.26 The fourth column contains the probabilities of obtaining *X* or more successes in *N* trials, where the probability of a chance success is equal to the values in P_SUCCES.

N	X	P_SUCCES	P_VALUE
5	0	0.1667	1.0000
5	1	0.1667	0.5982
5	2	0.1667	0.1963
5	3	0.1667	0.0355
5	4	0.1667	0.0033
5	5	0.1667	0.0001

Two-tailed (nondirectional) probabilities, symmetric distributions. A *p*-value in Table 2.26 corresponds to the probability of finding *x* or more successes in *n* trials. Sometimes, however, you may be interested in calculating the chance probability of finding any outcome "at least as improbable" as a particular number of successes *x*, regardless of the direction of the deviation of the outcome from chance. When the distribution is symmetric, that is, the probability of a success is 0.5, you can find this probability by simply multiplying the probability of finding x or more successes in n trials by two.

For example, suppose that you flipped a coin ten times in an attempt to determine whether it might be biased, but you did not make a strong prediction about whether it would be biased towards heads or tails. Suppose that you obtained eight heads. The probability of obtaining eight heads or more in ten coin tosses rounds to .055, which can be verified using the command from the previous section:

```
>LET P_VALUE = NCF( 10-8, 10, 0.5)
```

However, obtaining many tails might also be indicative of bias, and the probability of getting eight or more tails in ten coin tosses is also .055. So, to find the probability of obtaining any outcome at least as extreme as x successes, you simply multiply the probability of obtaining x or more successes by two, which gives $2 \times .055 = .110$. This suggests that outcomes that extreme are not terribly unusual.

Two-tailed (nondirectional) probabilities, asymmetric distributions. The computation of two-tailed p-values is considerably trickier when a binomial distribution is asymmetric, i.e., when $p \neq 0.5$. It it much easier (and probably wiser) just to find critical values for the desired level of significance (see the latter part of section 2.2.3.5.2) than to compute exact two-tailed probabilities for asymmetric distributions.

In asymmetric distributions you cannot simply multiply the probability of x or more successes by two to find the probability of obtaining any outcome at least as improbable as x. This is because there may not exist a number of failures y for which y or more failures is *exactly* as improbable as x or more successes. In fact, there may not exist a number of failures y for which y or more failures is even *at least* as improbable as x or more successes! Instead, you must *add* the probability of x or more successes to the probability of y or more failures, where the latter, if it exists, is at least as improbable as the former.

For example, suppose that you rolled a particular die 24 times to determine whether it might be biased for or against rolling a 1. If the die is fair, the probability of rolling a 1 is 1/6, so the expected number of 1's in 24 rolls is $24 \times 1/6 = 4$. Suppose instead that you rolled eight 1's. The probability of rolling eight or more 1's in 24 trials can be found with the command from section 2.2.3.5.1:

```
>LET P_VALUE = NCF( 24-8, 24, 1-1/6)
```

The resulting probability is .035.

How few 1's would be at least as improbable an outcome as eight or more 1's? Table 2.27 shows the probabilities of obtaining three or fewer, two or fewer, one or fewer, or zero 1's in 24 rolls of a fair die. These probabilities were found with the command

```
>LET P = NCF( X, 24, 1/6)
```

where X is the label of the column containing the number of 1's and P is the column in which SYSTAT will put the probabilities. Only zero or fewer 1's is "at least as improbable" as eight or more 1's (which was .035). Therefore, the two-tailed p-value for eight or more 1's, that is, the probability of obtaining any outcome at least as improbable as eight or more 1's, is .035 + .013 = .048. So an outcome of eight or more 1's meets the traditional two-tailed .05 significance level.

Table 2.27 The third column gives the probability of obtaining X or fewer 1's in 24 rolls of a fair die.

Number of 1's in 24 rolls	X	P
three or fewer 1's	3	0.416
two or fewer 1's	2	0.212
one or fewer 1's	1	0.073
zero 1's	0	0.013

What if you had instead rolled ten 1's? The probability of rolling ten or more 1's in 24 trials is about 0.003. None of the probabilities in Table 2.27 are this small. In other words, there is no number y of 1's for which y or fewer 1's is at least as improbable as ten or more 1's! Thus, the two-tailed p-value for ten or more 1's is .003 + 0 = .003. The one-tailed and two-tailed probabilities are equal! This is a curious property of binomial distributions that results from the fact that the outcomes are discrete and the lower tail ends at zero.

Finding the probability of exactly x successes. Sometimes you may be interested in finding the probability of exactly x successes exactly in n trials. This can be computed in a few different ways, but the easiest is to find the probability of x or fewer successes, and subtract from this the probability of $x - 1$ or fewer successes. The generic command is

```
>LET P_X = NCF(X,N,P) - NCF(X-1,N,P)
```

where P_X is the label of the column that will contain the probability of exactly x successes in n trials, and p is the probability of a chance success. You can either use the variables X, N, and P as column labels to compute more than one critical value at once, or replace them with the particular values that you want to use.

For example, to find the probability of obtaining exactly 14 heads in 20 coin tosses, you would use the command

```
>LET P_X = NCF(14,20,.5) - NCF(14-1,20,.5)
```

With this command SYSTAT will put the correct p-value of .037 into the column P_X.

2.2.3.5.2 Finding Critical Values of Binomial Outcomes from α

One-tailed (directional) critical values. Sometimes you may want to find the critical number of successes (or failures) necessary to meet a particular level of significance. This is easy in SYSTAT using NIF(), the *biNomial Inverse Function*. This function takes a proportion (probability), a number of trials, and the probability of a chance success, and calculates the value on the binomial distribution that would cut off that proportion of the distribution in the lower tail; that is, NIF() computes the number of successes x which falls at the upper end of the desired proportion of the distribution.

Therefore, if you want to compute the number of successes necessary to meet or exceed a particular significance level α, you must ask SYSTAT to compute the number of *failures* that cuts off the lower $1 - \alpha$ of the distribution and subtract this value from the number of trials n. This is done with the command

```
>LET X = N - NIF(ALPHA,N,1-P)
```

where N is the number of trials, ALPHA is the desired significance level, and P is the probability of a chance success. You can either use the variables N, ALPHA, and P as column labels to compute more than one critical value at once, or replace them with the particular values that you want to use.

For example, suppose you suspect that a particular coin is biased in favor of heads, and you toss it 20 times to test this suspicion. How many times would the coin have to come up heads to meet or exceed the .05 level of significance? SYSTAT will give the answer with the command

```
>LET X = 20 - NIF(.05,20,1-.5)
```

which puts the number 14 into the column labelled X. In 20 coin tosses, you would expect to see 14 or more heads less than 5% of the time due to chance.

Two-tailed (nondirectional) critical values, symmetric distributions. When the binomial distribution is symmetric, that is, the probability of a chance success is .5, finding values of binomial corresponding to two-tailed probabilities is similar to doing so for one-tailed probabilities, but you must remember to divide the desired alpha level by two before entering it into the NIF() function. For example, if you want to find the critical numbers of successes and failures that cut off the extreme $\alpha = 0.01$ of the distribution, you must ask SYSTAT to compute the values that cut off the extreme upper 0.005 and lower 0.005 of the distribution. The following command divides the α level by two within the argument to NIF():

```
>LET X = N - NIF(ALPHA/2,N,0.5)
```

where N is the number of trials, ALPHA is the desired significance level, and 0.5 is the probability of a chance success for symmetric distributions. You can either use the variables N and ALPHA as column labels to compute more than one critical value at once, or replace them with the particular values that you want to use. For example, for $\alpha = .01$ and $n = 20$, the command is

```
>LET X = 20 - NIF(.01/2,20,.5)
```

which puts the value 16 into the column labelled X. This means that you would need either 16 successes or 16 failures in 20 trials to meet the .01 level of significance, two-tailed.

Two-tailed (nondirectional) critical values, asymmetric distributions. When the binomial distribution is not symmetric, i.e., the probability of success is not 0.5, the number of successes and failures needed to cut off the upper $\alpha/2$ and lower $\alpha/2$ of the distribution may differ. Thus, you need to find them separately. We have already seen how to do this in the previous section: To find the critical number of successes, type

```
>LET SUCC = N - NIF(ALPHA/2,N,1-P)
```

and to find the critical number of failures type

```
>LET FAIL = N - NIF(ALPHA/2,N,P)
```

where N is the number of trials, ALPHA is the desired significance level, and P is the probability of a chance success. Suppose that the probability of a chance success is 0.3. For $\alpha = .01$ and $n = 20$, these two commands are

```
>LET SUCC = 20 - NIF(.01/2,20,1-.3)
>LET FAIL = 20 - NIF(.01/2,20,.3)
```

which give, respectively, 12 as the number of successes and 19 as the number of failures that would be needed to meet the .01 level of significance, two-tailed.

As usual, you can compute an entire column of probabilities with one pair of commands. The first three columns in Table 2.28 give the number of trials, the desired two-tailed α level, and the probability of a chance success, respectively, for four separate cases. The fourth and fifth columns were generated with the following commands

```
>LET SUCC = N - NIF(ALPHA/2,N,1-PS)
>LET FAIL = N - NIF(ALPHA/2,N,PS)
```

With N = 20 and the probability of success at 0.4, Table 2.28 shows how the numbers of successes and failures required for significance change as the desire α level decreases. For example, to meet $\alpha = .001$, two-tailed, you would either need to obtain 15 or more successes or 19 or more failures. These are the values that cut off the extreme 0.0005 of their respective tails.

Table 2.28 The fourth and fifth columns contain the numbers of successes and failures, respectively, needed to meet the two-tailed level of significance given in the column ALPHA.

N	ALPHA	PS	SUCC	FAIL
20	0.10	0.4	12	16
20	0.05	0.4	12	16
20	0.01	0.4	14	17
20	0.001	0.4	15	19

3
Describing Data in SYSTAT

This chapter shows how to compute summary statistics in SYSTAT and how to use SYSTAT's graphics procedures to plot data in various ways. To illustrate these procedures, this chapter uses two data sets, the SHOES data set, which will also be used to illustrate SYSTAT's regression procedures in Chapters 5, 6, and 7, and the TEST SCORES data set, which will be used for one-way ANOVA, Chapter 8. The first five cases of the SHOES data set are shown in Table 3.1 (the rest are shown in section 5.6). The TEST SCORES data set is shown in Table 3.2.

3.1 Summary Statistics

3.1.1 By Columns

SYSTAT offers 13 summary statistics for single variables, all of which can be generated with a single command (older versions of SYSTAT had only 12, and did not include the median). To

Table 3.1 The first 5 of 100 cases in the SHOES example data set (see section 5.6).

CASE	Family	Income (X)	Pairs of Shoes (Y)
1	Smith	18,244	9
2	Jones	21,883	7
3	Lincoln	15,439	11
4	Douglas	16,915	16
5	Gates	42,742	15
...			

Table 3.2 The TEST SCORES data set: test scores for students in drug dose conditions.

Group 1	Group 2	Group 3
94.4	95.3	98.1
75.7	117.2	101.2
88.1	97.9	120.1
108.7	82.7	77.5
94.8	105.3	124.7
130.6	85.2	136.1
121.1	86.7	132.6
82.9	104.8	130.5
112.0	67.9	130.0
85.2	106.1	138.8
98.7		105.8
50.1		70.4
86.1		
99.8		
121.8		

generate all 13 statistics for the shoes and income variables, open the SHOES data set and type the following commands:

```
>STATS
>STATISTICS SHOES INCOME/ALL
```

MACINTOSH

① Choose the menu item **Stats/Stats/Statistics...**
② Select SHOES and INCOME from the variable list.
③ Check the **All** option box.
④ Click **OK**.

MS-DOS

① Choose the menu item **Statistics/Stats/Statistics/Variables/**.
② Select SHOES and INCOME from the variable list.
③ Choose the menu item **Statistics/Stats/Statistics/Options/All**.
④ Choose the menu item **Statistics/Stats/Statistics/Go!**

The "all" at the end of the second command line tells SYSTAT to compute all 13 statistics. If you want only some of the statistics to be computed, you simply specify those that you want after the slash in the statistics command. All of the statistics are shown in the following command:

```
>STATISTICS SHOES INCOME/ CV KURTOSIS MAXIMUM ,
MEAN  MINIMUM  N RANGE SD  SEM  SKEWNESS  SUM ,
VARIANCE  MEDIAN
```

Note that a comma at the end of a command line allows the command to continue onto the next line. This works with all SYSTAT commands.

The output generated by these commands is shown in Table 3.3. Brief descriptions and formulas for the statistics in Table 3.3 are given in Table 3.4. The meaning of most of these statistics can be found in any statistics textbook, but skewness and kurtosis, which tell us about the shapes of distributions, are not as well known as the rest, so I will briefly describe them here.

Table 3.3 Summary statistics of shoes and income in the SHOES data set.

TOTAL OBSERVATIONS:	100	
	SHOES	INCOME
N OF CASES	100	100
MINIMUM	1.000	1533.000
MAXIMUM	26.000	122855.000
RANGE	25.000	121322.000
MEAN	10.140	20676.310
VARIANCE	22.768	.296682E+09
STANDARD DEV	4.772	17224.465
STD. ERROR	0.477	1722.446
SKEWNESS(G1)	0.890	2.775
KURTOSIS(G2)	1.170	11.746
SUM	1014.000	2067631.000
C.V.	0.471	0.833
MEDIAN	9.000	17111.000

Skewness is positive when the distribution is positively skewed and negative when the distribution is negatively skewed. For normally distributed data and sample sizes greater than about 150, skewness is approximately normally distributed with mean zero and standard deviation $\sqrt{6/n}$. By dividing the value of skewness by its standard deviation to obtain a Z, you can find an approximate probability of getting a skewness of that magnitude or larger due to chance (see section 2.2.3.1.1). For sample sizes less than 500 this gives an inaccurate value of the probability. More accurate values for sample sizes up to 500 can be found in Snedecor and Cochran (1980, Appendix 20i).

For the example data summarized in Table 3.3, skewness for the shoes variable is 0.89, with a standard deviation of $\sqrt{6/100}$ = 0.245. Dividing the former by the latter to obtain an approximate value of Z gives 3.63, which is very unlikely due to chance. For the income variable $Z = 11.3$, which is extremely high. Thus, the population distributions of both variables appear to be positively skewed.

Kurtosis is positive when a distribution has thicker tails than a normal distribution (t distributions, for example). Distributions with negative kurtosis have short tails and are "flat-topped." For normally distributed data and sample sizes greater than 1,000, kurtosis is approximately normally distributed

Table 3.4 Descriptions and formulas for the statistics computed by SYSTAT.

N of cases n
The number of nonmissing data points for each variable.

Range $max - min$
The difference between the *minimum* and *maximum* values for each variable.

Mean $\overline{X} = \Sigma X_i / n$
The sum divided by the *N of cases*.

Variance $\hat{\sigma}^2 = SS/(n-1)$
Of a sample, the sum-of-squares divided by *N of cases* minus 1.

Standard dev $\hat{\sigma} = \sqrt{SS/(n-1)}$
The "standard deviation" of a sample—the square-root of the sample variance.

Std error $\sqrt{\hat{\sigma}^2 / n}$
The "standard error of the mean"—equal to the square-root of the sample variance divided by the *N of cases*.

C.V. $\hat{\sigma} / \overline{X}$
The "coefficient of variation"—equal to the sample standard deviation divided by the *mean*.

Skewness (G1) $\dfrac{\Sigma(X_i - \overline{X})^3 / n}{\sigma^3}$

A measure of the symmetry of a distribution (see text).

Kurtosis (G2) $\dfrac{\Sigma(X_i - \overline{X})^4 / n}{\sigma^4}$

A measure of the "flatness" of a distribution (see text).

Median Q_2
The score that divides the data in half; the 50th centile.

Notes: $\sigma^2 = SS/n$; $\hat{\sigma}^2 = SS/(n-1)$; $SS = \Sigma(X_i - \overline{X})^2$.

with mean 0 and standard deviation $\sqrt{24/n}$ and can be compared to a Z distribution to find approximate probabilities (see section 2.2.3.1.1). For smaller sample sizes, probabilities obtained in this way will be somewhat inaccurate—more so for negative than positive values of kurtosis. More accurate values for small sample sizes can be found in Snedecor and Cochran (1980, Appendix 20ii). In the example data in Table 3.3, the value of kurtosis for shoes is 1.17, with a standard deviation of $\sqrt{24/100} = 0.490$. Dividing kurtosis by its standard deviation gives $Z = 2.39$, which is unlikely due to chance, suggesting that the shoes distribution is thick-tailed. For the income variable, kurtosis equals 11.746 with standard deviation 0.490, which gives the very unlikely value of $Z = 24$. Thus, the income distribution is also thick-tailed.

For the shoes and income variables, skewness and kurtosis do not tell us much that we could not see in the plots of those variables in the next section. However, they do allow a rough test of what we think we see in the plots, and in less clear circumstances they can be very informative.

3.1.2 By Groups

The previous section describes how to find statistics for entire columns of data, without regard to any groupings within those columns. To compute statistics, or perform any analysis in SYSTAT within groups in the data set, you must use the *by* command followed by one or more categorical grouping variables. Thus, the generic command looks like this:

```
>BY GROUPINGVARIABLE1 GROUPINGVARIABLE2 ...
```

After this command is issued, all calculations will be repeated within each group in the data set. To turn the groupings off, just type

```
>BY
```

3.2 Using Graphics to Describe Data

This section illustrates some of the graphics procedures in SYSTAT that will be used throughout this book. Each type of graph is described, along with some options that are particularly relevant to that type of graph. The final section, 3.2.4, gives some of the common options for controlling the layout of graphs. The procedures below will be demonstrated using the SHOES data set (Table 3.1), which contains two continuous

---

MACINTOSH

① Choose the menu item **Data/By Groups...**

② Select the grouping variable(s) from the variable list, and click **OK**.

MS-DOS

① Choose the menu item **Data/By Groups/Select**.

② Select the grouping variable(s) from the variable list.

MACINTOSH

① Choose the menu item **Data/By Groups...**

② Click on *By groups off*, and click **OK**.

MS-DOS

① Choose the menu item **Data/By Groups/Clear**.

variables, and the TEST SCORES data set (Table 3.2), which has one continuous and one categorical variable.

3.2.1 Plotting Single Variables

3.2.1.1 Histograms

A histogram shows the distribution of data points across a single variable. This distribution is constructed by dividing the variable scale into some number of categories and then counting the number of data points that fall into each of those categories. The height of a bar above each category represents the number of cases that fall into that category. Figure 3.1 shows histograms for the income variable from the SHOES example data set. This plot was created by typing

> ▶ >GRAPH
> >DENSITY INCOME /HIST

into the command window. The *hist* option after the slash tells SYSTAT to generate a histogram. Actually, you could have left this option off because histogram is the default density plot. Another useful type of density plot is the "dit" plot, which creates a stack of small circles (dits) rather than bars as in a histogram (see section 3.2.1.4 for an example).

SYSTAT allows you to control the bar width and the number of bars, but will automatically adjust the number of bars for you if you do not specify a number. To control the number of bars, simply append to the command line a slash and "bars=" followed by the number of bars you wish. For example, if you want 30 bars, type

> ▶ >DENSITY INCOME /BARS=30

But keep in mind that the number of bars includes bars of zero height (intervals with no cases). If you want to control the width of the *intervals* on the *X*-axis, append to the command line a slash and "bwidth=" (for *bar width*) and the size of the interval. In Figure 3.1 SYSTAT's default was to create bars with a width of $10,000. If you want to show bars representing intervals of $5,000 in the income histogram, type in the command window

---

MACINTOSH

① Choose the menu item **Graph/Density/Histogram**.

② Select INCOME from the variable list.

③ Click **OK**.

MS-DOS

① Choose the menu item **Graph/Density/Variables/Dependent/**.

② Select INCOME from the variable list.

③ Choose the menu item **Graph/Density/Type/Hist**.

④ Choose the menu item **Graph/Density/Go!**

MACINTOSH

① Choose the menu item **Graph/Density/Histogram**.

② Select INCOME from the variable list.

③ Click on **Bars**, enter "30" into the dialog box, and click **OK**.

④ Click **OK**.

MS-DOS

① Choose the menu item **Graph/Density/Variables/Dependent/**.

② Select INCOME from the variable list.

③ Choose the menu item **Graph/Density/Type/Hist**.

④ Choose the menu item **Graph/Density/Options/Bars** and type "30".

⑤ Choose the menu item **Graph/Density/Go!**

```
>DENSITY INCOME/BWIDTH=5000
```

From the heights of the bars you can read off the number of cases falling in a given income interval. For example, Figure 3.1 indicates that 37 families had incomes between $10,000 and $19,999. Because there are exactly 100 cases in this data set, this corresponds to 37% of the cases as shown by the left Y-axis in Figure 3.1.

WARNING: WHAT TO DO WITH NEW PLOTS

When you produce a second plot in SYSTAT, it gives you the option to **Overlay plot**, **Stack plot**, or **Next plot**. *Overlay plot* superimposes the new plot on top of the old plot. *Stack plot* creates a scrolling window and adds the new plot below the old plot; this allows you to scroll back and forth between the plots and save or print them as a group. *Next plot* deletes the old plot or plots and replaces them with the new plot. Do not select *Next plot* unless you are ready to delete the work you have already done.

One interesting feature of Figure 3.1 is that the distribution of incomes is very asymmetric, with a large positive skew and a sudden drop off at zero dollars. Therefore, the underlying income distribution may not be normally distributed, as

MACINTOSH
① Choose the menu item **Graph/Density/Histogram**.
② Select INCOME from the variable list.
③ Click on **Bwidth**, enter "5000" into the dialog box, and click **OK**.
④ Click **OK**.

MS-DOS
① Choose the menu item **Graph/Density/Variables/Dependent/**.
② Select INCOME from the variable list.
③ Choose the menu item **Graph/Density/Type/Hist**.
④ Choose the menu item **Graph/Density/Options/BWidth** and type "5000".
⑤ Choose the menu item **Graph/Density/Go!**

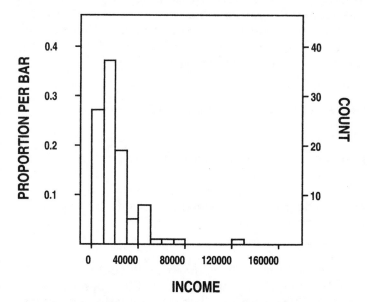

Figure 3.1 Histogram of income from the SHOES example data set.

suggested by the values of skewness and kurtosis for the income variable computed above in section 3.1. Another interesting feature of Figure 3.1 is that one of the cases, the family with an income between $120,000 and $130,000 stands apart from the rest of the group. You would want to keep an eye on this case in any further analysis to determine whether it has an undue influence on your results.

SYSTAT has a number of other density graphs: fuzzygrams, jitter plots, dit and dot plots, polygon plots, and stripe plots. Any of these may reveal something new to your eye that was not apparent in the simple histogram, but they will not be discussed further in this book.

HINT: AVOIDING COMMAND ERRORS

The error message

```
***ERROR*** ABOUT HERE SYSTAT EXPECTS A
SYSTAT COMMAND
```

is generated when you mistype a command, or when you type a proper command in the wrong place. SYSTAT originally was made up of a number of separate "modules," such as STATS and GRAPH, that each contained its own command set. Most STATS commands would not work in GRAPH, and vice versa. When using menus, this structure is implicit in the organization of the menu items, but when using the command interface, you must remember to switch to the module in which the desired command is located. So, to execute statistics commands, first type

```
> STATS
```

To execute graphics commands, first type

```
> GRAPH
```

To use SYSTAT's powerful data module, first type

```
> DATA
```

And, finally, to execute general commands, such as sorting a file, first type

```
> SYSTAT
```

Once you are "inside" a module you do not have to retype these before each command, so in this chapter they are usually shown only before the first example in each section.

3.2.1.2 Box Plots

Box plots (Tukey, 1977) give a different visual impression than do histograms, and contain more explicit information about the distribution of cases. The plots easily can be created using SYSTAT's default parameters with the commands

```
>GRAPH
>BOX INCOME
```

The resulting box plots for income and shoes are shown in Figure 3.2. The left and right sides of the box show the 25th and 75th centiles in the data, respectively. (Technically, the sides of the box fall at the boundaries of the upper and lower "fourths" of the data, which are defined slightly differently than quartiles; see Frigge, Hoaglin, and Iglewicz, 1989.) The entire box, therefore, contains the middle 50 percent of the data, and the difference between the 25th and 75th centiles is sometimes referred to as the "interquartile range." The vertical line through the box shows the 50th centile, or *median*. Extending from either side of the box are "whiskers," which represent 1.5 times the interquartile range beyond the edges of the box on either side. The ends of the whiskers mark the "inner fences." The "outer fences" represent three times the interquartile range beyond the edges of the box. Any data points falling outside the inner fences but within the outer fences are shown as asterisks. Data points falling outside the outer fences are shown as small circles. Data points falling outside fences are possible outliers, and you should examine them individually to determine whether they might exert undue influence on your data analyses.

Box plots are very useful for the detection of outliers because the existence of some outliers will rarely mask the existence of others. Up to about 25% of the data can be outliers without causing great difficulties with outlier detection, and with that many "outliers" it would be difficult to maintain that the data really came from a normal distribution anyway. For data sampled from a normal distribution, the expected proportion of data points falling outside the fences depends on the sample size. For samples of five, the smallest number that permits the construction of box plots, about 8.6% of data points fall beyond the inner fences and about 3.3% fall beyond the outer fences (Hoaglin, Iglewicz, and Tukey, 1986). However, these proportions decrease rapidly with sample size; for samples of size six, the rate of outliers beyond the inner fences drops to about 1 in 30 and beyond the outer fences drops to less than

---

MACINTOSH

① Choose the menu item **Graph/Box**.
② Select INCOME from the left variable list.
③ Click **OK**.

MS-DOS

① Choose the menu item **Graph/Box/Variables/Dependent/**.
② Select INCOME from the variable list.
③ Choose the menu item **Graph/Box/Go!**

1 in 100. With sample sizes of 20, the rate of outliers is only about 1 in 60 outside the inner fences and less than 1 in 1,000 beyond the outer fences. Finally, in a normal distribution with infinite sample size, only 1 in 143 of the cases falls beyond the inner fences and only about 1 in 435,000 falls beyond the outer fences! These numbers can serve as rough guides for determining whether your sample has an unusual number of outliers for data from a normal distribution.

Figure 3.2 shows the box plot for income. As in the histogram for this variable (Figure 3.1), the box plot indicates positive skew. Interestingly, the box plot suggests that data on income has four mild and two serious outliers. This is considerably more than we would expect in a sample of 100 cases from a normal distribution. However, further examination of the histogram (Figure 3.1) suggests that only the largest outlier is markedly separate from the rest of the data. The other outliers are consistent with the skewed shape of the rest of the data: Rather than being "outliers" per se, they simply may be large values expected in a non-normal distribution. This illustrates one reason why you might want to look at more than one type of plot of the variables to get a better feel for how your data is distributed.

3.2.1.3 Stem and Leaf Displays

Stem and leaf displays combine features of box plots and histograms, while presenting all of the data points in the data set in numeric form. Data points are grouped together on each line by the largest digits that they have in common (the "stem"). The final digits of each number are then abutted onto the end of the line (the "leaves"). Thus, the more data points in a category, the longer the line. This allows the stem and leaf display to take on the form of a histogram and provide a visual sense of the shape of the distribution. Like a box plot, the stem and leaf display can include information about medians,

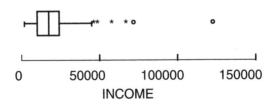

Figure 3.2 Box plot of income in the SHOES example data set.

hinges, and outliers. In SYSTAT, the median, hinges, and minimum and maximum values are listed in the analysis window, in addition to being represented in the stem and leaf display itself.

Figure 3.3 shows the stem and leaf displays of the income variable from the SHOES data set. The "stems" in this display along the left margin are in units of $10,000. This display was generated by the following two command lines:

```
>GRAPH
>STEM INCOME / HEIGHT=4IN WIDTH=2IN
```

The *height* and *width* options adjust the size of the display (see section 3.2.4 for other options). Because SYSTAT scales the font size to correspond to the size of the display, you may have to play around with the dimensions to create a readable display.

The stem that contains the median is marked with the letter M, and stems containing the upper and lower hinges (75th and 25th centiles) are marked with the letter H. The actual values of these numbers are given by the text in the analysis window. For example, the median for the income variable is $17,111, and this value is located in the fourth line from the top in the

```
STEM AND LEAF PLOT OF VARIABLE:   INCOME,   N = 100

MINIMUM IS:          1533.000
LOWER HINGE IS:      9728.000
MEDIAN IS:          17111.000
UPPER HINGE IS:     24153.000
MAXIMUM IS:        122855.000

    0       113444
    0     H 555566666677778999999
    1       00011222233444
    1     M 5555666777788888889999
    2     H 0011222333444
    2       556779
    3       04
    3       779
    4       002244
            ***OUTSIDE VALUES***
    4       68
    5       8
    6       7
    7       2
   12       2
```

Figure 3.3 Stem and leaf display of income variable in the SHOES data set.

MACINTOSH

① Choose the menu item **Graph/Stem**.
② Select INCOME from the variable list.
③ Click on the **ruler icon**, enter "4" in the **Height** box, "2" in the **Width** box, and click **OK**.
④ Click **OK**.

MS-DOS

① Choose the menu item **Graph/Stemleaf/ Variables/**.
② Select INCOME from the variable list.
③ Choose the menu item **Graph/Stemleaf/Options/ Height/**, type "4", and choose **/IN**.
④ Choose the menu item **Graph/Stemleaf/Options/ Width/**, type "2", and choose **/IN**.
⑤ Choose the menu item **Graph/Stemleaf/Go!**

stem and leaf display in the left panel of Figure 3.3. The six outliers that we found in the box plots for the income variable are also listed in the stem and leaf display. But here we can read the values of the numbers in thousands of dollars directly from the display: 46, 48, 58, 67, 72, and 122 thousand dollars. We know from the histogram (Figure 3.1) that only the $122,000 value represents a substantial departure from the rest of the data.

You can vary the number of stems (lines) in the display, but this is not as straightforward as you might expect. This is because not just any number of lines will make a difference in the display. For example, in the top two lines of the stem and leaf display in Figure 3.3, SYSTAT has divided the interval 00 to 09 into two parts: 00 to 04 and 05 to 09. To evenly divide this interval into a larger number of equally spaced groups, you would have to divide it into five new intervals: 00 to 01, 02 to 03, 04 to 05, 06 to 07, and 08 to 09. Therefore, although Figure 3.3 has only 14 lines, instructing SYSTAT to display 15 lines will make no difference in the display. To increase the number of stems, you would need to allow for five 0 stems, five 1 stems, five 2 stems, five 3 stems, and three 4 stems (the three before the outliers). Thus, the minimum number of lines you need to specify to change the display is approximately 23. (I say "approximately" because the actual number varies with the number of outliers and how the tails of the distribution are grouped.) You change the number of stems by adding "lines=" and the number of stems you wish after a slash at the end of the command line. For example:

> `>STEM INCOME/LINES=23 HEIGHT=4 IN WIDTH=2 IN`

The output from this command is shown in Figure 3.4.

3.2.1.4 Normal Probability Plots

Probability plots, also called *rankit* plots, provide one of the most powerful (and underused) methods of determining whether a given data set follows a particular underlying theoretical distribution. In these plots, the rank-ordered data values (from smallest to largest) are plotted against the centiles from a theoretical distribution corresponding to those ranks. When the data are drawn from a distribution that approximates the theoretical distribution, these data points will fall on a straight line in the plot. Because the normal distribution is of greatest importance to the types of analyses covered in this book, I will use it to illustrate probability plots. You should be aware that SYSTAT can also make probability plots for other

---

MACINTOSH

① Choose the menu item **Graph/Stem**.

② Select INCOME from the variable list.

③ Click on **Lines**, enter "23" into the dialog box, and click OK.

④ Click on the **ruler icon**, enter "4" in the **Height** box, "2" in the **Width** box, and click OK.

⑤ Click OK.

MS-DOS

① Choose the menu item **Graph/Stemleaf/Variables/**.

② Select INCOME from the variable list.

③ Choose the menu item **Graph/Stemleaf/Options/Lines/** and type "23" <ret>.

④ Choose the menu item **Graph/Stemleaf/Options/Height/**, type "4", and choose /IN.

⑤ Choose the menu item **Graph/Stemleaf/Options/Width/**, type "2", and choose /IN.

⑥ Choose the menu item **Graph/Stemleaf/Go!**

```
0    11
0    3
0    4445555
0    6666667777
0  H 8999999
1    00011
1    222233
1    4445555
1  M 66667777
1    88888889999
2    0011
2    222333
2  H 44455
2    677
2    9
3    0
3
3    4
3    77
3    9
4    00
4    22
4    44
     ***OUTSIDE VALUES***
4    68
5    8
6    7
7    2
12   2
```

Figure 3.4 Stem and leaf display of income variable in the SHOES data set, with the number of lines set at 23.

theoretical distributions: chi-squares, exponentials, gammas, uniforms, and Weibulls. Also, note that SYSTAT draws probability plots with the X- and Y-axes switched from the more common manner of drawing these plots (see section 3.2.4.10).

To illustrate normal probability plots, I generated 1,000 cases in a column labelled Z with the following command:

>LET Z = ZRN

Therefore, the random numbers in the column Z come from a normal distribution with mean zero and variance equal to 1.0. A "dit" plot (see section 3.2.1.1) of this variable is shown in the left panel of Figure 3.5. As you can see, this distribution looks quite normal.

Probability plots are generated in SYSTAT using the *pplot* command. This command should be followed by the variable(s) that you want to plot, a slash, and then the name of the theoretical distribution that you want to evaluate. To generate a normal probability plot of the variable labelled Z, type

>GRAPH
>PPLOT Z /NORMAL

MACINTOSH
① Choose the menu item **Graph/Prob'y/Normal...**
② Select Z from the variable list.
③ Click **OK**.

MS-DOS
① Choose the menu item **Graph/PPlot/Variables/**.
② Select Z from the variable list.
③ Choose the menu item **Graph/PPlot/Type/Norm**.
④ Choose the menu item **Graph/PPlot/Go!**

This plot is shown in the right panel of Figure 3.5. If the data points are drawn from a normal distribution, they should fall on a straight line in this plot. In fact, the data in Figure 3.5 fall almost exactly on a straight line, as we would expect given the way the data was generated. The slope of this line gives an estimate $1/\sigma$ for the data. In the right panel of Figure 3.5 the line through the data looks like it will pass through the coordinates [–4,–4] and [+4,+4], which means the slope is approximately 1.0. Recall that the standard deviation of this distribution is 1.0.

3.2.4.1.1 Detecting Skew

When the data in a normal probability plot comes from a distribution that is *positively skewed*, the shape of the plot tends to be bowed up (like a bow shooting an arrow towards the sky to the left). To illustrate this, I took the normal data in the column Z and squared all of the values greater than 1.0. This gives the distribution a clear positive skew, as you can see in the dit plot in the left panel of Figure 3.6. The normal probability plot of these data is shown in the right panel of Figure 3.6. This

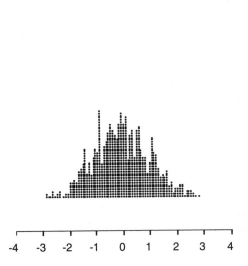
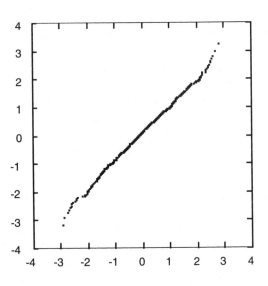

Figure 3.5 The left panel shows a dit plot of a normally distributed variable with mean zero and variance equal to 1.0. The right panel shows a normal probability plot of this variable, with the data closely falling on a straight line.

plot has an obvious upward bow to it, confirming the positive skew in the distribution.

Negatively skewed distributions tend to be bowed down (like a bow shooting an arrow towards the ground to the right). To illustrate this, I took the normal data in the column Z and squared all of the values less than –1.0 (and then reattached the minus sign). This gives the distribution a clear negative skew, as you can see in the dit plot in the left panel of Figure 3.7. The normal probability plot of these data is shown in the right panel of Figure 3.7. This plot has an obvious downward bow to it, indicative of the negative skew in the distribution.

3.2.4.2.2 Assessing Tail "Thickness"

Distributions with long, thick tails (such as *t* distributions with small *df*) tend to have an "ogive" shape, ∫. To illustrate this, I took the normal data in the column Z and squared all of the values greater than 1.0 and less than –1.0 (and then reattached minus signs to the latter). This gives the distribution thicker tails than a normal, as you can see in the dit plot in the left panel of Figure 3.8. The normal probability plot of these data is shown in the right panel of Figure 3.8. This plot does have a ∫ shape, indicative of the thickness of the tails in the distribution, although it also starts to turn again at the ends.

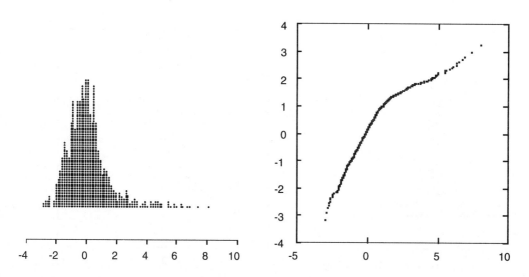

Figure 3.6 The left panel shows a dit plot of a positively skewed variable. The right panel shows a normal probability plot of this variable, with the data showing a clear upward bow.

Distributions with short, thin tails tend to be ∫ shaped in normal probability plots. To illustrate this, I used a transformation of the data in column Z that pulls in the tails of the distribution and gives the distribution the short, thin tails

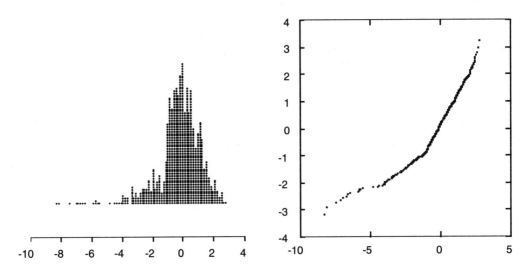

Figure 3.7 The left panel shows a dit plot of a negatively skewed variable. The right panel shows a normal probability plot of this variable, with the data showing a clear downward bow.

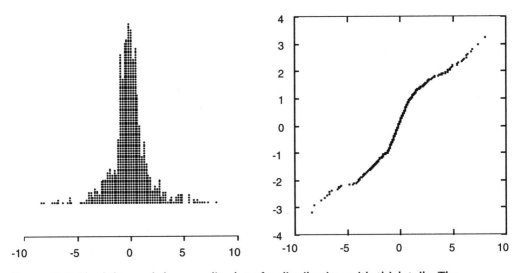

Figure 3.8 The left panel shows a dit plot of a distribution with thick tails. The right panel shows a normal probability plot of this variable, with the data showing an ogive ∫ shape.

shown in the dit plot in the left panel of Figure 3.9. (I took the normal data in the column Z, added 1 to the positive scores and subtracted 1 from the negative scores, took the square root of all of these values, and then subtracted 1 from the positive scores and added 1 to the negative scores to bring the two halves of the distribution back together.) The normal probability plot of these data is shown in the right panel of Figure 3.9. This plot has an obvious ∫ shape, indicative of the thinness of the tails in the distribution. In fact, the plot of the original Z data in Figure 3.5 tends to be a little "short-tailed."

3.2.4.3.3 Detecting Unusual Cases

When most of your data is normally distributed, the normal probability plot can be used to detect cases that are unusually large or small, and thus may not come from the same distribution as the rest of the data. Such cases can often be detected in normal probability plots because they will fall off the line formed by the rest of the data. To illustrate this, I took the original data, which ranges from roughly −3 to +3, and added the scores 4.0, 5.0, and 6.0. The normal probability plot of this data is shown in the left panel of Figure 3.10, along with the best-fitting line through the data. Note that the three new cases fall off to the right of the line, suggesting, as we know,

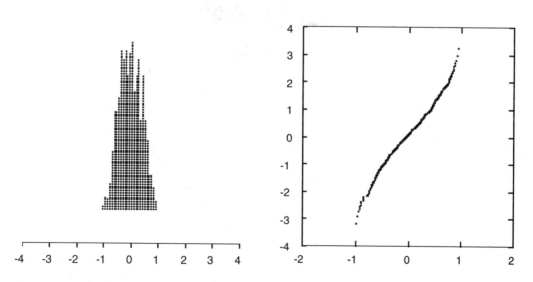

Figure 3.9 The left panel shows a dit plot of a distribution with short tails. The right panel shows a normal probability plot of this variable, with the data showing a ∫ shape.

that these cases may not come from the same distribution as the rest of the data.

To estimate the standard deviation of the distribution from the plot, you should first delete the three cases that do not belong (or just select all of the data except for those cases). Otherwise, these cases would pull the best-fitting line towards them, and for small n's they can alter the shape of the line. With these cases removed, you can recompute the normal probability plot and best-fitting line as shown in the right panel of Figure 3.10 (this figure is identical to the right panel of Figure 3.5 except that the line has been added). By noting where the line intersects the plot frame, you can estimate the standard deviation of the distribution. In the right panel of Figure 3.10, the line intersects the frame at about [–4,–4] and [+4,+4], which means that the slope is about equal to 1.0. Therefore, the standard deviation is approximately 1.0. Note that the remaining cases show some evidence of coming from a short-tailed distribution (see the previous section).

In real data sets the interpretation of unusual cases may not be so clear. Such cases may either come from a different distribution, or they may just indicate that the distribution in the plot is not normal. Looking at the trend in the non-unusual cases can help determine which situation seems most parsimonious. Consider the income variable in the SHOES data set. To

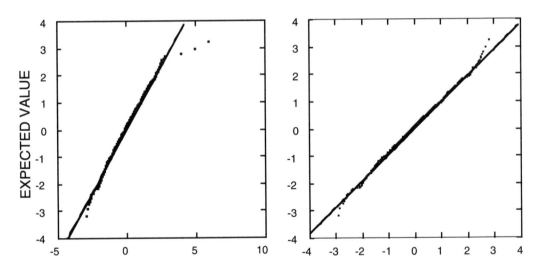

Figure 3.10 The left panel shows a normal probability plot—a normal distribution with three unusual cases. The right panel shows the same plot with those cases removed.

create a normal probability plot of this variable, open the SHOES data set and then type

```
>GRAPH
>PPLOT INCOME/NORMAL
```

This plot is shown in Figure 3.11. In this plot, if the data points are drawn from a normal distribution they should fall on a straight line: This is clearly not the case in Figure 3.11. The data in this plot are substantially bowed up, which indicates positive skew; this is in agreement with the large value of skewness computed for this variable in section 3.1. However, four of the cases fall considerably off the "line" formed by the rest of the data. Do these cases come from a different distribution? Probably not. The remainder of the cases are also bowed up, and the top four cases appear to fall on the curve formed by the rest of the cases. Cases that are unusually large for a normal distribution may be perfectly reasonable for a positively skewed distribution.

MACINTOSH
① Choose the menu item **Graph/Prob'y/Normal...**
② Select INCOME from the variable list.
③ Click **OK**.

MS-DOS
① Choose the menu item **Graph/PPlot/Variables/**.
② Select INCOME from the variable list.
③ Choose the menu item **Graph/PPlot/Type/Norm**.
④ Choose the menu item **Graph/PPlot/Go!**

3.2.1.5 Identifying Individual Cases in Plots

Both the box plots and stem and leaf displays locate outliers. The stem and leaf display will even give the actual value of the

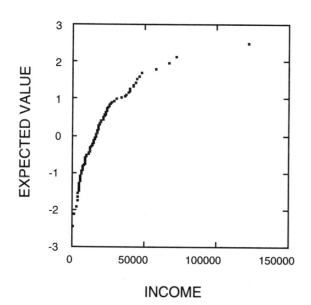

Figure 3.11 Normal probability plot of the income variable in the SHOES data set. The upward bow in the curve indicates positive skew.

outliers rounded off to some value. In order to identify which particular cases have outlying values on a variable, you must refer back to the original data set. To find individual cases that meet a specific criterion, you can use SYSTAT's find command. For example, to find the case with income over $100,000 in the SHOES data set, with the data editor open you can simply type

> ▶ >FIND INCOME > 100000

This will move the data window so that the desired case is shown at the top of the window. If there is more than one case in the data set that meets the criterion, type the find command again and SYSTAT will move to the next appropriate case.

One easy way to find multiple outliers is to sort the data set on the variable of interest and then count down the number of outliers. For example, the stem and leaf display of the income variable in the SHOES data set indicates that this variable has six outliers, and the box plot suggests that two of these fall outside the upper fence. Table 3.5 shows the last eight cases of this data set after it was sorted in ascending order by income (see section 2.1.1 on sorting data).

SHORTCUT: IDENTIFYING CASES IN PLOTS (MACINTOSH ONLY)

SYSTAT provides an easy way of identifying cases in scatterplots, which are described in section 3.2.6: Choose the arrow tool and move the cursor to the data point you wish to identify. Click on this point and the case will be highlighted in the data editor window. To actually *select* (use separately) one or more cases in the data set, use the marquee or lasso tool to select the cases in the plot. These cases will be bulleted in the data editor window, and any further analyses will use only those cases until the selection is removed by choosing **Data/Deselect** or by making a different selection.

3.2.1.6 Selecting Cases in the Data Set

After you have identified individual cases in a data set, you may want to perform some analyses excluding one or more cases. SYSTAT allows you to select subsets of cases on which to perform analyses using the select command, followed by some criteria (these criteria follow the same syntax as the conditions in if-then statements described in section 2.2.2). For example,

---

MACINTOSH

① Choose the menu item **Data/Find Case...**

② Enter the desired variable into the **Find Test Variable** box, e.g., INCOME.

③ Select the desired relation, e.g., ">".

④ Type the test criterion into the **Variable or expression** box, e.g., "100000".

⑤ Click **OK**.

⑥ (To find the next case meeting the criterion simply choose **Data/Find Next**.)

MS-DOS

① Choose the menu item **Data/Find Case...**

② Enter the desired variable into the **Find Test Variable** box, e.g., INCOME.

③ Select the desired relation, e.g., ">".

④ Type the test criterion into the **Variable or expression** box, e.g., "100000".

⑤ Click **OK**.

⑥ (To find the next case meeting the criterion simply choose **Data/Find Next**.)

Table 3.5 Last eight cases from the SHOES data set, sorted in ascending order of income.

Case	Income	Shoes
93	44,805.000	24.000
94	44,915.000	16.000
95	46,765.000	12.000
96	48,822.000	4.000
97	58,499.000	11.000
98	67,397.000	15.000
99	72,463.000	19.000
100	122,855.000	26.000

if you wished to exclude case 93 from your analysis, you could type the command

```
>SELECT CASE <> 93
```

Any further analyses will exclude this case. To turn the selection off, simply type

```
>SELECT
```

with no arguments and SYSTAT will use all of the data in subsequent analyses.

The criteria can be as complicated as you like. For example, in the SHOES data set, to use only those cases with income less than $5,000 or greater than $50,000 except for case number 93, type

```
>SELECT (INCOME<5000 OR INCOME>50000) AND,
CASE <> 93
```

3.2.2 Plotting Two Variables

3.2.2.1 Scatterplots

3.2.2.1.1 Simple Scatterplots

In SYSTAT, scatterplots can be used to show the joint distribution of variables in two or three dimensions. A two-dimensional scatterplot of shoes versus income for the SHOES data set is shown in Figure 3.12. This plot was created by typing

```
>GRAPH
>PLOT SHOES*INCOME
```

MACINTOSH
① Choose the menu item **Data/Select Cases...**
② Enter the selection criteria into the box, clicking on **And** or **Or** if needed.
③ Click **OK**.

MS-DOS
① Choose the menu item **Data/Select/Select**.
② Select a variable name, then select a criterion, or hit <esc> and type a criterion.
③ Repeat selecting variables and criteria until you complete the selection criterion.

MACINTOSH
① Choose the menu item **Data/Deselect**.

MS-DOS
① Choose the menu item **Data/Select/Clear**.

MACINTOSH
① Choose the menu item **Graph/Plot/Plot**.
② Select SHOES from the *Y variables* list and INCOME from the *X variable* list.
③ Click **OK**.

MS-DOS
① Choose the menu item **Graph/Plot/Data/2-D/Variables/Y**.
② Select SHOES from the variable list.
③ Choose the menu item **Graph/Plot/Data/2-D/Variables/X**.
④ Select INCOME from the variable list.
⑤ Choose the menu item **Graph/Plot/Data/2-D/Go!**

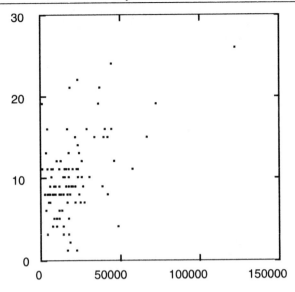

Figure 3.12 Scatterplot of shoes versus income in the SHOES data set.

This plot reveals that both variables thin out as the values increase. One case contains the largest value on *both* variables, and such a case may have a large influence on a regression analysis, for example. The outlier case in Figure 3.12 is, of course, the family with the highest income, which was found in Table 3.5 with the data set sorted on the income variable. This family also has the largest number of shoes at 26.

3.2.2.1.2 Influence Plots

In scatterplots, SYSTAT offers an option of displaying the influence of single cases on the correlation between the two variables, and hence on a regression line, by varying the size of the points to correspond with their influence. The influence plot for shoes versus income is shown in Figure 3.13. This plot was created by typing into the command window

▶ `>PLOT SHOES*INCOME/INFLUENCE`

The case with the largest value on both variables, case 100 in Table 3.5, has a very large influence on the correlation between these variables, as suspected from the scatterplot in Figure 3.13. This case will "pull" a regression line, for example, towards it to a much greater extent than the other cases. The remaining cases are very similar in their influence.

MACINTOSH
① Choose the menu item **Graph/Plot/Influence**.
② Select SHOES from the *Y variables* list and INCOME from the *X variable* list.
③ Click **OK**.

MS-DOS
① Choose the menu item **Graph/Plot/Data/2-D/ Variables/Y**.
② Select SHOES from the variable list.
③ Choose the menu item **Graph/Plot/Data/2-D/ Variables/X**.
④ Select INCOME from the variable list.
⑤ Choose the menu item **Graph/Plot/Data/2-D/ Options/Influence**.
⑥ Choose the menu item **Graph/Plot/Data/2-D/Go!**

HINT: "BORDER PLOTS" (MACINTOSH ONLY)

SYSTAT offers Macintosh users an easy way to generate "border plots," plots that combine scatterplots and box plots into a single display. A border plot for the shoes and income variables in the SHOES data set is illustrated in Figure 3.12a below. This plot was created as easily as a plain scatterplot using the menu selection **Graph/Plot/Border**, and then selecting shoes and income from the Y and X variable lists, respectively. The box plot at the top of this display is identical to Figure 3.2, and the scatterplot is identical to that in Figure 3.12. The same plot can be generated in DOS using the following lengthy set of commands:

```
>BEGIN
>ORIGIN 15.00 8.00
>PLOT SHOES*INCOME/HEIGHT=2.54 IN,
WIDTH=2.54 IN
>ORIGIN 15.00 2.86 IN
>BOX INCOME/HEIGHT=1.27 IN WIDTH=2.54 IN,
SCALE=0 AXES=0 XLABEL=' '
>ORIGIN 3.14 IN 8.00
>BOX SHOES/HEIGHT=2.54 IN WIDTH=1.27 IN,
TRANSPOSE SCALE=0 AXES=0 XLABEL=' '
>ORIGIN 15.0 8.0
>END
```

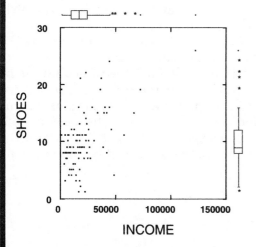

Figure 3.12a Border plot for shoes and income in the SHOES data set.

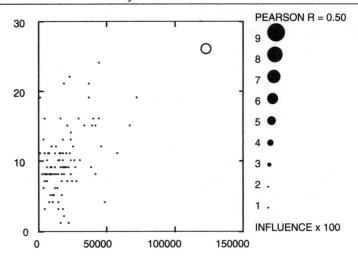

Figure 3.13 Influence plot of shoes versus income in the SHOES data set.

3.2.2.1.3 Fitting Ellipses

SYSTAT offers a large number of scatterplot options. I will just discuss three more that are particularly nice in regression contexts (see Chapters 5 and 6). First, SYSTAT can draw ellipses over the scatterplot that show regions estimated to cover certain proportions of the data, assuming a bivariate normal distribution. For example, for the default 50% region, if the joint distribution of the X and Y variables is normal in both dimensions, then the ellipse should contain approximately 50% of the cases. The orientation of the ellipse is determined by the covariance of X and Y. The length and width of the ellipse are determined by the sample standard deviations of the X and Y variables. Figure 3.14 shows the 50% ellipse for the shoes versus income scatterplot in the SHOES data set. In fact, this ellipse contains 65% of the cases, which suggests that the bivariate distribution has more cases near the mean of both variables than one would expect in a normal distribution. The ellipse in the scatterplot can be generated by the following command:

```
>GRAPH
>PLOT SHOES*INCOME/ELL=.5
```

MACINTOSH

① Choose the menu item **Graph/Plot/Plot**.
② Select SHOES from the *Y variables* list and INCOME from the *X variable* list.
③ Click on **Ellipse**.
④ Type ".50" into the dialog box (if it is not already there) and click **OK**.
⑤ Click **OK**.

MS-DOS

① Choose the menu item **Graph/Plot/Data/2-D/Variables/Y**.
② Select SHOES from the variable list.
③ Choose the menu item **Graph/Plot/Data/2-D/Variables/X**.
④ Select INCOME from the variable list.
⑤ Choose the menu item **Graph/Plot/Data/2-D/Options/Ell** and type ".5".
⑥ Choose the menu item **Graph/Plot/Data/2-D/Go!**

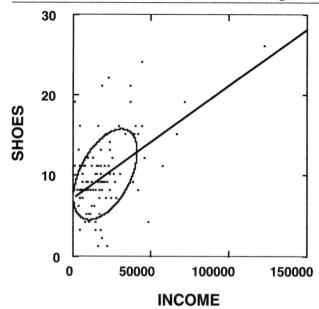

Figure 3.14 Scatterplot of shoes versus income in the SHOES data set, with 50% bivariate ellipse and best-fitting least squares regression line.

3.2.2.1.4 Fitting Lines

In addition to ellipses, SYSTAT will draw the best-fitting line (least squares) for the scatterplot. Figure 3.14 shows this line in the same plot as the ellipse. Keep in mind that the best-fitting line is heavily influenced by one case (the highest value on both variables in Figure 3.14). The plot shown in Figure 3.14 was created by the following command line:

```
>PLOT SHOES*INCOME/ELL=.5 SMOOTH=LINEAR
```

3.2.2.1.5 Scatterplots and Categorical Variables

When your independent variable is categorical, it is still sometimes useful to create a scatterplot of the data within each category. This plot can be used, for example, to examine the validity of some of the assumptions underlying an ANOVA (see section 9.6). These plots are constructed in precisely the same manner as any other scatterplot. For example, to create a scatterplot of the TEST SCORES across groups in the TEST SCORES data set, open that file and then type

MACINTOSH
① Choose the menu item **Graph/Plot/Plot**.
② Select SHOES from the *Y variables* list and INCOME from the *X variable* list.
③ Click on **Ellipse**.
④ Type ".50" into the dialog box (if it is not already there) and click **OK**.
⑤ Click on **Smooth**.
⑥ Check **Linear** and click **OK**.
⑦ Click **OK**.

MS-DOS
① Choose the menu item **Graph/Plot/Data/2-D/Variables/Y**.
② Select SHOES from the variable list.
③ Choose the menu item **Graph/Plot/Data/2-D/Variables/X**.
④ Select INCOME from the variable list.
⑤ Choose the menu item **Graph/Plot/Data/2-D/Options/Ell** and type ".5".
⑥ Choose the menu item **Graph/Plot/Data/2-D/Options/Smooth/Method/Linear**.
⑦ Choose the menu item **Graph/Plot/Data/2-D/Go!**

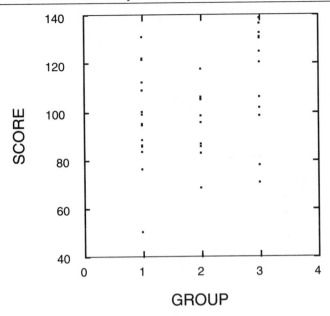

Figure 3.15 Scatterplot of the test scores across groups in the TEST SCORES data set.

```
> GRAPH
> PLOT SCORES * GROUP
```

This plot is shown in Figure 3.15.

3.2.2.2 Kernel Density Plots

Although both the income and shoes distributions are positively skewed (see Table 3.3 and Figure 3.1, for example), we still cannot see clearly what the joint density distribution of these two variables looks like. SYSTAT offers a view of this distribution using three-dimensional kernel density plots. These plots are like histograms in 3-D: The height of the plot shows the proportion of cases falling into each two-dimensional interval defined by combinations of the Y and X variables. Unlike histograms, the variables are treated as continuous, and are not partitioned into intervals. The result is a 3-D surface as shown in Figure 3.16. In this figure you can see that the joint distribution of shoes and income is piled up rather heavily in the center, trailing off quickly in the tails.

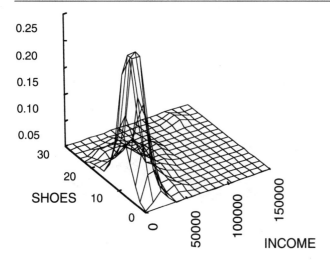

Figure 3.16 Kernel density plot of shoes versus income in the SHOES data set.

This is consistent with the observation in Figure 3.14 that more than 50% of the cases fall within the predicted 50% ellipse, which assumed that the distribution was multivariate normal. The kernel density plot in Figure 3.16 was created with the following command:

```
>PLOT SHOES*INCOME/SMOOTH=KERNEL
```

Unfortunately, SYSTAT does not allow you to change the viewing angle in Figure 3.16. However, you can replot the data with the Y- and X-axes switched to give another view of the kernel density plot. Such a plot is shown in Figure 3.17. It is important to plot the data both ways so you can see what is going on behind the "hump" in the distribution. For example, you know from Table 3.5 that there is a small "bump" in this distribution out near income=125,000 and shoes=26. This bump can be seen in the upper center portion of Figure 3.16, but it is barely visible behind the hump in Figure 3.17. To

MACINTOSH

① Choose the menu item **Graph/Plot/Plot**.
② Select SHOES from the *Y variables* list and INCOME from the *X variable* list.
③ Click on **Smooth**.
④ Check **Kernel** and click OK.
⑤ Click **OK**.

MS-DOS

① Choose the menu item **Graph/Plot/Data/2-D/Variables/Y**.
② Select SHOES from the variable list.
③ Choose the menu item **Graph/Plot/Data/2-D/Variables/X**.
④ Select INCOME from the variable list.
⑤ Choose the menu item **Graph/Plot/Data/2-D/Options/Smooth/Kernel**.
⑥ Choose the menu item **Graph/Plot/Data/2-D/Go!**

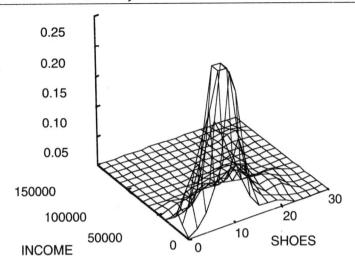

Figure 3.17 Kernel density plot of income versus shoes in the SHOES data set.

MACINTOSH
① Choose the menu item **Graph/Plot/Plot**.
② Select INCOME from the *Y variables* list and SHOES from the *X variable* list.
③ Click on **Smooth**.
④ Check **Kernel** and click **OK**.
⑤ Click **OK**.

MS-DOS
① Choose the menu item **Graph/Plot/Data/2-D/Variables/Y**.
② Select INCOME from the variable list.
③ Choose the menu item **Graph/Plot/Data/2-D/Variables/X**.
④ Select SHOES from the variable list.
⑤ Choose the menu item **Graph/Plot/Data/2-D/Options/Smooth/Kernel**.
⑥ Choose the menu item **Graph/Plot/Data/2-D/Go!**

create the plot with the axes switched, simply reverse the order of the variable names in the command:

> `>PLOT INCOME*SHOES/SMOOTH=KERNEL`

3.2.2.3 Multiple Box Plots

When the number of data points within a category gets large, it can become difficult to distinguish one case from another in scatterplots of those groups (see section 3.2.2.1.5). Fortunately, one of the box plot options allows you to display several box plots representing different groups in the same display. For example, in the TEST SCORES data set there are three groups of subjects, and you can display the individual box plots for each group in the same graph. To do this, after the box command you simply type the name of the continuous variable (the *Y*-axis), type an asterisk, and then type the name of the grouping variable (the *X*-axis). For example, open the TEST SCORES data set and type

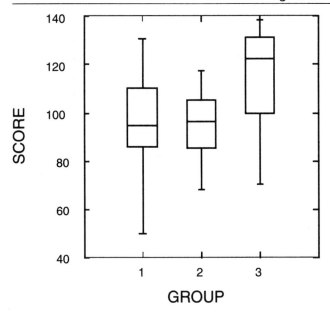

Figure 3.18 Box plots for each group in the TEST SCORES data set. (Table 3.2)

```
>GRAPH
>BOX SCORES * GROUP
```

This command generates the three box plots shown in Figure 3.18.

3.2.2.4 Bar Graphs

When one of your independent variables is categorical and is not on a meaningful scale, that is, the ordering of the categories is arbitrary, it is often useful to plot the means of the categories using a bar graph. This can be done easily with the commands

```
>GRAPH
>BAR SCORE*GROUP
```

This command computes the means within each category and displays them in the bar graph shown in Figure 3.19.

Frequently, you will want to plot the means along with some type of error bars to indicate the amount of variability in the data. SYSTAT can automatically generate error bars showing ±1 standard error of the mean for each group using that

---

MACINTOSH

① Choose the menu item **Graph/Box**.

② Select SCORES from the left variable list and GROUP from the right variable list.

③ Click **OK**.

MS-DOS

① Choose the menu item **Graph/Box/Variables/ Dependent/** and select SCORES from the variable list.

② Choose the menu item **Graph/Box/Variables/ Grouping/** and select GROUP from the variable list.

③ Choose the menu item **Graph/Box/Go!**

MACINTOSH

① Choose the menu item **Graph/Bar/Bar**.

② Select SCORE from the left variable list and GROUP from the right variable list.

③ Click **OK**.

MS-DOS

① Choose the menu item **Graph/Bar/Variables/ Dependent/** and select SCORE from the variable list.

② Choose the menu item **Graph/Bar/Variables/ Grouping/** and select GROUP from the variable list.

③ Choose the menu item **Graph/Bar/Go!**

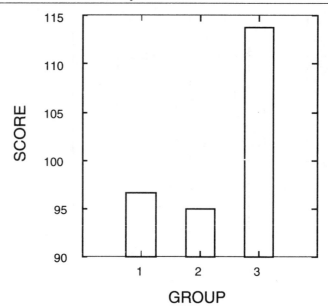

Figure 3.19 Bar graph of the test scores means for each group in the TEST SCORES data set.

MACINTOSH

① Choose the menu item **Graph/Bar/Bar**.
② Select SCORES from the left variable list and GROUP from the right variable list.
③ Click on the **Error** icon, click on **SError**, and click **OK**.
④ Click on the **Axes** icon, type 80 into the "Y min" box and 130 into the "Y max" box, and click **OK**.
⑤ Click **OK**.

MS-DOS

① Choose the menu item **Graph/Bar/Variables/Dependent/** and select SCORES from the variable list.
② Choose the menu item **Graph/Bar/Variables/Grouping/** and select GROUP from the variable list.
③ Choose the menu item **Graph/Bar/Options/SError** and choose the variable SCORES.
④ Choose the menu item **Graph/Bar/Options/Ymin/** and type "80".
⑤ Choose the menu item **Graph/Bar/Options/Ymax/** and type "130".
⑥ Choose the menu item **Graph/Bar/Go!**.

group's own variance (section 8.7). For the TEST SCORES data set, type

> ▶ >GRAPH
> >BAR   SCORE * GROUP/YMIN=80 YMAX=130 SERROR

The *ymin* and *ymax* values were chosen so that the error bars fit completely on the graph. The option *SERROR* tells SYSTAT to include error bars based on the standard error of the mean for each group. The resulting graph is shown in Figure 3.20.

Often, it is more informative to the reader to show the *confidence intervals* around the group means, rather than just the standard errors (see section 8.7). In general, you can generate any error bars you wish by creating a new column in your data set that contains the magnitude of the error bars for each group. Section 8.7 shows how to find the 95% confidence limits around each group mean. To plot confidence intervals around the means in the bar graph for each group in the TEST SCORES data set, first create a new column of data that contains those confidence limits by typing

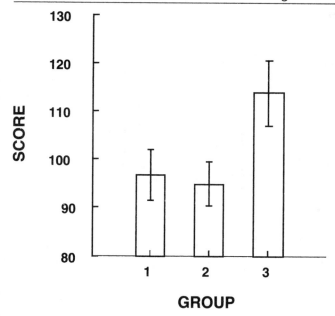

Figure 3.20 Bar graph showing the mean of the three groups in the TEST SCORES data set, along with error bars of ±1 SE of the mean, where the standard errors are based on each group's separate variance.

```
>IF GROUP=1 THEN LET CL=11.4
>IF GROUP=2 THEN LET CL=10.3
>IF GROUP=3 THEN LET CL=14.6
```

where "CL" is just my label for the new column of confidence limits, and 11.4, 10.3, and 14.6 are the confidence limits for groups 1, 2, and 3, respectively. To put these confidence limits around the means in the bar graph, type

```
>BAR SCORE*GROUP/YMIN=80 YMAX=130 ERROR=CL
```

The option ERROR=CL tells SYSTAT to look in the data column CL to find the magnitudes of the error bars in the graph. The resulting graph is shown in Figure 3.21. In general, you can use this procedure to set the error bars to anything you want.

3.2.2.5 Line Plots

3.2.2.5.1 Equally Spaced Groups

When the independent variable is on a meaningful scale, it is standard to plot the means of the categories using a line graph.

MACINTOSH
① Choose the menu item **Graph/Bar/Bar**.
② Select SCORES from the left variable list and GROUP from the right variable list.
③ Click on the **Error** icon, type the column label "CL" into the box, and click **OK**.
④ Click on the **Axes** icon, type 80 into the "Y min" box and 130 into the "Y max" box, and click **OK**.
⑤ Click **OK**.

MS-DOS
① Choose the menu item **Graph/Bar/Variables/Dependent/** and select SCORES from the variable list.
② Choose the menu item **Graph/Bar/Variables/Grouping/** and select GROUP from the variable list.
③ Choose the menu item **Graph/Bar/Options/Error** and choose the column CL.
④ Choose the menu item **Graph/Bar/Options/Ymin/** and type "80".
⑤ Choose the menu item **Graph/Bar/Options/Ymax/** and type "130".
⑥ Choose the menu item **Graph/Bar/Go!**

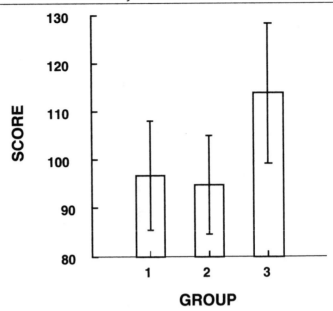

Figure 3.21 Bar graph showing the mean of the three groups in the TEST SCORES data set, along with error bars showing the 95% confidence intervals around each mean, with standard errors based on the pooled variance.

MACINTOSH

① Choose the menu item **Graph/Category/Line**.

② Select SCORES from the left variable list and GROUP from the right variable list.

③ Click on the **Error** icon, type the column label "CL" into the box, and click **OK**.

④ Click on the **Axes** icon, type 80 into the "Y min" box and 130 into the "Y max" box, and click **OK**.

⑤ Click **OK**.

MS-DOS

① Choose the menu item **Graph/CPlot/Variables/Dependent/** and select SCORES from the variable list.

② Choose the menu item **Graph/CPlot/Variables/Grouping/** and select GROUP from the variable list.

③ Choose the menu item **Graph/CPlot/Options/Line** and choose line type #1.

④ Choose the menu item **Graph/CPlot/Options/Error** and choose the column CL.

⑤ Choose the menu item **Graph/CPlot/Options/Ymin/** and type "80".

⑥ Choose the menu item **Graph/CPlot/Options/Ymax/** and type "130".

⑦ Choose the menu item **Graph/CPlot/Go!**

The options for choosing error bars are the same as those discussed in section 3.2.2.4 and will not be repeated here. To create a line graph for the means in the TEST SCORES data set showing the 95% confidence intervals around the means, type

```
>GRAPH
>CPLOT SCORE * GROUP/YMIN=80 YMAX=130 ERROR=CL
>LINE
```

where "CL" is the column in the data set that contains the values of the confidence limits, as discussed in section 3.2.2.4. The *ymin* and *ymax* values were chosen so that the error bars fit completely on the graph. The option *LINE* tells SYSTAT to draw the line connecting the means of the groups. This graph is shown in Figure 3.22.

3.2.2.5.2 Unequally Spaced Groups

Sometimes the independent variable is on a meaningful (interval or ratio) scale, and in such a case the groups may not be equally spaced on the X-axis. To change this spacing in the

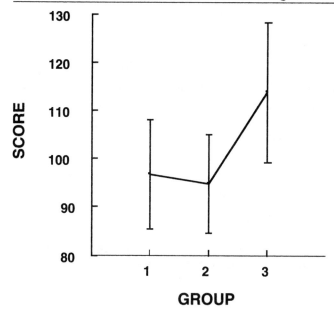

Figure 3.22 Line graph showing the mean of the three groups in the TEST SCORES data set, along with error bars showing the 95% confidence intervals around each mean, with standard errors based on the pooled variance.

graph, it is necessary to open a new data worksheet and create a new column containing the values of the independent variable, and a second column with the means of the groups. If you desire, you can create a third column containing the values of the error bars for each of the means. These columns of data can then be used to generate a graph using SYSTAT's *plot* command.

For example, suppose that the three groups of subjects in the TEST SCORES data set were each administered different quantities of adrenalin to assess its effects on test-taking performance. Suppose that the dosages were 100 units, 200 units, and 600 units for groups 1, 2, and 3, respectively. If dosage is used on the *X*-axis, these groups should not be equally spaced. To plot the groups using dosage as the *X*-axis, create a new file containing the three dosage levels in one column, the three group means in another column, and the magnitudes of the confidence intervals in a third column. The entries in this new file are as follows:

GROUP	SCORE	CL
100	96.7	11.4
200	94.8	10.3
600	113.8	14.6

MACINTOSH

① Choose the menu item **Graph/Plot/Plot**.

② Select SCORES from the left variable list and GROUP from the right variable list.

③ Click on the **Error** icon, type the column label "CL" into the box, and click **OK**.

④ Click on the **Line** icon, choose the line type you want, and click **OK**.

⑤ Click on the **Axes** icon, type "80" into the **Y min** box and "130" into the **Y max** box, and click **OK**.

⑥ Click **OK**.

MS-DOS

① Choose the menu item **Graph/Plot/Data/2-D/ Variables/Y/** and select SCORES from the variable list.

② Choose the menu item **Graph/Plot/Data/2-D/ Variables/X/** and select GROUP from the variable list.

③ Choose the menu item **Graph/Plot/Data/2-D/ Options/Line** and choose line type #1.

④ Choose the menu item **Graph/Plot/Data/2-D/ Options/Error** and choose the column CL.

⑤ Choose the menu item **Graph/Plot/Data/2-D/ Options/Ymin/** and type "80".

⑥ Choose the menu item **Graph/Plot/Data/2-D/ Options/Ymax/** and type "130".

⑦ Choose the menu item **Graph/Plot/Data/2-D/Go!**

The column labelled "CL" contains the confidence limits for each group based on the pooled error variance from the ANOVA (see section 8.7). To generate the plot of the means and error bars, type

```
>GRAPH
>PLOT SCORE*GROUP/YMIN=70 YMAX=130 ERROR=CL,
LINE=1
```

The *ymin* and *ymax* options were set to ensure that the error bars would fit completely on the graph. The *error=CL* option tells SYSTAT to use the values in the column CL as the magnitudes of the error bars. Finally, the option *line=1* causes SYSTAT to draw a solid line connecting the means of the

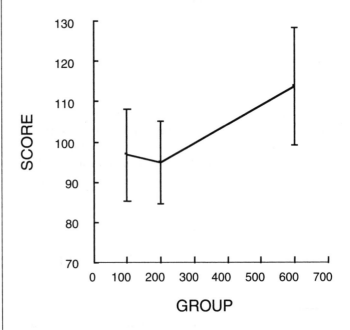

Figure 3.23 Line graph showing the means of the three groups in the TEST SCORES data set spaced according to three hypothetical drug dosage levels, along with error bars showing the 95% confidence intervals around each mean, with standard errors based on the pooled error variance.

groups (the ten line types are numbered 1 through 10). The resulting graph is shown in Figure 3.23. This graph preserves the proper spacing between the dosage levels on the X-axis.

3.2.3 Plotting Three or More Variables

3.2.3.1 Bar Graphs with Two Grouping Variables

SYSTAT allows you to plot more than one dependent variable in the same bar graph. Therefore, if you want to construct a bar graph with two or more subgroups within each group on the X-axis, you can do so by redefining a new dependent variable for each of those subgroups. For example, suppose that each group in the TEST SCORES data set was made up by approximately half men and half women, and the gender of each subject is coded in a column called GENDER$. To create separate bars for men and women in this plot, you must create two new columns containing test scores of men and women, respectively. This can be done easily using the commands

```
>IF GENDER$ = "MALE" THEN LET M = SCORE
>IF GENDER$ = "FEMALE" THEN LET F = SCORE
```

Each of these columns will have missing data, but this won't affect the graph. With these new columns you can now create the bar graph by typing, for example

```
>GRAPH
>BAR M F * GROUP /YMIN=80 YMAX=120 FILL=7,4 ,
LLABEL= "males" "females" YLAB = "TEST SCORE"
```

The resulting bar graph is shown in Figure 3.24. The *ymin* and *ymax* options were set by trial and error to make the graph look good. The *fill* option specifies the numbers corresponding to the desired patterns used to fill the bars. The *llabel* option allows you to specify the names of the subgroups to be used in the legend. Make sure that the order of the labels corresponds to the order of the subgroups at the beginning of the bar command. Finally, the *ylab* option allows you to specify the label of the Y-axis. Because SYSTAT thinks you have two dependent variables in this plot, it does not automatically know what to call the Y-axis.

3.2.3.2 Line Graphs with Two Grouping Variables

3.2.3.2.1 Equally Spaced Groups

When one of your grouping variables is on a meaningful scale, it is standard to plot the means of the categories using a line

MACINTOSH
① Choose the menu item **Graph/Bar/Bar**.
② Select M and F from the left variable list and GROUP from the right variable list.
③ Click on the **Fill** icon, type "7,4" into the box, and click **OK**.
④ Click on the **Axes** icon, type "80" into the **Y min** box and "120" into the **Y max** box, type "TEST SCORE" into the **Y label** box, and click **OK**.
⑤ Click on the **Legend** icon, type "males" into the top box and "females" into the second box, and click **OK**.
⑥ Click **OK**.

MS-DOS
① Choose the menu item **Graph/Bar/Variables/Dependent** and select M and F from the variable list.
② Choose the menu item **Graph/Bar/Variables/Grouping** and select GROUP from the variable list.
③ Choose the menu item **Graph/Bar/Options/Fill**, hit <esc>, and type "7,4" into the box.
④ Choose the menu item **Graph/Bar/Options/Ymin/** and type "80".
⑤ Choose the menu item **Graph/Bar/Options/Ymax/** and type "120".
⑥ Choose the menu item **Graph/Bar/Options/Ylabel/** and type "TEST SCORES".
⑦ Choose the menu item **Graph/Bar/Options/Llabel/** and type "males" on the first line and "females" on the second line and hit <esc>.
⑧ Choose the menu item **Graph/Bar/Go!**

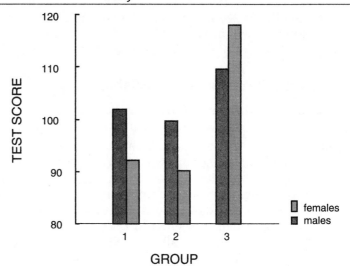

Figure 3.24 Bar graph showing the means of male and female subgroups in the TEST SCORES data set.

MACINTOSH

① Choose the menu item **Graph/Category/Line**.

② Select M and F from the left variable list and GROUP from the right variable list.

③ Click on the **Line** icon, type "1,6" into the box, and click **OK**.

④ Click on the **Axes** icon, type "80" into the **Y min** box and "120" into the **Y max** box, type "TEST SCORE" into the **Y label** box, and click **OK**.

⑤ Click **OK**.

MS-DOS

① Choose the menu item **Graph/CPlot/Variables/ Dependent/** and select M and F from the variable list.

② Choose the menu item **Graph/CPlot/Variables/ Grouping/** and select GROUP from the variable list.

③ Choose the menu item **Graph/CPlot/Options/ Line**, hit <esc>, and type "1,6" into the box.

④ Choose the menu item **Graph/CPlot/Options/ Ymin/** and type "80".

⑤ Choose the menu item **Graph/CPlot/Options/ Ymax/** and type "120".

⑥ Choose the menu item **Graph/CPlot/Options/ Ymin/** and type "TEST SCORE".

⑦ Choose the menu item **Graph/CPlot/Go!**

graph. As with bar graphs, if you want to construct a line graph with two or more subgroups within each group on the X-axis, you can do so by redefining a new dependent variable for each of those subgroups. Again, suppose that each group in the TEST SCORES data set is made up of approximately half males and half females, and the gender of each subject is coded in a column called GENDER$. To create separate lines for men and women in this plot, you must create two new columns containing test scores of men and women, respectively. This can be done using the commands

```
>IF GENDER$ = "MALE"   THEN LET M = SCORE
>IF GENDER$ = "FEMALE" THEN LET F = SCORE
```

Each of these columns will have missing data, but this won't affect the plot. With these new columns you can now create the line plot by typing

```
>GRAPH
>CPLOT M F * GROUP / YMIN=80 YMAX=120 LINE=1,6,
YLAB="TEST SCORE"
```

The resulting bar graph is shown in Figure 3.25. The *ymin* and *ymax* options were set by trial and error to make the graph look good. The *line* option specifies the numbers corresponding to the desired line pattern used in the plot. For example, the number '1' corresponds to the solid line. Next, the *ylab* option

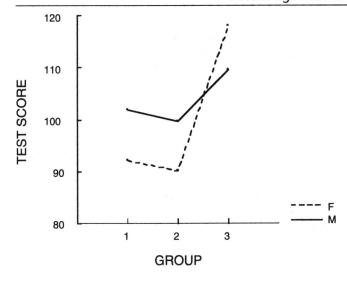

Figure 3.25 Line plot showing the means of male and female subgroups in the TEST SCORES data set.

allows you to specify the label of the *Y*-axis. Because SYSTAT thinks you have two dependent variables in this plot, it does not automatically know what to call the *Y*-axis. You must specify it yourself. Finally, unlike the bar graph in the previous section, this time the *llabel* option was omitted, and SYSTAT used the column labels to denote the subgroups in the legend.

3.2.3.2.2 Unequally Spaced Groups

Sometimes one of your groupings variables may be on an interval or ratio scale, and in such a case the groups may not be equally spaced on the *X*-axis. To change this spacing in the graph, it is necessary to open a new data worksheet and create a new column containing the values of the grouping variable, and additional columns with the means of the subgroups. These columns of data can then be used to generate a graph using SYSTAT's *plot* command (see also section 3.2.2.5.2).

For example, suppose that the three groups of subjects in the TEST SCORES data set were each administered different quantities of adrenalin to assess its effects on test-taking performance. Suppose that the dosages were 100 units, 200 units, and 600 units for groups 1, 2, and 3, respectively. If dosage is used on the *X*-axis, these groups should not be equally spaced.

Furthermore, suppose that each group in the TEST SCORES data set is made up of approximately half males and half females. To create separate lines for men and women in this plot, you must find the means for men and women separately within each group, and for this the gender of each subject must be coded in a numerical column. Suppose that you have created such a column labelled GENDER. Then the means can be found using SYSTAT's statistics procedures (see section 3.1).

```
>BY GROUP GENDER
>STATS
>STATISTICS SCORE / MEAN
```

The output from these commands is shown in Table 3.6. You can now enter these means into columns of a new data set (or at the rightmost column of your current data set): Put the three means for the males into one column and the three means for the females into another column. You also must have a column indicating the dosage level for each group. These three columns are shown in Table 3.7.

With this new file you can generate the line plots of the means for males and females by typing

```
>GRAPH
>PLOT MALES FEMALES * DOSAGE/YMIN=80 YMAX=120,
YLAB="TEST SCORE" LINE=1,6
```

The *ymin* and *ymax* options were set to ensure that the error bars would fit completely on the graph. The *ylab* option allows you to specify the label of the Y-axis: Because SYSTAT thinks you have two dependent variables in this plot, it does not automatically know what to call the Y-axis, so you must specify it yourself. The resulting graph is shown in Figure 3.26. This graph preserves the proper spacing between the dosage levels on the X-axis.

3.2.4 Common Graph Layout Options

This book is not about graphics, but graphs are used so extensively in many of the following chapters that I thought it would be useful to give brief summaries of some of the layout options available in SYSTAT. Usually, you won't need to bother with these settings because SYSTAT will make intelligent default choices for you. However, once in a while a graph may not look just the way you like it, so it is important to know that you do have control over just about every aspect of the graph. If there is something about your graph that you want to change but you don't find how to do it here, don't assume that it can't be done. This section just covers the basics. For the hard stuff, see Chapter 19 of SYSTAT's *Graphics* manual.

MACINTOSH

① Choose the menu item **Graph/Plot/Plot**.

② Select MALE and FEMALE from the left variable list and DOSAGE from the right variable list.

③ Click on the **Line** icon, type "1,6" into the box, and click **OK**.

④ Click on the **Axes** icon, type "80" into the **Y min** box and "120" into the **Y max** box, type "TEST SCORE" into the **Y label** box, and click **OK**.

⑤ Click **OK**.

MS-DOS

① Choose the menu item **Graph/Plot/Data/2-D/ Variables/Y/** and select MALE and FEMALE from the variable list.

② Choose the menu item **Graph/Plot/Data/2-D/ Variables/X/** and select DOSAGE from the variable list.

③ Choose the menu item **Graph/Plot/Data/2-D/ Options/Line**, hit <esc>, and type "1,6".

④ Choose the menu item **Graph/Plot/Data/2-D/ Options/Ymin/** and type "80".

⑤ Choose the menu item **Graph/Plot/Data/2-D/ Options/Ymax/** and type "120".

⑥ Choose the menu item **Graph/Plot/Data/2-D/ Options/Ylabel/** and type "TEST SCORE".

⑦ Choose the menu item **Graph/Plot/Data/2-D/Go!**

Table 3.6 SYSTAT output with the means of males and females within each of the three dosage level groups in the TEST SCORES data set.

```
THE FOLLOWING RESULTS ARE FOR:
          GROUP    =    1.000
          GENDER   =    1.000
TOTAL OBSERVATIONS:       7
                    SCORE
  N OF CASES                7
  MEAN              101.914

THE FOLLOWING RESULTS ARE FOR:
          GROUP    =    1.000
          GENDER   =    2.000
TOTAL OBSERVATIONS:       8
                    SCORE
  N OF CASES                8
  MEAN               92.075

THE FOLLOWING RESULTS ARE FOR:
          GROUP    =    2.000
          GENDER   =    1.000
TOTAL OBSERVATIONS:       5
                    SCORE
  N OF CASES                5
  MEAN               99.680

THE FOLLOWING RESULTS ARE FOR:
          GROUP    =    2.000
          GENDER   =    2.000
TOTAL OBSERVATIONS:       5
                    SCORE
  N OF CASES                5
  MEAN               90.140

THE FOLLOWING RESULTS ARE FOR:
          GROUP    =    3.000
          GENDER   =    1.000
TOTAL OBSERVATIONS:       6
                    SCORE
  N OF CASES                6
  MEAN              109.617

THE FOLLOWING RESULTS ARE FOR:
          GROUP    =    3.000
          GENDER   =    2.000
TOTAL OBSERVATIONS:       6
                    SCORE
  N OF CASES                6
  MEAN              118.017
```

Table 3.7 A new data file containing a column with the dosage levels and columns with the means of males and females within the groups in the TEST SCORES data set.

DOSAGE	MALE	FEMALE
100	101.914	92.075
200	99.680	90.140
600	109.617	118.017

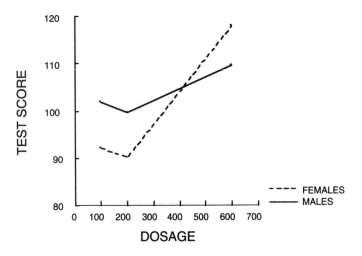

Figure 3.26 Line plot showing the means of male and female subgroups in the TEST SCORES data set, preserving the unequal spacing between dosage levels on the X-axis.

Most of the options below use commands that follow the slash in the command line for the graph procedure. Generically, these take the form

>GRAPHIC *VARIABLE1 VARIABLE2* ... / *OPTION1, OPTION2* ...

In the Macintosh menus, the options can be found by clicking on the icons in the graphics dialog boxes. In DOS menus, the options are listed under the /Options/ heading in the menus. For brevity's sake, only the command versions are shown in the following sections.

3.2.4.1 Size: /height = 2.5in width = 2.5in

You can control the height and width of the graph. SYSTAT's defaults usually center the graph on the screen in a square frame, 2.5 × 2.5 inches. (If you change either the height or width of the plot frame, you may make the plot off-center, which you can correct with the *origin* command; see the next section.) Sometimes these defaults are not appropriate. For example, in the stem and leaf display of the income variable in section 3.2.1.3, the default display size made the lines in the display too small to read. To correct this I increased the height of the display to make more room for the lines with the command

```
>STEM INCOME / HEIGHT=4IN WIDTH=2IN
```

3.2.4.2 Controlling Position: Origin x y

Unlike most of the options in this section, the position of a graph on the screen is controlled by a separate command: *origin*, followed by a pair of X-Y coordinates. These coordinates are in percentage units of the total size of the View window, and indicate the position of the lower left corner of the graph frame. SYSTAT's usual default is to center the graph on the screen, but if you change the height or width of the display (see previous section), you may also wish to recenter the graph. (I don't worry about this one because I always copy the SYSTAT graph and paste it into another program, which is the way to go if you have the appropriate software.)

To exactly center a graph, you can (a) divide the height of the graph by the height of the View window to determine what proportion of the height of the window the graph covers, (b) divide one minus this proportion by two to find the percentage of the window that should be shown below the graph, (c) enter this proportion into the X argument to the origin command, and (d) repeat a–c for the width of the graph. I just use trial and error.

3.2.4.3 Controlling Axes: /Xmin=0 Xmax=10 Ymin=0 Ymax=10

SYSTAT gives options for controlling the range of the X- and Y-axes. The *Xmin* and *Ymin* options set the minimum value of their respective axes, in units defined by the data. The *Xmax* and *Ymax* options do the same for the maximum values of those axes. (When you are plotting a 3-D graph, you may also control the maximum and minimum values in the third dimension.) Values in the data that are outside the specified range are treated as missing, so be aware that fitted lines will not be the same in a graph that excludes some of the data points as the line fitted to all the data points.

For example, in plotting the bar chart in Figure 3.20, I wanted to ensure that the standard error bars would fit within the range of the Y-axis. By trial and error I determined that a range of 80 to 130 on the Y-axis produced a good-looking graph, so I generated the graph with the command

```
>BAR SCORE * GROUP/YMIN=80 YMAX=130 SERROR
```

(See sections 3.2.2.4, 8.7, and 8.8 for more on the *SERROR* option.) By not specifying values for *xmin* or *xmax* I allowed SYSTAT to use its default values, which seem perfectly appropriate in this case.

SYSTAT also enables you to transform your X- and/or Y-axes using log or power transformations. These options are discussed further in Chapter 7.

3.2.4.4 Filling Bars: /Fill 1,8

Bars, symbols, and some other features in SYSTAT graphs can be filled with patterns that help differentiate them visually. For example, in the bar graph in Figure 3.24, males and females are represented by separate bars within each group. To help distinguish the male and female bars I filled the male bars with a gray pattern and the female bars with a lighter, hashed pattern. SYSTAT has eight patterns available, which are designated by the numbers 1 through 8 after the 'fill=' option. To select more than one type of pattern, simply separate the numbers by commas. For example, to select the two patterns in Figure 3.24, I used the option *fill=7,4* in the command

```
>BAR M F * GROUP /YMIN=80 YMAX=120 FILL=7,4,
LLABEL= "MALES" "FEMALES" YLAB = "TEST SCORE"
```

This option filled the male bars (the first set) with pattern #7 and the female bars (the second set) with pattern #4.

3.2.4.5 Legend Controls: /Llabel = "var1" "var2" Legend= 3in 2in

The legend option, *llabel*, allows you to specify legend labels other than the names of the variables used in the graph. When *llabel* is not specified, SYSTAT uses as a default the names of the variables in the order listed before the slash in the command line. For example, in Figure 3.24 the two subgroup variables were the columns labelled 'M' and 'F', but I chose to use the legend labels "males" and "females" in the command

```
>BAR M F *GROUP/YMIN=80 YMAX=120 FILL=7,4,
LLABEL="males" "females" YLAB = "TEST SCORE"
```

You can also control the position of the legend using the option 'legend=' followed by the lower left corner of the legend in X and Y coordinates. For example, to put the lower left corner of the legend three inches over and two inches up in the previous graph, type

```
>BAR M F * GROUP /YMIN=80 YMAX=120 FILL=7,4,
LLABEL= "males" "females" YLAB = "TEST SCORE",
LEGEND=3IN 2IN
```

3.2.4.6 Setting Line Types: / Line=1,11

When you have SYSTAT connect points in a graph with lines, you can control the line type with the option 'line=' followed by the number(s) corresponding to the type(s) of line you want. If you want to use more than one type, separate the numbers with commas. SYSTAT has 11 line types, ranging from solid to dots in decreasing dash length. For example, in Figure 3.25 I wanted to have a solid line representing the males and a dashed line representing females, so I chose line numbers 1 and 6, respectively. The complete command was

```
>CPLOT M F * GROUP / YMIN=80 YMAX=120 LINE=1,6,
YLAB="TEST SCORE"
```

3.2.4.7 Controlling Category Order on the X-Axis: / Nsort

For categorical variables, SYSTAT's default is to sort the categories along the X-axis in alphabetical or numerical order. For example, with the default 'sort' option on, SYSTAT would place "females" to the left of "males" on the X-axis. The other option is to have SYSTAT place categories from left to right in the order in which they are encountered in the data set. So, if you want to have males to the left of females, make sure the first male case in the data set occurs before the first female case, and turn sorting off with the option *nsort*. For example, the command

```
>BAR GENDER / NSORT
```

would create a bar chart of the gender variable, with the first gender encountered in the data set in the bar on the left and the other gender in the bar on the right.

3.2.4.8 Controlling Plot Symbols: / Symbol= 1 Fill=2 Size=3

Some plots, especially category plots, can be made more readable by controlling the symbols used to represent points on the graph. SYSTAT offers a number of symbol types (the number depends on the type of printer you are using), different sizes, and eight different fill patterns. For example, suppose I wanted to use a different symbol for males and females in the line graph in Figure 3.25. The following command gives one possibility:

```
>CPLOT M F * GROUP/YMIN=80 YMAX=130 YLAB=,
"TEST SCORE" SYMBOL=20,21 FILL=1,7 SIZE= 4
```

The resulting graph is shown in Figure 3.27. Symbols 20 and 21 correspond to the male and female symbols, respectively, as shown in the graph. I filled these symbols with the patterns 1 (black) and 7 (gray), respectively. Finally, both types of symbols were set to size 4.

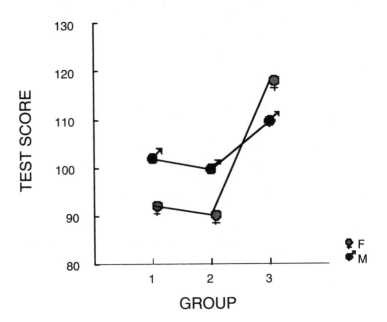

Figure 3.27 Line graph from Figure 3.25, shown here with solid lines and two types of filled symbols.

If you want no symbols in your graph, an easy way to eliminate them is to set the size option to zero.

3.2.4.9 Title: / Title = "This Is the Title of My Graph."

You can easily attach a title to the top of your graph with the 'title=' option. You simply follow this option by the title you want, in quotes. For example, the command

```
>CPLOT M F * GROUP/YMIN=80 YMAX=130 YLAB=,
"TEST SCORE" SYMBOL=20,21 FILL=1,7 SIZE=4,
TITLE="MEANS OF MALES AND FEMALES"
```

would put the title "MEANS OF MALES AND FEMALES" at the top of the graph in Figure 3.27.

3.2.4.10 Transposing X- and Y-Axes

Sometimes you may want to switch the X-axis with the Y-axis in your graph. SYSTAT will do this automatically with the *transpose* option. Just turn this option on and SYSTAT will reverse the axes.

For example, normal probability plots are usually graphed with the expected values on the X-axis and the ranked data on

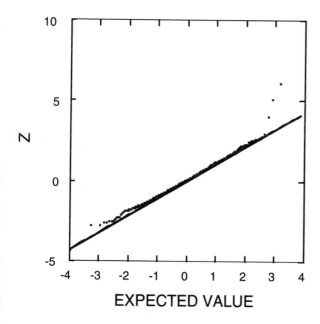

Figure 3.28 The normal probability plot from the left panel of Figure 3.10 with the axes transposed.

the Y-axis (Cook & Weisberg, 1982). SYSTAT graphs these axes the other way around (it really doesn't make a difference which way you do it but make sure your interpretation of patterns matches the way the axes are oriented). If you want to put the axes in the more usual locations, just type *transpose* at the end of the command line. Figure 3.28 shows the normal probability plot from the left panel of Figure 3.10 with the axes switched. This plot was generated with the command

```
>PPLOT Z / NORMAL SMOOTH=LINEAR TRANSPOSE
```

3.2.4.11 Formats and Global Settings

SYSTAT has many other options for choosing fonts, styles and sizes, line thickness, scale resizing, color, depth in 3-D plots, printer resolution, and so on. Because the appropriate values of these settings depend on both the type of plot and your personal taste, and because their appearance is constrained by the type of computer, printer, and monitor that you are using, I will not even attempt to cover them here. You should be aware that just about any feature of a graph in SYSTAT can be changed.

4
Meta-Analysis

Meta-analysis refers to a relatively new family of procedures that is used to combine or compare the results of multiple studies. The details of the procedures that you will actually use depend on the type of information that you have available about the studies, but the general approach can be illustrated with a single example. The steps in any meta-analysis are, roughly: (1) get the results of each study into some common metric, for example, the effect-size r; (2) combine the results of the studies to find overall summary statistics; and (3) compare the set of studies to see whether the results are "homogeneous," i.e., whether they appear to be estimating the same thing. Fortunately, each step can be accomplished using SYSTAT procedures from Chapters 1–3 of this book. For more information on meta-analysis, see Rosenthal (1991). For brevity's sake, menu equivalents for commands are not shown in this chapter, but references are given to the relevant sections in previous chapters.

4.1 Converting Results to a Common Metric

4.1.1 Computing Effect Sizes

Depending on the number of studies that you are attempting to meta-analyze, it may or may not be worth your while to transform the results of those studies in SYSTAT. The following example is a case in which it would be worth your while. Suppose that you have dug up 20 studies on a particular topic, each of which reports a significance test of the same effect. Ten of these studies used a t test of the significance of the effect, and ten of the studies used a chi-square test with one degree of freedom. The first five columns of Table 4.1 show these data as you might enter them into SYSTAT: The first column, labelled

STUDY$, is a character column containing the study labels A through T; the second column, TYPE$, is a character column indicating whether the study used a *t* test or chi-square test of the effect; the third column, RESULT, contains the value of the significance test statistic; the fourth column, N, contains the number of subjects used in each study; and the fifth column, DIR$, codes the direction of the result, positive or negative.

Table 4.1 A SYSTAT data set with the results of 20 hypothetical studies.

STUDY$	TYPE$	RESULT	N	DIR$	R	SIGNED_R	P
A	t	−0.051	8	neg	.021	−.021	0.480
B	t	0.341	6	pos	.168	.168	0.375
C	t	0.184	10	pos	.065	.065	0.429
D	t	0.484	30	pos	.091	.091	0.316
E	t	0.835	10	pos	.283	.283	0.214
F	t	0.880	50	pos	.126	.126	0.192
G	t	0.101	30	pos	.019	.019	0.461
H	t	−0.230	10	neg	.081	−.081	0.412
I	t	1.126	15	pos	.298	.298	0.141
J	t	1.630	35	pos	.273	.273	0.057
K	chi-sq	0.592	8	pos	.172	.172	0.221
L	chi-sq	0.390	6	pos	.114	.114	0.266
M	chi-sq	0.050	10	neg	.071	−.071	0.411
N	chi-sq	0.014	30	neg	.037	−.037	0.454
O	chi-sq	0.857	10	pos	.138	.138	0.177
P	chi-sq	0.015	50	neg	.032	−.032	0.451
Q	chi-sq	2.302	30	pos	.277	.277	0.065
R	chi-sq	2.268	10	pos	.213	.213	0.066
S	chi-sq	0.691	15	pos	.240	.240	0.203
T	chi-sq	0.151	35	neg	.087	−.087	0.348

The first step in meta-analysis is to get the results of the studies into a common metric. Usually, this will be some type of effect-size measure. In Table 4.1 the results are of two types, *t* statistics and $\chi^2_{(1)}$, and both of these can easily be transformed into the Pearson product-moment correlation *r* (although *r* is usually called "ϕ" when it comes from chi-square). This is an effect-size measure with many desirable properties: It ranges from −1 to +1, most people are familiar with it, and so on. The formula for transforming *t* into *r* is

$$r = \sqrt{\frac{t^2}{t^2 + df}} \qquad (4.1)$$

and the formula for transforming $\chi^2_{(1)}$ into r is

$$r = \sqrt{\frac{\chi^2_{(1)}}{n}} \qquad (4.2)$$

To transform those studies that use t as a significance test into r, you can use SYSTAT's conditional recoding procedures (section 2.2.2) to transform with Equation 4.1 only those studies that used t. For the example data, this can be done by typing

```
>IF TYPE$="T" THEN LET R = SQR(RESULT^2/
(RESULT^2 + N-2))
```

where N–2 is the degrees of freedom for t. To transform those studies that use $\chi^2_{(1)}$ as a significance test into r, you can employ a similar command using Equation 4.2:

```
>IF TYPE$="CHI-SQ" THEN LET R = SQR(RESULT/N)
```

These two transformations place the proper magnitudes of the effect-size r in the column R, the third column from the right in Table 4.1.

However, the values of six of these r's have the wrong sign. Note that the results for studies A and H in Table 4.1 were in the opposite direction of the other studies, but the minus sign was removed when we transformed t into r. Similarly, chi-squares are always positive, but we know from the column DIR$ that four of the studies using chi-square had results in the direction opposite the rest. One of the most crucial steps in a meta-analysis is making sure that the directions of all of the effects are coded properly. Assuming that the signs are correct in column DIR$, and that "pos" for studies using t indicate effects in the same direction as "pos" for a study using chi-square, you can reattach the minus sign using the following pair of transformations:

```
>IF DIR$="POS" THEN LET SIGNED_R = R
>IF DIR$="NEG" THEN LET SIGNED_R = -R
```

These transformations finally get all of the effects on the same scale with signs appropriate to the direction of the effects in the studies. The results are shown in the column SIGNED_R in Table 4.1. Now the data are ready for meta-analysis.

Some of the studies may have estimates of *r* that are significantly different from zero; others certainly will not. Out of curiosity, let's begin by seeing which of the studies taken separately would have been considered statistically significant at the .05 level, two-tailed. To do this, you need one command to transform the *t*'s into one-tailed *p*-values using the TCF() function (see section 2.2.3.2.1) and another to transform the chi-squares into one-tailed (i.e., directional) *p*-values using the XCF() function (see section 2.2.3.4.1 for a discussion of "one-tailed" chi-squares). These commands are

```
>IF TYPE$="T" THEN LET P=1-TCF(ABS(RESULT),N-2)
>IF TYPE$="CHI-SQ" THEN LET P=(1-XCF,
(RESULT,1))/2
```

In the first command, ABS() is the absolute value function, which converts all of the *t*'s in the column RESULT into their absolute values. The expression "1 – TCF()" takes the values of *t* and their degrees of freedom given by N–2 and computes the probability of obtaining *t* greater than or equal to the magnitude specified, due to chance. Because we first took the absolute value of the *t*'s, this command provides one-tailed probabilities, whether the original *t*'s were positive or negative.

In the second command, the expression "1 – XCF()" takes the value of the chi-square with degrees of freedom equal to 1 and computes the probability of obtaining a chi-square greater than or equal to the value specified due to chance. Because a chi-square is a squared statistic, it is "two-tailed" (i.e., non-directional; see section 2.2.3.4.1) and the probability must be divided by two to obtain the "one-tailed" value. This is the function of the "/2" at the end of the command line.

The resulting one-tailed probabilities are shown in the last column of Table 4.1. None of the studies found statistically significant effects! Is there any point in continuing with the meta-analysis in such a circumstance? Emphatically *yes*, as we shall later see.

4.1.2 Fisher's Z_r Transformation

An *r* distribution in general is not a normal distribution. This is obvious from the fact that *r*'s are bounded between –1 and +1. To make the *r*'s look more normal, Fisher suggested a transformation that has come to be called Fisher's Z_r, which is not bounded and is approximately normally distributed. So before you proceed with any meta-analytic procedures using

r's, you should first transform them into Z_r's. Conveniently, this transformation is equivalent to the ATH() ("arc-tangent hyperbolic") function available in SYSTAT. For example, to transform the signed r's in Table 4.1 into Z_r's, simply type

```
>LET ZR = ATH(SIGNED_R)
```

The resulting Fisher's Z_r's are shown in the fifth column, labelled ZR, of Table 4.2 (page 109). Note that the changes from the original r's (shown in the fourth column of Table 4.2) are relatively small, but increase with the magnitude of the r. For large r's this transformation can make a huge difference.

4.1.3 Displaying the Data

4.1.3.1 Stem and Leaf Displays

Before proceeding with the meta-analytic procedures, take a moment to look at the r's and Z_r's in stem and leaf displays (see section 3.2.1.3), box plots, and graphs of the Z_r's versus n. The stem and leaf displays are shown in Figure 4.1. (Note: Don't believe SYSTAT if it says that the minimum, maximum, median, and upper and lower hinges are all equal to zero! You may need to adjust the number of decimal places shown in the output or display numbers near zero in exponential notation as done in Figure 4.1.) The r's range from $-.087$ to $+.298$, with a median of $+.120$. Neither the r's nor the Z_r's look normally distributed, but this is not very surprising given the small number of data points. Importantly, none of the studies really stand apart from the rest.

4.1.3.2 Box Plots

With so few data points, the stem and leaf displays in Figure 4.1 do not provide a very compelling sense of the distribution. Another way to examine the distribution of data is through a box plot (see section 3.2.1.2). Box plots of both the r's and Z_r's in Figure 4.2 suggest that the data are fairly symmetrically distributed around the median. Again, there are no outliers in these plots.

4.1.3.3 Plotting Z_r versus n

There is one more plot that you might want to examine. Because the Z_r's will be weighted in the following analyses according to the sizes of the studies, you might want to look at a plot of Z_r versus n to see which studies will be influencing the results the most. To generate this plot, you can use SYSTAT's usual scatterplot commands (see section 3.2.2.1). The resulting

```
STEM AND LEAF PLOT OF VARIABLE:   SIGNED_R,
N =    20

MINIMUM IS:         -.870000E-01
LOWER HINGE IS:     -.265000E-01
MEDIAN IS:           .012000E+01
UPPER HINGE IS:      .022650E+01
MAXIMUM IS:          .029800E+01

       -0     887
       -0   H 332
        0     1
        0     69
        1   M 123
        1     67
        2   H 14
        2     7789


STEM AND LEAF PLOT OF VARIABLE:            ZR,
N =    20

MINIMUM IS:         -.872205E-01
LOWER HINGE IS:     -.265070E-01
MEDIAN IS:           .012059E+01
UPPER HINGE IS:      .023054E+01
MAXIMUM IS:          .030732E+01

       -0     887
       -0   H 332
        0     1
        0     69
        1   M 123
        1     67
        2   H 14
        2     889
        3     0
```

Figure 4.1 Stem and leaf displays of the signed r's (top panel) and Fisher Z_r's (bottom panel) from the data in Table 4.1.

plot is shown in Figure 4.3 with the best-fitting, least-squares line. This plot suggests that large studies tend to have somewhat larger effect sizes, on average, than small studies. Also, the plot has the shape of a megaphone opening to the left, suggesting that the variance of the effect-size estimates may be smaller for larger studies. This is what you would expect if the

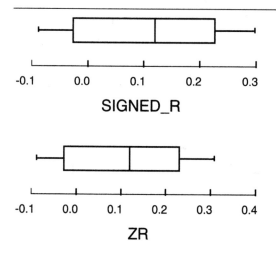

Figure 4.2 Box plots of the signed r's (top panel) and Fisher Z_r's (bottom panel) from the data in Table 4.1.

larger studies yielded better estimates of the same population effect size.

The points toward the right on the X-axis in Figure 4.3 are going to have the largest influence in the analyses that follow. It is a good idea to identify those cases in the data set (see section 3.2.1.5) and examine those studies carefully to ensure that they are of high quality. Meta-analyses can be weighted by study quality, although this will not be covered here.

4.2 Combining Studies
4.2.1 Combining Effect Sizes
The mean Z_r in Table 4.2 (column ZR) can be found easily using SYSTAT's statistics command (see section 3.1). The mean, minimum, and maximum computed with this command are shown in Table 4.3. However, for the purpose of finding a combined effect-size estimate from a number of studies, you will usually want to weight the effect-size estimates by the sizes of the studies that produced them, giving greater weight to effect-size estimates from larger studies. This is because the larger studies themselves should give better estimates of the true population effect size (assuming no true differences among studies and that these studies are all of equal quality). To find the weighted mean Z_r of the Fisher's Z-transformed effect-size estimates, you can use the following formula (Rosenthal, 1991, p. 74):

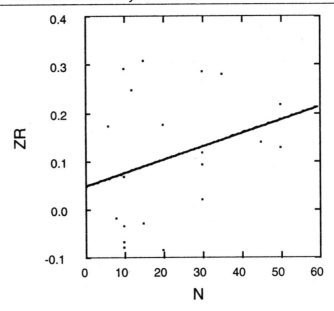

Figure 4.3 Scatterplot of Z_r versus n for the twenty studies in Table 4.1.

$$\bar{Z}_r = \frac{\Sigma(n_j - 3)Z_{rj}}{\Sigma(n_j - 3)} \qquad (4.3)$$

where n_j is the number of subjects in study j. This can be computed with the help of SYSTAT in four steps. First, create a new column in the data set containing $n-3$ for each study. This can be done, for example, with the command

>LET N_3 = N-3

The result of this command is shown in the sixth column, N_3, of Table 4.2. (I used an underscore rather than a minus sign because the minus sign is illegal in variable labels.)

Second, create a column called something like WEIGHTZ that is equal to $(n-3) \times Z_r$ for each study: These are the values to be summed in the numerator of Equation 4.3. They can be computed with the command

>LET WEIGHTZ = N_3 * ZR

The result of this command is shown in the seventh column, WEIGHTZ, of Table 4.2.

Third, use the *statistics* command to find the sums of the

Table 4.2 The first four columns are from Table 4.1. Columns five through eight show steps in the computation of the weighted mean *r* of the twenty studies.

STUDY$	N	DIR$	SIGNED_R	ZR	N_3	WEIGHTZ	MEAN_R
A	8	neg	−.021	−0.021	5	−0.105	0.137
B	6	pos	.168	0.170	3	0.509	0.137
C	10	pos	.065	0.065	7	0.456	0.137
D	30	pos	.091	0.091	27	2.464	0.137
E	10	pos	.283	0.291	7	2.037	0.137
F	50	pos	.126	0.127	47	5.954	0.137
G	30	pos	.019	0.019	27	0.513	0.137
H	10	neg	−.081	−0.081	7	−0.568	0.137
I	15	pos	.298	0.307	12	3.688	0.137
J	35	pos	.273	0.280	32	8.963	0.137
K	8	pos	.172	0.174	17	2.953	0.137
L	6	pos	.114	0.114	27	3.091	0.137
M	10	neg	−.071	−0.071	7	−0.498	0.137
N	30	neg	−.037	−0.037	7	−0.259	0.137
O	10	pos	.138	0.139	42	5.833	0.137
P	50	neg	−.032	−0.032	12	−0.384	0.137
Q	30	pos	.277	0.284	27	7.680	0.137
R	10	pos	.213	0.216	47	10.167	0.137
S	15	pos	.240	0.245	9	2.203	0.137
T	35	neg	−.087	−0.087	17	−1.483	0.137

Table 4.3 Minimum, maximum, and mean of the Z_r's in Table 4.2.

```
TOTAL OBSERVATIONS:      20

                         ZR

N OF CASES               20
MINIMUM                  -0.087
MAXIMUM                  0.307
MEAN                     0.110
```

columns just created: WEIGHTZ and N_3. These sums are the numerator and denominator, respectively, of Equation 4.3. To find these sums, type

```
>STATS
>STATISTICS WEIGHTZ N_3 / SUM
```

The output of this command is shown in Table 4.4.

Table 4.4 Sum of the column of weighted Z_r's and their weights, n–3, for the data in Tables 4.1 and 4.2.

TOTAL OBSERVATIONS: 20

	WEIGHTZ	N_3
N OF CASES	20	20
SUM	53.21288	386

The fourth and final step of the procedure is to divide the sum of the column of weighted Z_r's by the column of n–3 using a hand calculator. This gives 53.213/386 = .138 as the weighted mean Z_r for the 20 studies in Table 4.2. To find the weighted mean r, this weighted mean Z_r can be converted back using the following formula (Rosenthal & Rosnow, 1991, p. 599):

$$\bar{r} = \frac{e^{2Z} - 1}{e^{2Z} + 1} \tag{4.4}$$

This can be computed either on a hand calculator ($e \cong 2.718$) or in SYSTAT with the command

```
>LET MEAN_R = (2.718^(2*.138)-1)/,
(2.718^(2*.138)+1)
```

where the value 0.138 is the value of the mean Z_r from above. This command fills the new column MEAN_R with the value 0.137, as shown in the last column of Table 4.2, which is the proper value of the weighted mean r for the 20 studies.

This value of 0.137 is slightly higher than the unweighted mean in Table 4.3, partly because of Study R in Table 4.2, a study with a large effect size and large n. If you wish to see what the weighted mean would be without Study R, you could exclude Study R from the computation before repeating the statistics command:

```
>SELECT STUDY$ <> "R"
>STATISTICS WEIGHTZ N_3 / SUM
```

This gives the values 43.046 and 339, respectively, for the numerator and denominator of Equation 4.3. This gives 43.046/339 = .127 as the weighted mean Z_r of the studies, and .126 as the weighted mean r of the studies. This value is still slightly greater than the unweighted mean value in Table 4.3. Study R does not dramatically influence the mean effect size of the twenty studies.

4.2.2 Combining Probabilities

Is the weighted mean effect size, $\bar{r} = .137$, significantly different from zero? There are a few different methods for determining whether a number of studies taken together produce an effect-size estimate that is statistically different from zero (see Rosenthal, 1991, p. 90). The simplest of these methods is the *Stouffer* method, which converts the *p*-values of each study into a standard normal deviate Z, and then combines these Z's (these are not Fisher's Z_r's). The *p*-values for each study must be one-tailed, as are those in Table 4.1, and you will need to keep track of which tail each one comes from, i.e., whether the results are positive or negative. To find the Z's corresponding to the *p*-values in the fourth column of Table 4.5, use SYSTAT's *Z inverse function* (see section 2.2.3.1.2). For example, type

```
>IF DIR$="NEG" THEN LET Z_FROM_P = ZIF(P)
>IF DIR$="POS" LET Z_FROM_P = ZIF(1-P)
```

The two separate commands are necessary to take the signs of the effects into account. This new column of Z's, called Z_FROM_P, is shown in the fifth column of Table 4.5. These Z's can now be combined with the formula

$$Z = \frac{\Sigma Z_j}{\sqrt{k}} \qquad (4.5)$$

where *k* is the number of studies. To compute this, you simply find the sum of the column of Z's using the command

```
>STATS
>STATISTICS Z_FROM_P / SUM
```

which gives the value 10.451, and divide this by the square root of the number of studies: $10.451/\sqrt{20} = 2.34$. Finally, you can convert this Z into a two-tailed *p*-value, for example, using the following command (see section 2.2.3.1.1)

```
>LET P_FROM_Z = 2*(1-ZCF(Z_FROM_P))
```

(You would leave off the "1–" if the Z were negative.) This command fills the column P_FROM_Z with the value .01, as shown in the last column of Table 4.5, which is significant at the traditional .05 level. Thus, you may conclude that the combined effect size for the twenty studies is significantly different from zero. This shows one of the great virtues of meta-analysis: Although no single study in Table 4.1 found a significant result, this was probably because they lacked the power to find one, not because the effect did not exist. The meta-analysis takes advantage of the relative consistency of the results across the studies.

Table 4.5 The first four columns are from Table 4.1. Column five shows the Z's corresponding to the one-tailed p-values in column four. Column six is the two-tailed p-value for the combined effect size from the 20 studies.

STUDY$	N	DIR$	P	Z_FROM_P	P_FROM_Z
A	8	neg	0.480	–0.049	0.010
B	6	pos	0.375	0.319	0.010
C	10	pos	0.429	0.179	0.010
D	30	pos	0.316	0.479	0.010
E	10	pos	0.214	0.793	0.010
F	50	pos	0.192	0.872	0.010
G	30	pos	0.461	0.099	0.010
H	10	neg	0.412	–0.222	0.010
I	15	pos	0.141	1.078	0.010
J	35	pos	0.057	1.585	0.010
K	8	pos	0.221	0.769	0.010
L	6	pos	0.266	0.625	0.010
M	10	neg	0.411	–0.225	0.010
N	30	neg	0.454	–0.117	0.010
O	10	pos	0.177	0.925	0.010
P	50	neg	0.451	–0.124	0.010
Q	30	pos	0.065	1.518	0.010
R	10	pos	0.066	1.506	0.010
S	15	pos	0.203	0.831	0.010
T	35	neg	0.348	–0.389	0.010

4.3 Comparing Studies

4.3.1 Three or More Studies

Are there significant differences among the effect sizes in Table 4.1? To assess the heterogeneity of effect sizes you can use a test statistic with a chi-square distribution on $k-1$ degrees of freedom, where k is the number of studies (Rosenthal, 1991, p. 74). The formula for this chi-square test statistic is

$$\chi^2_{(k-1)} = \Sigma[(n_j - Z_{rj} - \bar{Z}_r)^2] \tag{4.6}$$

where $\bar{Z}_r$ is the weighted mean of the Z_r's as found above in section 4.2.1. This is equal to 0.138. To compute the ingredients of Equation 4.6, you must create one more new column in the data set corresponding to the values on the right side of Equation 4.6 that are to be summed; that is, for each study you need to compute the weighted deviation $(n_j - 3)(Z_{rj} - \bar{Z}_r)^2$. You can do this with the following command

```
>LET WEIG_DEV = (N-3)*(ZR-.138)^2
```

where WEIG_DEV is the new column of weighted, squared deviations from the mean Z_r of 0.138. This column is shown in Table 4.6.

The sum of this WEIG_DEV column in Table 4.6 is distributed as χ^2 with $k-1$ degrees of freedom. You can find the sum using SYSTAT's statistics command (see section 3.1), the output of which is shown in Table 4.7. This sum indicates that χ^2 = 4.840 with 19 df, which is nowhere near statistical significance, $p = .999$ (see section 2.2.3.4.1 for computing exact significance levels). Thus, it appears that the 20 effect sizes are statistically homogeneous. (In fact, this degree of homogeneity would be very unusual, and suspect, in real data!) However, remember that a chi-square with more than 1 df that is not significant can still contain significant 1 df effects. For example, perhaps studies conducted on the East Coast yield different effect sizes than studies conducted on the West Coast. Such differences can be tested using contrasts on the effect sizes from those studies (Rosenthal, 1992).

Table 4.6 The first four columns are from Table 4.2. Column five shows the weighted, squared deviations of the values in the fourth column around their weighted mean.

STUDY$	N	DIR$	ZR	WEIG_DEV
A	8	neg	−0.021	0.126
B	6	pos	0.170	0.003
C	10	pos	0.065	0.037
D	30	pos	0.091	0.059
E	10	pos	0.291	0.164
F	50	pos	0.127	0.006
G	30	pos	0.019	0.382
H	10	neg	−0.081	0.336
I	15	pos	0.307	0.344
J	35	pos	0.280	0.646
K	8	pos	0.174	0.022
L	6	pos	0.114	0.015
M	10	neg	−0.071	0.306
N	30	neg	−0.037	0.214
O	10	pos	0.139	0.000
P	50	neg	−0.032	0.347
Q	30	pos	0.284	0.579
R	10	pos	0.216	0.288
S	15	pos	0.245	0.103
T	35	neg	−0.087	0.862

4.3.2 Comparing Two Studies

You can also compare the results of two studies to see whether they differ. This can be done on a hand calculator using the equation

$$Z = \frac{Z_{r_1} - Z_{r_2}}{\sqrt{\frac{1}{n_1 - 3} + \frac{1}{n_2 - 3}}} \qquad (4.7)$$

where the Z on the left side is approximately normally distributed (it is not a Fisher's Z_r). To illustrate, suppose you were interested in comparing the smallest effect size in Table 4.1, −0.087, with the largest effect size in that table, 0.298. The n's of these two studies are 20 and 15, respectively. Plugging these numbers into Equation 4.7, you find

Table 4.7 Sum of the weighted, squared deviations in column WEIG_DEV in Table 4.6.

```
TOTAL OBSERVATIONS:        20

                       WEIG_DEV

   N OF CASES                20
   SUM                  4.84032
```

$$\frac{.298 - (-.087)}{\sqrt{\dfrac{1}{20-3} + \dfrac{1}{15-3}}} = 1.02 \qquad (4.8)$$

which is approximately normally distributed. The one-tailed p-value corresponding to this Z is .154, which is not significant at the traditional level. Thus, even the two most extreme effect sizes in Table 4.1 do not differ significantly.

5

Simple Linear Regression

The SYSTAT command for estimating and testing regression coefficients is simple, and in itself would not require an entire chapter of discussion. But a complete regression analysis is much more than estimating coefficients: It involves computing summary statistics, graphing the data in a variety of ways, testing the assumptions underlying the regression, and possibly transforming the data in one or more ways. Aside from transformation, which is treated separately in Chapter 7, this chapter pulls together all of the procedures for conducting a complete regression analysis with a single predictor variable. Most regression procedures can be illustrated with a single predictor variable; additional issues that arise only with two or more variables, such as collinearity, are covered in Chapter 6. As in the chapters that follow, this chapter will analyze a single data set from start to finish. The analysis proceeds in three overlapping stages: (1) looking at the raw data, (2) estimating the regression coefficients, and (3) checking the assumptions underlying the regression. Various confidence interval calculations are shown at the end of the chapter.

The Example Data Set

The next three chapters use the SHOES data set to illustrate SYSTAT's regression procedures. In this chapter, assume that you have two measurements on 100 families: their total family income and the number of pairs of shoes that the family owns. Such a data set might be of interest in marketing research, for example. The first five cases are shown in Table 5.1. (The complete data set is given in section 5.6, and can be reconstructed by Mac users using SYSTAT functions as described in section 5.6.) Suppose that you want to use family income to predict the number of pairs of shoes that a family owns. Following convention, I will call family income the X, or *predictor*, variable and number of pairs of shoes the Y, or *outcome*, variable. The goal of this chapter is to carry out a complete regression analysis of Y on X for this data set.

5.1 Preliminary Diagnostics: Look at the Raw Data

The first step in data analysis almost always should be to look at plots of the data along with simple summary statistics. SYSTAT's graphics capabilities make this step quite easy. This section proceeds by looking at summary statistics, plotting variables separately to examine their distributions, and then plotting the joint (two-dimensional) distributions of the variables. Some people think that looking at the data is a form of cheating, because the data may suggest relationships that were not expected beforehand, and thus permit the analyst to capitalize on chance patterns in the data. However, looking at the data is only cheating if you present what you find as a *prediction* and fail to take into account the possibility of chance results. Rather than missing potentially important patterns in the data, it is better to go ahead and snoop, and then warn the reader by saying "an examination of the data *suggests* ...," or by formally adjusting the levels of the significance tests, or both. You should not sacrifice to p-values the valuable information about your data that only plotting it can provide.

5.1.1 Summary Statistics

SYSTAT offers 13 summary statistics for single variables, all of which can be generated with a single command (see section 3.1). To generate all 13 statistics for the shoes (Y) and income (X) variables, open the SHOES data set and type the following commands:

▶ >STATS
>STATISTICS SHOES INCOME/ALL

The "all" option at the end of the second command line tells SYSTAT to compute all 13 statistics. The output generated by these commands is shown in Table 5.2.

The statistics in Table 5.2 are described in section 3.1. In

MACINTOSH

① Choose the menu item **Stats/Stats/Statistics...**
② Select SHOES and INCOME from the variable list.
③ Check the **All** option box.
④ Click **OK**.

MS-DOS

① Choose the menu item **Statistics/Stats/Statistics/Variables/**.
② Select SHOES and INCOME from the variable list.
③ Choose the menu item **Statistics/Stats/Statistics/Options/All**.
④ Choose the menu item **Statistics/Stats/Statistics/Go!**

Table 5.1 The first five cases of the SHOES example data set in section 5.6. The names are fictitious; the numbers were generated in SYSTAT.

CASE	Family	Income (X)	Pairs of Shoes (Y)
1	Smith	18,244	9
2	Jones	21,883	7
3	Lincoln	15,439	11
4	Douglas	16,915	16
5	Gates	42,742	15
...			

Table 5.2, skewness for the shoes variable is 0.89, with a standard deviation of

$$\sqrt{6/100} = 0.245$$

Dividing the former by the latter to obtain an approximate value of Z gives 3.63, which is very unlikely due to chance. For the income variable $Z = 11.3$, which is extremely high. Thus, the population distributions of both variables appear to be positively skewed.

Table 5.2 Summary statistics of shoes and income in the SHOES data set.

```
TOTAL OBSERVATIONS:      100

                     SHOES            INCOME

N OF CASES             100               100
MINIMUM              1.000          1533.000
MAXIMUM             26.000        122855.000
RANGE               25.000        121322.000
MEAN                10.140         20676.310
VARIANCE            22.768       .296682E+09
STANDARD DEV         4.772         17224.465
STD. ERROR           0.477          1722.446
SKEWNESS(G1)         0.890             2.775
KURTOSIS(G2)         1.170            11.746
SUM               1014.000       2067631.000
C.V.                 0.471             0.833
MEDIAN               9.000         17111.000
```

The value of kurtosis for shoes is 1.17, with a standard deviation of

$$\sqrt{24/100} = 0.490$$

Dividing kurtosis by its standard deviation gives $Z = 2.39$, which is unlikely due to chance, suggesting that the shoes distribution is long-tailed. For the income variable, kurtosis equals 11.746 with standard deviation 0.490, which gives the very unlikely value of $Z = 24$. Thus, the income distribution is also long-tailed.

For the shoes and income variables, skewness and kurtosis do not tell us much that we could not see in the plots of those variables below. However, they do allow for a rough test of what we think we see in the plots, and in less clear circumstances they can be very informative.

5.1.2 Histograms

A histogram shows the distribution of data points across a single variable (see section 3.2.1.1). Figure 5.1 shows histograms for the income and shoes variables from the SHOES example data set. These plots were created by typing

```
>GRAPH
>DENSITY INCOME SHOES
```

From the heights of the bars you can read off the number of cases falling in a given income or shoe interval. For example, the top panel of Figure 5.1 indicates that 37 families had incomes between $10,000 and $19,999. Because we had exactly 100 cases in this data set, this corresponds to 37% of the cases as shown by the left Y-axis in the top panel of Figure 5.1.

For our purposes, the interesting feature of the top panel of Figure 5.1 is that the distribution of incomes is very asymmetric, with a large positive skew and a sudden drop off at zero dollars. Therefore, the underlying income distribution may not be normally distributed, as suggested by the values of skewness and kurtosis computed above. Another feature of the top panel of Figure 5.1 is that one of the cases, the family with an income between $120,000 and $130,000 stands apart from the rest of the group. We will want to keep an eye on this case to see if it has an undue influence on the regression analysis.

The bottom panel of Figure 5.1 indicates that the distribution of numbers of pairs of shoes also may not be normal. This distribution is more peaked than we would like, although it is fairly symmetric. The value of skewness computed above indicates that the distribution is positively skewed, but you can see in the histogram that the skewness is not dramatic.

5.1.3 Box Plots

Box plots (Tukey, 1977) give a different visual impression than do histograms and contain more explicit information about the distribution of cases (see section 3.2.1.2). These plots can

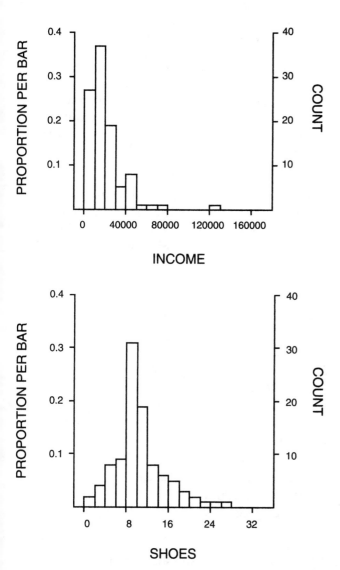

Figure 5.1 Histograms of income and shoes from the SHOES example data set.

easily be created using SYSTAT's default parameters with the commands

```
>GRAPH
>BOX INCOME SHOES
```

The top panel of Figure 5.2 shows the box plot for income. As in the histogram for this variable (top panel of Figure 5.1), the box plot indicates positive skew. Interestingly, the box plot suggests that data on income has four mild and two serious outliers. However, an examination of the histogram suggests that only the largest outlier is markedly separate from the rest of the data. The other outliers are consistent with the skewed shape of the rest of the data: Rather than being "outliers" per se, they may simply be large values expected in a skewed distribution. This illustrates one reason why you might want to look at more than one type of plot of the variables to get a better feel for how your data is distributed.

5.1.4 Stem and Leaf Displays

Stem and leaf displays combine features of box plots and histograms, while presenting all of the data points in the data set in numeric form (see section 3.2.1.3). Figure 5.3 shows the stem and leaf displays of the INCOME and SHOES variables.

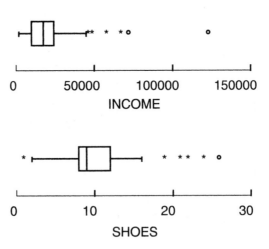

Figure 5.2 Box plots of income and shoes in the SHOES example data set.

These displays were generated by the following two commands:

```
>GRAPH
>STEM INCOME SHOES/ HEIGHT=4IN WIDTH=2IN
```

The line that contains the median is marked with the letter *M*, and lines containing the upper and lower hinges (75th and 25th centiles) are marked with the letter *H*. The actual values of these numbers are given by the text in the analysis window. For example, the median for the income variable is $17,111, and this value is located in the fourth line from the top in the stem and leaf display in the left panel of Figure 5.3. The six outliers that we found in the box plots for the income variable are also listed in the stem and leaf display. But here we can read the values of the numbers in thousands of dollars directly from the display: 46, 48, 58, 67, 72, and 122 thousand dollars. We know from the histogram (top panel of Figure 5.1) that only the $122,000 value represents a substantial departure from the rest of the data.

5.1.5 Summary: Plotting Single Variables

Plotting the individual predictor and outcome variables allows you to do three important things: (1) check the shape of the distributions, (2) identify outliers that may represent troublesome cases in the regression analysis, and (3) identify grossly misrecorded data points. If the data is non-normal, you may want to consider whether the variable is on an appropriate scale, and whether you should transform one or more variables to make them more normal. The normality of variables taken individually is not assumed by regression, but non-normal distributions may have more outliers that heavily influence the regression estimates. If outliers cause problems in the analysis, you may want to delete them to see how the results of the analysis change. Both of these topics are treated in more detail below.

In the SHOES data set we found that the income variable was positively skewed and contained six outliers, two of which were more than three times the interquartile range beyond the upper hinge (the 75th centile). However, given the positive skew of this distribution, only the largest of these outliers, $122,000, looks out of line with the rest of the data. A transformation that removed the positive skew will probably bring the rest of the outliers back into the normal range of the data.

```
STEM AND LEAF PLOT OF VARIABLE:
INCOME     , N =   100

MINIMUM IS:         1533.000
LOWER HINGE IS:     9728.000
MEDIAN IS:         17111.000
UPPER HINGE IS:    24153.000
MAXIMUM IS:       122855.000

   0      113444
   0    H 555566666677778999999
   1      00011222233444
   1    M 5555666677778888889999
   2    H 0011222333444
   2      556779
   3      04
   3      779
   4      002244
          ***OUTSIDE VALUES***
   4      68
   5      8
   6      7
   7      2
  12      2


STEM AND LEAF PLOT OF VARIABLE:
SHOES      , N =   100

MINIMUM IS:            1.000
LOWER HINGE IS:        8.000
MEDIAN IS:             9.000
UPPER HINGE IS:       12.000
MAXIMUM IS:           26.000

   1      00
          ***OUTSIDE VALUES***
   2      0
   3      000
   4      0000
   5      0000
   6      000
   7      000000
   8    H 000000000000000
   9    M 0000000000000000
  10      0000000
  11      000000000000
  12    H 00000
  13      000
  14      0
  15      00000
  16      00000
          ***OUTSIDE VALUES***
  19      000
  21      00
  22      0
  24      0
  26      0
```

Figure 5.3 Stem and leaf displays of income and shoes variables in the SHOES data set.

The shoes variable is more symmetrically distributed, but tends to pile up a little too much in the middle, giving it a sharp peak in the center. The shoes variable also had outliers, as shown in the box plot and stem and leaf display. But in this case, even the most serious of these, the family with 25 pairs of shoes, does not fall very far beyond the rest of the data. In the analyses below, each of these points will be examined to determine whether they represent catagorically different kinds of values than the remainder of the data.

5.1.6 Identifying Individual Cases

Both the box plots and stem and leaf displays help locate outliers. For example, the stem and leaf display of the income variable in the SHOES data set indicates that this variable has six outliers, and the box plot suggests that two of these fall outside the upper fence. Table 5.3 shows the last eight cases of this data set after it was sorted in ascending order by income (see section 2.1.1). The six outliers found above are now labelled cases 95 through 100. Cases 99 and 100 lie beyond the outer fence. You should pay special attention to such cases in your regression analyses. Note that these cut-offs are somewhat arbitrary: Case 95 does not differ much from case 94 on the income variable. Outlier cut-offs should be used as hueristic devices for bringing cases to our attention, not as rigid criteria for treating some points differently than others.

5.1.7 Plotting More Than One Variable

Outliers do not necessarily affect a regression analysis. The influence that a case has on the regression line is measured by its "leverage," which I shall return to below. To further help identify cases that may cause trouble for the regression, you can look at the joint distribution of the variables by creating

Table 5.3 Last eight cases from the SHOES data set, sorted in ascending order of income.

Case	Income	Shoes
93	44,805.000	24.000
94	44,915.000	16.000
95	46,765.000	12.000
96	48,822.000	4.000
97	58,499.000	11.000
98	67,397.000	15.000
99	72,463.000	19.000
100	122,855.000	26.000

scatterplots of the outcome variable Y on the predictor variable X.

A scatterplot of shoes versus income for the SHOES data set is shown in Figure 5.4 (see section 3.2.2.1). This plot reveals that both variables thin out as the values increase (which you already knew from the plots of the individual variables). What we did not know from previous plots is that one case contains the largest value on *both* variables. This case may have a large influence on the regression. The outlier case in Figure 5.4 is, of course, the family with the highest income, which was found in Table 5.3 when we sorted the data set on the income variable. This family also has the largest number of shoes at 26.

Other plots that may be of interest are influence plots and plots with fitted ellipses and lines. The influence plot for shoes versus income is shown in Figure 5.5 (see section 3.2.2.1.2). The *influence* of each point is the amount by which the correlation between the variables would change if that case were eliminated. The case with the largest value on both variables, case 100 in Table 5.3, has a very large influence on the correlation between these variables, as suspected from the scatterplot in Figure 5.4. This case will *pull* the regression line towards it to a much greater extent than the other cases. The remaining cases are very similar in their influence. Whether

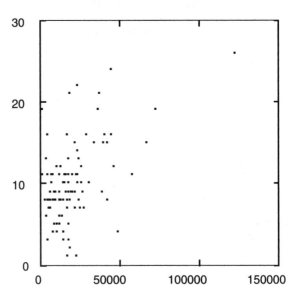

Figure 5.4 Scatterplot of shoes versus income in the SHOES data set.

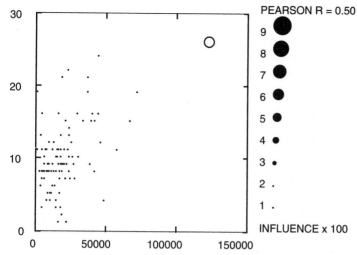

Figure 5.5 Influence plot of shoes versus income in the SHOES data set.

this point makes a difference depends on where the line would be without it; this is addressed in section 5.3, "Cook's D."

Figure 5.6 shows the 50% ellipse and best fitting line for the shoes versus income scatterplot in the SHOES data set (see sections 3.2.2.1.3 and 3.2.2.1.4). When the data is bivariate normal, the ellipse should cover approximately 50% of the cases. In fact, this ellipse contains 65% of the cases, which suggests that the bivariate distribution has more cases near the mean of both variables than one would expect in a bivariate normal distribution. The line in Figure 5.6 is the best fitting (least squares) line for the scatterplot. However, keep in mind that the best fitting line is heavily influenced by one case (the highest value on both variables in Figure 5.6).

Finally, although the income distribution is positively skewed, and the shoes distribution is more "peaked" than a normal, we still cannot see clearly what the joint density distribution of these two variables looks like. SYSTAT offers a view of this distribution using three-dimensional kernel density plots (see section 3.2.2.2). These plots are like histograms in 3-D: The height of the plot shows the proportion of cases falling into each two-dimensional interval defined by combinations of the Y and X variables. Unlike histograms, the variables are treated as continuous and are not partitioned into intervals. The result is a 3-D surface like that shown in Figure 5.7. In this figure you can see that the joint distribution of shoes and income is piled up rather heavily in the center,

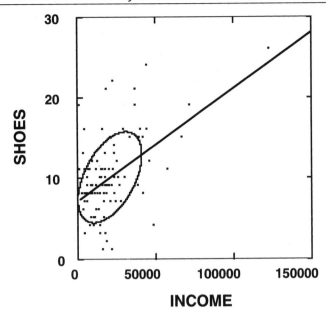

Figure 5.6 Scatterplot of shoes versus income in the SHOES data set, with 50% bivariate ellipse and best fitting least squares regression line.

trailing off quickly in the tails. This is consistent with the observation in Figure 5.6 that more than 50% of the cases fall within the predicted 50% ellipse, which assumed that the distribution was multivariate normal.

Unfortunately, SYSTAT does not allow you to change the viewing angle in Figure 5.7. However, you can replot the data with the Y and X axes switched to give another view of the kernel density plot. Such a plot is shown in Figure 5.8. It is important to plot the data both ways so you can see what is going on behind the "hump" in the distribution. For example, you know from the foregoing analyses that there is a small "bump" in this distribution near income=125,000 and shoes=26. This bump can be seen in the upper center portion of Figure 5.7, but it is barely visible behind the hump in Figure 5.8.

5.2 Simple Regression

The length or the foregoing section is an indication of how important viewing the data is to regression analysis and any other analysis for that matter. Regression gives you a view of

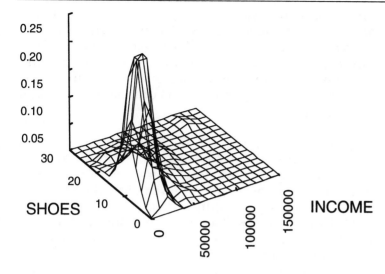

Figure 5.7 Kernel density plot of shoes versus income in the SHOES data set.

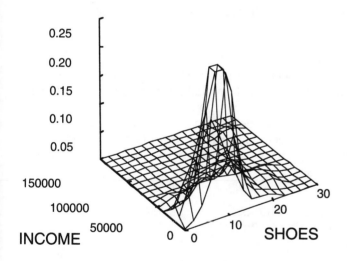

Figure 5.8 Kernel density plot of income versus shoes in the SHOES data set.

your data through a very specific mathematical lens, and you cannot determine whether that lens is appropriate if you do not look at the data from any other perspective. If you skipped the previous sections on plotting data, proceed at your own risk!

To perform a regression analysis, you first have to specify a linear model. This model tells SYSTAT what variable is to be treated as the outcome and which variables are the predictors whose regression coefficients you wish to estimate. In simple regression this will include your single predictor variable and a constant. Recall that simple linear equations take the form

$$Y = mX + b \tag{5.1}$$

For our purposes, Y is the outcome variable, X is the predictor variable, m is the slope of the line relating Y to X, and b is the Y-intercept. In linear regression, values of m and b are chosen that minimize the squared deviations of the actual data points from the line, and if the data is bivariate normal the expected values of these two parameters are the "right answers" in the actual relationship between the two variables. Of course, estimates are not perfect because there will be noise in real data, and we have probably omitted relevant variables. Any important variables that have been left out of the model, any error in recording data, or other noise in the system are lumped together into the term "residual variation" or "error variation." Thus, the linear model for simple regression is

$$Y = mX + b + e \tag{5.2}$$

where e represents error in the system.

When you enter a model into SYSTAT, include the outcome, constant, and any predictor variables that you want, but leave off the term for the error component. Anything that you leave out of the model will be treated by SYSTAT as error variance (see the warning below). Without further ado, the commands for computing a simple regression of shoes on income from the SHOES data set is

```
>MGLH
>MODEL SHOES = CONSTANT + INCOME
>SAVE SHOES.RES /MODEL
>ESTIMATE
```

The *model* command tells SYSTAT which variables to include in the regression. The *save* command will save to a file the residuals and other diagnostic information that you can later use to assess the validity of the assumptions underlying regression (more on this in the following section). The *model* option after the slash in the *save* command asks SYSTAT to include the original variables in the new file with the residuals. The output of the regression analysis is shown in Table 5.4.

MACINTOSH

① Choose the menu item **Stats/MGLH/Regression...**
② With the cursor in the *Dependent* box, select SHOES from the variable list.
③ With the cursor in the *Independent* box after CONSTANT, select INCOME from the variable list.
④ Check **Save residuals**.
⑤ Click **OK**.
⑥ Type a file name, e.g., "shoes.res" into the dialog box and click **Save**.
⑦ Check **Save model and diagnostics**.
⑧ Click **OK**.

MS-DOS

① Choose the menu item **Statistics/MGLH/Regression/Model/Variables/**.
② Select SHOES from the variable list.
③ Select INCOME from the variable list.
④ Choose the menu item **Statistics/MGLH/Regression/Estimate/Save/**, hit <esc>, type, e.g., "shoes.res", and then choose **/Model**.
⑤ Choose the menu item **Statistics/MGLH/Regression/Estimate/Go!**

WARNING: SPECIFYING "ERROR"

The convention in SYSTAT for designating error variance in a linear model is to leave out of the model all sources of variance that you wish to be treated as error. So any variables or sources of variation that you *leave out* of the model will be lumped together as "error." Too many terms in the model relative to the number of cases may result in the model being overspecified, with no residual variance left for error. If important variables are left out of the model, the error variance will include variation due to those variables, and be too large.

The first two lines of the regression output give information about the overall fit of the model. The correlation between Y and X is given in two places in simple regression output, as

Table 5.4 Simple regression output from shoes on income in the SHOES data set.

```
DEP VAR: SHOES      N: 100   MULTIPLE R: 0.50   SQUARED MULTIPLE R: 0.255
ADJUSTED SQUARED MULTIPLE R: 0.247      STANDARD ERROR OF ESTIMATE: 4.140

VARIABLE     COEFFICIENT    STD ERROR    STD COEF  TOLERANCE      T     P(2 TAIL)

CONSTANT        7.249         0.649        0.000        .       11.173   .10E-14
INCOME     .139820E-03   .241578E-04       0.505     1.000       5.788   .86E-07

                         ANALYSIS OF VARIANCE

SOURCE       SUM-OF-SQUARES    DF    MEAN-SQUARE      F-RATIO         P

REGRESSION        574.199      1        574.199       33.498  .857598E-07
RESIDUAL         1679.841     98         17.141

WARNING: CASE    9 HAS LARGE LEVERAGE                (LEVERAGE =        0.084)
WARNING: CASE   21 IS AN OUTLIER (STUDENTIZED RESIDUAL =       2.913)
WARNING: CASE   24 IS AN OUTLIER (STUDENTIZED RESIDUAL =       2.879)
WARNING: CASE   39 HAS LARGE LEVERAGE                (LEVERAGE =        0.365)
WARNING: CASE   51 IS AN OUTLIER (STUDENTIZED RESIDUAL =       2.785)
WARNING: CASE   84 IS AN OUTLIER (STUDENTIZED RESIDUAL =       2.649)
WARNING: CASE   99 HAS LARGE LEVERAGE                (LEVERAGE =        0.101)

DURBIN-WATSON D STATISTIC     1.959
FIRST ORDER AUTOCORRELATION   0.020

RESIDUALS HAVE BEEN SAVED
```

multiple R (line 1) and as the standardized coefficient (*std coef*) for the *X* variable (line 5). The square of this correlation, *squared multiple R*, gives the proportion of the variance in *Y* that is predictable by *X*. Thus, the correlation between shoes and income is a substantial .505, which means that 25.5% of the variance in shoes is predictable from income, as shown in line 1 of Table 5.4. The *adjusted squared multiple R* is an estimate of what the proportion of variance accounted for would be in a new sample of data from the same population and will always be somewhat lower than the unadjusted value. The *standard error of estimate*, 4.140, is equal to the square root of the *mean square* for the residual, 17.141.

The fourth and fifth lines of the output give the best-fitting estimates of the coefficients in the linear equation (Equation 5.1) relating *X* and *Y*, along with their standard errors (*std error*) and significance tests of their differences from zero. In Table 5.4, the coefficient for the constant (the *Y*-intercept) is 7.249 and the coefficient for the income variable (the slope) is about 0.00014. Using these values you can show the function relating shoes to income and compute predicted values of the shoes variable by plugging the coefficient estimates into Equation 5.1:

$$\text{shoes} = (0.00014 \times \text{income}) + 7.249 \qquad (5.3)$$

For a family with an income of $25,000, this regression line predicts that they will own 10.9, or approximately 11 pairs of shoes.

Dividing the coefficients by their standard errors, you obtain the *t*'s shown in lines 4 and 5 of the output. Note that for simple regression the *t* for the test of the slope is equal to the square root of the *F* test for the regression, in this case $5.788^2 \approx 33.498$. *Tolerance* is not relevant in simple regression and will always equal 1.0.

The analysis of variance table provides a test of the overall fit of the regression. The fit of the regression line in the SHOES data set is highly reliable, $F_{(1,98)} \approx 33.5$, $p = .86 \times 10^{-7}$. Dividing the sum-of-squares for the regression by the total sum-of-squares gives the *squared multiple R* in line 1 of the output.

The remainder of the output in Table 5.4, the warnings, Durbin-Watson, and autocorrelation, is discussed below.

5.3 Case Diagnostics

In the first section of this chapter, a number of procedures were presented for examining the sample distributions of variables and checking for extreme cases. As noted above, such extreme cases are always of interest, but they may or may not have an impact on the regression coefficient estimates. To determine the impact that individual cases have on the regression coefficients, we can employ three of the diagnostics saved by SYSTAT when a regression is computed: studentized residuals, leverage, and Cook's D. In the example in the previous section, these variables were saved into the file SHOES.RES.

5.3.1 Studentized Residuals

When you ask SYSTAT to save the residuals from a regression analysis, it saves the studentized residuals in a column labelled STUDENT in the file that you specify. To examine those residuals, open that file in the data editor (see section 1.1.4 for opening files). The studentized residuals are "externally studentized," which means they are approximately distributed as t with $N-p-1$ degrees of freedom. The regression output in Table 5.4 indicates that four cases have unusually large residuals: cases 21, 24, 51, and 84. All four of these cases have absolute values greater than 2.627, which we would only expect to find about 1 time in 100 due to chance in a t distribution with 98 degrees of freedom (see section 2.2.3.2.2 for how to find critical values of t). This many large residuals suggests that the residuals may not be normally distributed. However, you should not stop with the warnings provided by SYSTAT. A good way to display studentized residuals is with a stem and leaf display and box plots, as shown in Figure 5.9. The stem and leaf display and box plot in Figure 5.9 were created by typing the following commands (see sections 3.2.1.2 and 3.2.1.3):

```
>GRAPH
>STEM STUDENT/HEIGHT=3.0 IN WIDTH=2.5 IN
>BOX STUDENT
```

The four large studentized residuals listed in the regression output in Table 5.4 are bunched up near the left side of the box plot in Figure 5.9 (the two biggest overlap). What the regression output did not mention are the three large *negative* studentized residuals near the right side of the box plot in Figure 5.9. The top line in the stem and leaf display indicates that these values are between approximately –2.1 and –2.5.

Perhaps the easiest way to identify the individual cases corresponding to large and small residuals is to sort the file by STUDENT and then reopen the newly sorted file (see section 2.1.1). Table 5.5 shows the first six and last six cases in this newly sorted file, containing the six largest negative and positive studentized residuals. Eight of these studentized re-

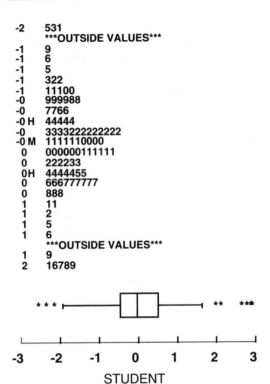

```
STEM AND LEAF PLOT OF VARIABLE:    STUDENT,   N = 100

MINIMUM IS:          -2.548
LOWER HINGE IS:      -0.477
MEDIAN IS:           -0.036
UPPER HINGE IS:       0.501
MAXIMUM IS:           2.913

    -2      531
            ***OUTSIDE VALUES***
    -1      9
    -1      6
    -1      5
    -1      322
    -1      11100
    -0      999988
    -0      7766
    -0 H    44444
    -0      3333222222222
    -0 M    1111110000
     0      000000111111
     0      222233
     0 H    4444455
     0      666777777
     0      888
     1      11
     1      2
     1      5
     1      6
            ***OUTSIDE VALUES***
     1      9
     2      16789
```

Figure 5.9 Stem and leaf display and box plot of studentized residuals from the regression of shoes on income in the SHOES data set.

siduals have absolute values greater than 2.0, which is not far from the five we would (roughly) expect due to chance in a *t* distribution with 98 degrees of freedom. However, five of these residuals are greater than 2.5, which is nearly four times as many as one would expect due to chance in a corresponding *t* distribution. You should examine such cases individually to determine whether they are unusual in some other respect and whether the data should be transformed in some manner, which I will return to below.

Table 5.5 The six largest negative and positive studentized residuals from the regression of shoes on income in the SHOES data set.

Case	Student
1	−2.548
2	−2.338
3	−2.128
4	−1.941
5	−1.659
6	−1.538
⋮	
95	1.987
96	2.106
97	2.649
98	2.785
99	2.879
100	2.913

5.3.2 Leverage

The regression output in Table 5.4 lists three cases, 9, 39, and 99, as having large values of *leverage*. Usually denoted symbolically as h_i or h_{ii}, the metaphorical name is appropriate: Imagine the regression line as a lever across a pivot at the mean of X and Y. Each case in the regression "pulls" the line toward itself, and h_i is a measure of how much "leverage" each case has in pulling on that line. The further a case is away from the mean on the X-axis, the more leverage that case will have. It is important to note that the values of Y do not enter into the computation of leverage at all: Leverage is entirely a function of the case's position in the predictor variable space, which, in simple regression, is just the X-axis. Thus, leverage helps identify outlying X values that have unusual "mechanical advantage" on the line. With a constant in the model, the bounds on h_i are

$$1/N \leq h_i \leq 1.0 \tag{5.4}$$

and the sum of the h_i is equal to $p+1$, where p is the number of predictor variables. To verify this, you can compute summary statistics on the leverage values, which are found in the same file as the studentized residuals (see section 3.1) by typing

```
>STATS
>STATISTICS LEVERAGE/N MINIMUM MAXIMUM MEAN SUM
```

The SYSTAT output with these statistics is shown in column 2 of Table 5.6. As expected, the values of leverage sum to 2. Dividing this value by N gives 0.02 as the mean value of h_i; this will be the mean value in any data set with $p+1=2$ coefficients and $N=100$ cases. A rule of thumb is to take any h_i greater than twice the mean as "large," that is, when

$$h_i > 2(p+1)/N \tag{5.5}$$

Such cases should be scrutinized to see if they are unusual in any way. For a data set with two coefficients and 100 cases, this value is 0.04. This is the criterion used by SYSTAT to bring the cases 9, 39, and 99 to our attention in Table 5.4.

Table 5.6 Summary statistics for leverage and Cook's D for the regression of shoes on income in the SHOES data.

TOTAL OBSERVATIONS: 100

	LEVERAGE	COOK
N OF CASES	100	100
MINIMUM	0.010	0.000
MAXIMUM	0.365	0.118
MEAN	0.020	0.010
SUM	2.000	0.955

For small values of p, this criterion tends to bring too many cases to our attention. The following transformation of h_i is approximately distributed as F when (and only when) the X values are normally distributed:

$$F(p, N-p-1) = \frac{(N-p-1)[h_i - (1/N)]}{(1-h_i)(p)} \tag{5.6a}$$

Solving for h_i we find:

$$h_i = \frac{pF + (N-p-1)/N}{pF + N-p-1} \tag{5.6b}$$

The $F_{(1,98)}$ value corresponding to $\alpha=.05$ is 3.94 (see section 2.2.3.3.2), and from Equation 5.6b this gives $h_i = 0.048$ as the criterion value, which means that you should only expect about five cases in 100 to have $h_i > .048$ due to chance in such a data set. Clearly, the heuristic used by SYSTAT approximates this value closely for a data set with 100 cases.

A note of warning, however: It is not clear what "due to chance" means in this context. In regression we assume that the X values are fixed, not randomly sampled; and high leverage is high leverage, whether the problem case was drawn from the same distribution as the rest of the X values or not. In general, leverage should only be used as a heuristic for locating potential problem cases and for understanding the cause of large Cook's Ds (discussed in the next section)—not as a basis for deleting cases. In any case, in the present example only three cases fall outside the criterion, and we would have expected about five "due to chance."

5.3.3 Cook's D

Cases with large studentized residuals increase the residual variance in the data, but do not necessarily have much impact on the regression line itself. Similarly, cases with high leverage may have little influence on the regression line if they fall near the regression line that would be produced by the remainder of the cases. However, cases with high leverage may have such tremendous influence on the slope of the regression line that they "artificially" decrease their own residuals. Cook's D provides a measure that takes into account both leverage and the residuals and indicates how much actual influence each case has on the slope of the regression line.

Values of Cook's D are saved as "COOK" in the file along with leverage and the studentized residuals. Summary statistics are shown in the last column of Table 5.6, which was produced by typing (see section 3.1)

```
>STATS
>STATISTICS COOK/N MINIMUM MAXIMUM MEAN SUM
```

As a heuristic, examine cases with Cook's D ≥ 1.0. Deletion of a case with D = 1 would move the estimate of the slope in the regression approximately to the edge of its 50% confidence interval based on all of the data, a potentially large change. In general, the probability of finding a value of D *or smaller* on an F distribution with $p+1$ and $N-p-1$ degrees of freedom gives the size of the confidence interval to the edge of which the estimate of the slope will be moved if the case is deleted. The p value for $F_{(2,98)} = 1.0$ is .63 (see section 2.2.3.3.1), which means that deleting a case with D = 1.0 would move the estimate of the slope to the edge of the 63% confidence interval based on all of the data. This is close to the 50% value assumed by the heuristic.

MACINTOSH

① Choose the menu item **Editor/Select...**
② Select COOK from the variable list (or type "cook" directly) and then type into the *Select expression* box "< .1".
③ Click **OK**.
④ Choose the menu item **Stats/MGLH/Regression...**
⑤ With the cursor in the *Dependent* box, select SHOES from the variable list.
⑥ With the cursor in the *Independent* box after CONSTANT, select INCOME from the variable list.
⑦ Check **Save residuals** and Click **OK**.
⑧ Type a file name, e.g., "newshoes.res" into the dialog box and click **Save**.
⑨ Check **Save model and diagnostics**.
⑩ Click **OK**.

MS-DOS

① Choose the menu item **Data/Select/Select/**.
② Select COOK from the variables list, choose /</, and type ".1" <ret>.
③ Choose the menu item **Statistics/MGLH/Regression/Model/Variables/**.
④ Select SHOES from the variable list.
⑤ Select INCOME from the variable list.
⑥ Choose the menu item **Statistics/MGLH/Regression/Estimate/Save/**, hit <esc>, type, e.g., "newshoes.res", and then choose /**Model**.
⑦ Choose the menu item **Statistics/MGLH/Regression/Estimate/Go!**

In the regression of shoes on income in the SHOES data set, the largest value of Cook's D is .118 (shown in Table 5.6), which is well within the acceptable range. The corresponding p value is .11, which indicates that deleting this case would only move the slope of the regression line to the edge of its 11% confidence region, a very small change. Thus, some cases in this data set have high leverage and others have large residuals, but no cases have high influence on the regression coefficients. To illustrate this, the regression excluding the two cases with the largest values of Cook's D (0.102 and 0.118) is shown in Table 5.7. It is always a good idea to check whether your conclusions would change with such cases deleted. The commands for producing Table 5.7 are

```
>SELECT COOK < .1
>MGLH
>SAVE NEWSHOES.RES
>MODEL SHOES=CONSTANT+INCOME
>ESTIMATE
```

Deleting the two cases barely improved the variance accounted for in the regression from 25.5% to 27.4% and, more importantly, only changed the slope of the regression line from about 0.000140 to about 0.000141. Thus, as the moderate values of Cook's D indicated, there simply is no good reason to drop any cases from the analysis.

Do not place too much confidence in Cook's D alone for finding influential cases. Cook's D only considers changes in the coefficients resulting from deletion of single cases; if two highly influential cases lie near one another, they may mask each other's effects in the calculation of Cook's D because deleting either of them alone will have little impact on the regression. Graphs are indispensable for diagnosing such situations.

5.4 Assumption Diagnostics

Linear regression makes four assumptions about the distribution of errors: They are normally distributed, independent, have constant variance, and are unbiased. Each of these assumptions is examined below, and SYSTAT procedures are illustrated for assessing their validity. The importance of checking these assumptions cannot be overemphasized: No regression results should be taken seriously until the validity of the assumptions has been assessed.

Table 5.7 Regression output for shoes on income in the SHOES data set, excluding the two cases with values of Cook's D greater than 0.1.

```
DEP VAR: SHOES        N: 98   MULTIPLE R: 0.524   SQUARED MULTIPLE R: 0.274
ADJUSTED SQUARED MULTIPLE R: 0.267     STANDARD ERROR OF ESTIMATE: 3.911

VARIABLE      COEFFICIENT    STD ERROR    STD COEF  TOLERANCE     T       P(2 TAIL)

CONSTANT         7.223         0.615       0.000         .       11.749    .10E-14
INCOME       .140904E-03   .233849E-04     0.524       1.000      6.025    .31E-07

                          ANALYSIS OF VARIANCE

SOURCE       SUM-OF-SQUARES   DF   MEAN-SQUARE     F-RATIO         P

REGRESSION       555.302       1      555.302      36.306    .310937E-07
RESIDUAL        1468.331      96       15.295

WARNING: CASE   9 HAS LARGE LEVERAGE               (LEVERAGE =             0.090)
WARNING: CASE  21 IS AN OUTLIER        (STUDENTIZED RESIDUAL =             3.110)
WARNING: CASE  24 IS AN OUTLIER        (STUDENTIZED RESIDUAL =             3.067)
WARNING: CASE  39 HAS LARGE LEVERAGE               (LEVERAGE =             0.387)
WARNING: CASE  51 IS AN OUTLIER        (STUDENTIZED RESIDUAL =             2.966)
WARNING: CASE  99 HAS LARGE LEVERAGE               (LEVERAGE =             0.108)

DURBIN-WATSON D STATISTIC      2.004
FIRST ORDER AUTOCORRELATION   -0.002

RESIDUALS HAVE BEEN SAVED
```

5.4.1 Normality

The normality assumption states that, for any value of X, the errors have a normal distribution. The values of the regression coefficients do not depend on this assumption, but their F and t tests, and corresponding confidence intervals, do. Fortunately, F's and t's are fairly robust to violations of the normality assumption, but serious violations may call for a transformation of some sort to make the errors look more normal. It turns out that in real data, transformations that make the errors more normal also often make them have more constant variance (see section 5.4.3 below).

5.4.1.1 Stem and Leaf Displays and Box Plots

We do not get to see the real errors in our data, but must instead use the residuals from our regression analysis. It turns out that in looking for cases with large studentized residuals in the previous section, I have already discussed two of the graphical methods used in examining the normality assumption: stem and leaf displays and box plots. These plots were shown in Figure 5.9. Both plots show a large number of outliers: nine cases, or almost 10% of the data, beyond the inner fences. In a normal distribution, only about 1 in 20 cases will fall beyond the inner fences. (Under the normality assumption, the studentized residuals are actually distributed as t. But for N greater than about 30, they will closely approximate the normal.) Furthermore, six of the nine cases are in the positive tail, suggesting the possibility of some positive skew.

5.4.1.2 Skewness and Kurtosis

Skewness and kurtosis can be computed on the studentized residuals to help assess their normality. These statistics are shown in Table 5.8 along with a few other statistics that were generated by the following commands (see section 3.1):

```
>STATS
>STATISTICS STUDENT/N MINIMUM MAXIMUM MEAN
SKEWNESS KURTOSIS
```

The standard deviation for skewness for 100 cases (see section 3.1) is 0.245, so the value of skewness in Table 5.8 divided by its standard deviation gives $Z = 1.95$, corresponding to p (two-tailed) = .051 (see section 2.2.3.1.1). For sample sizes less than about 100 this probability is somewhat conservative. Thus, as it appeared to the eye in Figure 5.9, the distribution of

Table 5.8 Summary statistics for the studentized residuals from the regression of shoes on income in the SHOES data set.

TOTAL OBSERVATIONS:	100
	STUDENT
N OF CASES	100
MINIMUM	-2.548
MAXIMUM	2.913
MEAN	0.003
SKEWNESS(G1)	0.477
KURTOSIS(G2)	1.207

studentized residuals is positively skewed to a degree expected by chance at most only about 1 in 20 times.

The standard deviation for kurtosis for 100 cases (see section 3.1) is 0.490. Dividing the value of kurtosis in Table 5.8 by this standard deviation gives $Z = 2.46$, p (two-tailed) $= .014$. This probability is only an approximation: the more accurate tables in Snedecor and Cochran (1982, Appendix 20ii) indicate that this value of kurtosis would be expected a little more than 1 in 50 times due to chance. Both approximations suggest that the distribution of studentized residuals may be longer-tailed than normal. (As noted above, under the normality assumption, the studentized residuals are actually distributed as t, but for N greater than about 30 they will closely approximate normal.)

5.4.1.3 Normal Probability Plots

One of the clearest methods of determining the normality of the studentized residuals is by displaying them in a *normal probability plot*, also called a *rankit* plot (see section 3.2.1.4). These plots can be generated easily in SYSTAT using the following commands:

```
>GRAPH
>PPLOT RESIDUAL/NORMAL
```

In this plot, if the residuals are drawn from a normal distribution they should fall on a straight line. The systematic deviation from a straight line in Figure 5.10 indicates that the residuals are not normally distributed. You would not usually worry about a couple of points falling off the end of the line, but in Figure 5.10 five points on the low end and eleven points on the high end are systematically off of the line formed by the rest of the data points. The plot in Figure 5.10 shows a ∫ shape, which is one sign that the distribution has short, thin tails (see section 3.2.1.4). However, this disagrees with the value of kurtosis in Table 5.8, 1.207, which indicates that the distribution is longer-tailed than normal.

This discrepancy is probably due to the outliers in this plot. The top three and bottom two points in Figure 5.10 fall off the lines formed by their nearest neighbors. You can select all of the data excluding these points using the command

```
>SELECT  RESIDUAL>-9 AND RESIDUAL<11
```

and then redraw the normal probability plot to see what the data looks like without those cases. The new plot is shown in Figure 5.11. In this plot, the data falls fairly closely on a straight line, with a slight bend just above 3 on the X-axis indicating some positive skew. What are the values of skewness and

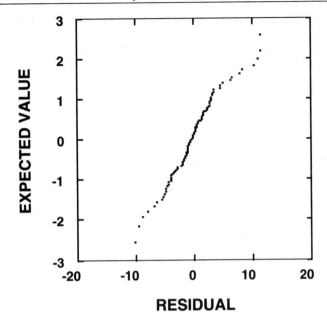

Figure 5.10 Normal probability plot of the residuals from the regression of shoes on income using the SHOES data set. Curvature at the ends of the straight line suggests non-normality.

kurtosis excluding those five cases? Skewness is now .310 and kurtosis is .804, neither of which is significant (Snedecor & Cochran, 1980, Appendix 20i and ii). This is consistent with (but certainly should not be weighted heavily as evidence for) the possibility that most of the cases come from a normal distribution, and that the five extreme cases are outliers.

5.4.1.4 Remedies

The distinction between a "non-normal distribution" and a "normal distribution with outliers" is usually not clear and should be determined by factors apart from the data itself: Do the cases in question represent different kinds of subjects in some way? Was their data collected in the same manner?—and so on. When the evidence is overwhelming that some cases do not belong, it is probably still safest just to report the results with and without those cases. Readers become suspicious when you start deleting data. There are other options: Transformations that correct positive skew, for example, may also pull in outliers so that they no longer make much difference.

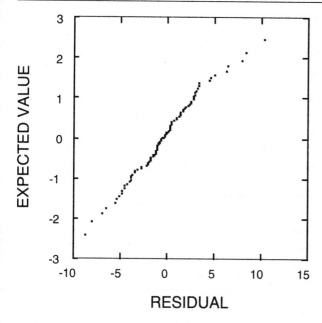

Figure 5.11 Normal probability plot of the residuals from the regression of shoes on income using the SHOES data set excluding the three top and two bottom cases iin Figure 5.10.

If the residuals are not normally distributed, the *p*-values from the *F* tests and *t* tests in the regression output may be too small. Fortunately, *F*'s and *t*'s are fairly robust to violations of the normality assumption, so moderate departures from normality can often be ignored if the other assumptions are met. In such a case, the reader should be warned that the *p*-values are not accurate. If the deviation from normality is substantial (and you must rely on your experience or consult an applied textbook for what should count as "substantial"), the most common remedies are: (1) to transform the outcome (*Y*) variable, or (2) do robust regression, which is not covered in this book. Transforming variables is covered in Chapter 7, but one word of warning is offered here: If you transform the *Y* variable to meet the normality assumption, you may find that the data then violates the linearity assumption. To recover linearity, you may have to transform one or more of the *X* variables. This in turn may cause new violations of normality ... and so on. Usually a satisfactory solution can be found, but you should be aware that remedies for violations of one assumption may produce violations of others.

5.4.2 Equal Variance

The second assumption underlying regression is that the residuals have the same variance at every value of X (or combination of X's in multiple regression). Equal variance is sometimes called *homoscedasticity*. In real data, the residuals often increase in magnitude as the value of X increases. The reverse can also happen in theory, but is rarely seen in practice. Violations of this assumption can often be seen in a plot of the studentized residuals versus the predicted values, and formal tests, such as the *score test* discussed below, are also available.

5.4.2.1 Plotting Studentized Residuals Versus Predicted Values

To assess the equal variance assumption graphically in simple regression, plot the studentized residuals against the estimated (also "predicted" or "fitted") Y values from the regression line. This can be accomplished in SYSTAT by opening the file containing the diagnostics saved from the regression and using the following commands (see section 3.2.2.1):

```
>GRAPH
>PLOT STUDENT*ESTIMATE
```

This scatterplot is shown in Figure 5.12. As the estimated values increase up to approximately 15 on the X-axis in Figure 5.12, the variance of the residuals may also be increasing: The range is about the same, but there are fewer cases bunched up near zero. Beyond estimated values of 15, the magnitudes of the residuals markedly decrease, but this may be due to the sparsity of data with large estimated values. The status of the equal variance assumption cannot be assessed with much confidence in Figure 5.12.

5.4.2.2 Plotting Studentized Residuals Versus Predictor Variables

A second type of plot that may help locate deviations from equal variance are plots of the studentized residuals versus the predictor (X) variables. In simple regression this plot will look almost identical to the plot of studentized residuals versus predicted values (Figure 5.12). This can be seen by comparing the plot in Figure 5.13 with that in Figure 5.12. In Figure 5.13 it is a little easier to see that, excluding the four largest income values, the variance of the residuals may be increasing—the range is about the same, but there are fewer values bunched near zero. As in Figure 5.12, however, violation of the equal

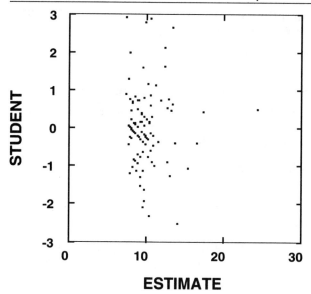

Figure 5.12 Plot of the studentized residuals versus the estimated values from the regression of shoes on income in the SHOES data set.

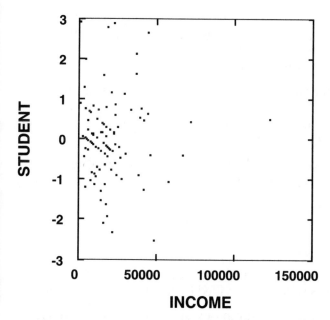

Figure 5.13 Plot of the studentized residuals versus income from the regression of shoes on income in the SHOES data set.

MACINTOSH

I. Regress Y on X.

① Choose the menu item **Stats/MGLH/Regression...**

② With the cursor in the *Dependent* box, select SHOES from the variable list.

③ With the cursor in the *Independent* box after CONSTANT, select INCOME from the variable list.

④ Check **Save residuals** and click **OK**.

⑤ Type a file name, e.g., "shoes.res" into the dialog box and click **Save**.

⑥ Check **Save model and diagnostics** and click **OK**.

II. Divide SS_{reg} by N (see text).

III. Create the column of scaled squared residuals.

① Choose the menu item **File/Open**.

② Choose the menu item **Editor/Math**.

③ Set *u* equal to (residual^2)/16.79841.

IV. Regress u on the estimated values of Y.

① Choose the menu item **Stats/MGLH/Regression...**

② With the cursor in the *Dependent* box, select U from the variable list.

③ With the cursor in the *Independent* box after CONSTANT, select ESTIMATE from the variable list.

④ Check **Save residuals** and click **OK**.

⑤ Type a file name, e.g., "u.res" into the dialog box and click **Save**.

⑥ Check **Save model and diagnostics** and click **OK**.

V. Compute the chi-square score test (see text).

continued ...

---

variance assumption, if any, is not *obvious* in Figure 5.13. The plot in Figure 5.13 was created using the command

```
>PLOT STUDENT*INCOME
```

5.4.2.3 Score Test for Non-Constant Variance

A formal test of the assumption of equal variance is provided by the "score test" (Cook & Weisberg, 1983; Weisberg, 1985, pp. 135–136). This test has five simple steps:

I. Regress *Y* on *X* and save the raw residuals, $\hat{e}_i$, into a file.

II. Divide the $SS_{residual}$ from the regression by *N* to get the quantity $\tilde{\sigma}^2$.

III. In the file with the residuals, create a new column of *scaled squared residuals* (u_i) by squaring the residuals and dividing by $\tilde{\sigma}^2$, that is: $u_i = \dfrac{\hat{e}_i^2}{\tilde{\sigma}^2}$.

IV. Regress the column of u_i on the variable or variables that you think the variance may be a function of (the estimated values of *Y*, the predictor variables, etc.).

V. Compute the score test $\chi^2_{(p)} = SS_{reg} / 2$, where SS_{reg} is the sum of squares for the regression of u_i on the other variables, and *p* is the number of regressors not including the constant.

The commands for computing each step of the score test in SYSTAT for the SHOES data set are as follows (the menu options to execute this procedure are shown in the margins beginning at the top of this page):

I. Regress Y on X (the output from this regression is shown in Table 5.4):

```
>MGLH
>MODEL SHOES = CONSTANT + INCOME
>SAVE "SHOES.RES"/ MODEL
>ESTIMATE
```

II. The $SS_{residual}$ from the regression output in Table 5.4 is 1679.841, which, divided by *N* = 100, equals 16.79841.

III. To create the column of scaled squared residuals, type

```
>EDIT SHOES.RES
>LET U = (RESIDUAL^2)/16.79841
```

IV. Regress *u* on the estimated values of *Y* using, for example, the commands

```
>MGLH
>MODEL U = CONSTANT + ESTIMATE
>SAVE   U.RES/MODEL
>ESTIMATE
```

(The output for this regression is shown in Table 5.9.)

V. The SS$_{regression}$ from the output in Table 5.9 is 1.522. The score test, then, is equal to $\chi^2_{(1)} = 1.522 / 2 = 0.761$, $p = .38$.

Therefore, the score test does not indicate a statistically significant departure from equal variance.

5.4.2.4 Remedies

As a rough rule, if you can't see unequal variance in the plot of the studentized residuals versus the estimated values (e.g., Figure 5.12), you don't have a serious problem. If you do not trust your eyes, try the score test. *F*'s and *t*'s are fairly robust to moderate departures from equal variance, but when the departure from equality is substantial the most common remedies are: (1) to perform a *variance stabilizing transformation* on the outcome (*Y*) variable (Snedecor & Cochran, 1982, pp. 282–292; Weisberg, 1985, pp. 133–135), or (2) do weighted least-squares regression, which is not covered in this book but is discussed in the SYSTAT *Statistics* manual. Transforming variables is covered in Chapter 7, but as mentioned above, you should be aware that transforming the *Y* variable to meet the equal variance assumption may produce violations of the linearity assumption. To recover linearity, you may have to transform one or more of the *X* variables. These issues are discussed in Chapter 7.

5.4.3 Independence

The third assumption underlying regression is that the errors are uncorrelated with one another. When the normality assumption is met, uncorrelated errors are also *independent*, so this assumption is often called the independence assumption. In practice, violations of this assumption are difficult to distinguish graphically from violations of linearity, and whether you should attribute the violation to independence or linearity largely depends on the design of the study and the way the data was collected.

MS-DOS

I. Regress Y on X.

① Choose the menu item **Statistics/MGLH/ Regression/Model/ Variables/**.

② Select SHOES from the variable list.

③ Select INCOME from the variable list.

④ Choose the menu item **Statistics/MGLH/ Regression/Estimate/ Save/**, hit <esc>, type, e.g., "shoes.res", and then choose **/Model**.

⑤ Choose the menu item **Statistics/MGLH/ Regression/Estimate/Go!**

II. Divide SS$_{reg}$ by N (see text).

III. Create the column of scaled squared residuals.

① Choose the menu item **Files/Use**, hit <esc>, and type "shoes.res".

② Choose the menu item **Data/Edit!**

③ Hit <esc> and type at the prompt:

>let u = (residual^2)/, 16.79841

IV. Regress u on the estimated values of Y.

① Choose the menu item **Statistics/MGLH/ Regression/Model/ Variables/**.

② Select U from the variable list.

③ Select ESTIMATE from the variable list.

④ Choose the menu item **Statistics/MGLH/ Regression/Estimate/ Save/**, hit <esc>, type, e.g., "u.res", and then choose **/Model**.

⑤ Choose the menu item **Statistics/MGLH/ Regression/Estimate/Go!**

V. Compute the chi-square score test (see text).

Table 5.9 Regression of the scales squared residuals u_i on the estimated values from the regression of shoes on income.

```
DEP VAR: U      N: 100     MULTIPLE R: 0.071    SQUARED MULTIPLE R: 0.005
ADJUSTED SQUARED MULTIPLE R: 0.000      STANDARD ERROR OF ESTIMATE: 1.761

VARIABLE   COEFFICIENT  STD ERROR   STD COEF  TOLERANCE     T      P(2 TAIL)

CONSTANT      0.478       0.766       0.000        .      0.624      0.534
ESTIMATE      0.051       0.073       0.071      1.000    0.701      0.485


                          ANALYSIS OF VARIANCE

    SOURCE     SUM-OF-SQUARES    DF    MEAN-SQUARE    F-RATIO      P

   REGRESSION       1.522         1        1.522       0.491      0.485
   RESIDUAL       303.984        98        3.102
```

5.4.3.1 Examine the Design

Is there any reason that one observation in your data set might be related to, or be predictable from, another? The most obvious way that errors might be correlated is in time: Perhaps you made more positive data-recording errors later in the day, or perhaps you collected multiple data points from the same subject over time. If so, you should plot the residuals versus time and look for "snakiness" in the plot—positive errors tending to follow positive errors, and negative errors tending to follow negative errors. If more than one of your data points were collected from the same person or group, their errors may be correlated, and you should adopt some type of *repeated measures* design, including a variable for person, family, etc. If there is any reason that errors could be correlated with each other across one or more of the predictor variables, then you should plot the residuals against those predictors and examine those plots for snakiness.

If, after you have considered in full the design of your study, you can find no reason that the errors would be correlated, then they probably aren't. Patterns in the residual plots are most likely due to violations of linearity, which often can be corrected with transformations (see below). For example, in the SHOES data set we have supposed that we have data on the number of pairs of shoes that each family owns and each family's income. If the families in the study do not influence each other, and are not influenced by the study itself (for example, if wealthier families heard about the study and rushed out to buy more shoes), then we are probably safe to assume that the errors are uncorrelated.

5.4.3.2 Durbin-Watson

The Durbin-Watson statistic is a measure of the correlation among the errors when the cases are equally spaced in time. A value less than about 1.4 or greater than about 2.6 indicates a possible violation or the independence assumption. In the SHOES data set the cases are not spaced in time at all as far as we know, so the Durbin-Watson statistic is not of much interest. The value of 1.959 in Table 5.4 is very near the middle of this range, which is what we would expect.

5.4.3.3 Autocorrelation and the ACF Plot

The first-order autocorrelation shown in Table 5.4 is the correlation between the column of residuals for case 1 through $N-1$ with the column of residuals for case 2 through N. In other words, it's the correlation of the residual for each case with the residual for the next case. The *order* of the autocorrelation is determined by the size of the "lag," or how many cases down you would shift the second column before computing the correlation. For example, the fourth order autocorrelation between residuals (four cases apart) has a lag of 4.

If you suspect that your errors may be correlated over some variable other than case, such as time or estimated Y value, you can sort the data file by that variable and compute the autocorrelation using the *ACF* plot. Higher-order autocorrelations can also be computed using this plot. In the SHOES data set the data are not ordered in time, but it can sometimes be helpful to examine higher-order autocorrelations in the residuals sorted by their estimated Y values. By quantifying how much residuals that are near to each other in their estimated values tend also to be similar in size, this can help uncover non-independence in plots such as Figure 5.12. This can be done in SYSTAT using the following commands:

```
>EDIT "SHOES.RES"
>SYSTAT
>SAVE "ESTIMATE.SRT"
>SORT ESTIMATE
>EDIT "ESTIMATE.SRT"
>SERIES
>ACF RESIDUAL/LAG=15
```

MACINTOSH

① Choose the menu item **File/Open...**
② Select the file, e.g., "shoes.res", from the list and click **Open**.
③ Choose **Data/Sort**. Select ESTIMATE from the variable list and click **OK**.
④ Type a file name, e.g., "estimate.srt" into the dialog box and click **Save**.
⑤ Check **Single precision** or **Double precision** and click **OK**.
⑥ Choose **File/Open...** Select the file, e.g., "estimate.srt" from the list and click **Open**.
⑦ Choose **Stats/Series/ACF plot...**
⑧ Select RESIDUAL from the variable list.
⑨ Type "15" into the *lag* box, if it is not there already.
⑩ Click **OK**.

DOS MENUS

① Choose the menu item **Files/Use**, hit <esc>, and type "shoes.res".
② Choose **Data/Sort/ Variables/**.
③ Select ESTIMATE from the variables list.
④ Choose **Data/Sort/Save/**, hit <esc>, and type, e.g., "estimate.srt".
⑤ Choose **Data/Sort/Go!**
⑥ Choose **Files/Use**, hit <esc>, and type "estimate.res".
⑦ Choose **Statistics/Series/ ACF/Variable/**.
⑧ Select RESIDUAL from the variables list.
⑨ Choose **Statistics/Series/ ACF/Lag**, and then type "15"<ret>.
⑩ Choose **Statistics/Series/ ACF/Go!**

The option *lag* allows you to select the maximum order of the autocorrelations computed by SYSTAT. The default is 15.

The ACF plot output with the first through fifteenth order autocorrelations among the residuals is shown in Table 5.10. The degree of lag is shown in the first column and the autocorrelation is shown in the second column. The third column contains the standard errors of the autocorrelations, which are used to put confidence intervals around the values in the plot itself. In general, autocorrelation values that do not extend beyond the parentheses in the plot are probably due to chance. In Table 5.10, none of the autocorrelations fall outside these intervals, and most are very near zero.

5.4.3.4 Plot of Residuals Versus Estimated Values

Correlated errors may show up as patterns in scatterplots of the residuals versus the estimated Y values. This plot is shown in Figure 5.12. When errors are positively correlated, you would expect to see a "snakey" pattern, with positive residuals tending to follow positive residuals and negative residuals tending to follow negative residuals. Such a pattern can be seen in Figure 5.12 to the right of 12 on the X-axis, but this is likely a chance pattern owing to the small number of data points in that region. To get a closer look at particular regions of a scatterplot, you can limit the range of X and Y values to be plotted. Figure 5.14 shows the same plot as in Figure 5.12, but with the X values limited between 6 and 12. There is a slight hint of downward bowing in this plot, but again this is mostly due to a few data points. The bulk of the data shows no discernible pattern. Figure 5.14 was generated in SYSTAT by typing (see section 3.2.2.1)

```
>GRAPH
>PLOT   STUDENT  *  ESTIMATE/XMIN=6 XMAX=12
```

5.4.3.5 Plotting Residuals Versus Other Variables

In general, you should plot the residuals versus the variable or variables over which you think they may be correlated. If you think they are correlated in time, then plot the residuals versus time. If you think they are correlated in one or more of the predictors, then plot the residuals versus those predictors. In simple regression there is only a single predictor, and it is perfectly correlated with the estimated values, so the plot of the residuals versus the predictor will look almost identical to the plot of the residuals versus the estimated values. This plot is shown in Figure 5.13. As in Figures 5.12 and 5.14, the patterns in this plot are due to a small number of data points,

---

MACINTOSH

① Choose the menu item **Graph/Plot/Plot**.

② Select STUDENT from the *Y variables* list and ESTIMATE from the *X variable* list.

③ Click on **Axes**, type "6" into the *X min* box and "12" into the *X max* box, and click **OK**.

④ Click **OK**.

MS-DOS

① Choose the menu item **Graph/Plot/Data/2-D/Variables/Y**.

② Select STUDENT from the variable list.

③ Choose the menu item **Graph/Plot/Data/2-D/Variables/X**.

④ Select ESTIMATE from the variable list.

⑤ Choose the menu item **Graph/Plot/Data/2-D/Options/Xmin** and type "6" <ret>.

⑥ Choose the menu item **Graph/Plot/Data/2-D/Options/Xmax** and type "12" <ret>.

⑦ Choose the menu item **Graph/Plot/Data/2-D/Go!**

Table 5.10 ACF plot of autocorrelations for the residuals from the regression of shoes on income in the SHOES data set.

```
PLOT OF RESIDUAL
NUMBER OF CASES =   100
MEAN OF SERIES =         0.000
STANDARD DEVIATION OF SERIES =        4.099

PLOT OF AUTOCORRELATIONS

LAG    CORR      SE -1.0   -.8   -.6   -.4   -.2    .0    .2    .4    .6    .8   1.0
                    |----|----|----|----|----|----|----|----|----|----|
  1   0.081    0.100                              (    |••  )
  2  -0.085    0.101                              (  ••|    )
  3  -0.167    0.101                              (••••|    )
  4   0.032    0.104                              (    |    )
  5  -0.024    0.104                              (    |    )
  6   0.148    0.104                              (    |••• )
  7  -0.061    0.106                              (   •|    )
  8   0.049    0.107                              (    |•   )
  9  -0.148    0.107                              ( •••|    )
 10   0.069    0.109                              (    |•   )
 11   0.030    0.109                              (    |    )
 12   0.177    0.109                              (    |••••)
 13  -0.048    0.112                              (   •|    )
 14   0.024    0.112                              (    |    )
 15  -0.055    0.112                              (   •|    )
```

and therefore do not suggest any substantial violation of the independence assumption.

5.4.3.6 Remedies for Correlated Errors

Correlated errors are a result of the "system" that produced the data, so the correlation either exists or it doesn't. Transformations of the data won't make it go away, but there are ways of modelling the correlation within the regression framework that can be found in very advanced textbooks. Fortunately, even when the independence assumption is violated, the coefficients in the regression are still unbiased estimates of the population values. Only the standard errors, significance tests, and confidence intervals are wrong. When the errors are positively correlated, p-values and confidence intervals will be too small, and readers should be warned of this.

When the residuals are correlated over time, a time series analysis will often be appropriate. Such procedures are available with SYSTAT's *Series* command, but they are beyond the scope of this book.

5.4.4 Linearity

The fourth assumption underlying regression is that the mean value of Y at each value of X is a straight line function of X. In other words, all the members of the population are described by the same linear model. Linearity is often described as an assumption that the errors are *unbiased*, with an expected value of zero, which would not be true, for example, if you have left out an important X variable. In reality, the linear model is almost always only an approximation of the true relationship between variables, but it is often close enough not to be misleading, at least within the restricted range of the data. When this assumption is violated, the least-squares estimates do not give unbiased estimates of coefficients or estimated values. The diagnostic procedures below are graphical, but formal tests do exist (e.g., Weisberg, 1985, pp. 89–95).

5.4.4.1 Plotting Residuals Versus Estimated Values

Plots of the residuals versus the estimated Y values from the regression are usually used as a diagnostic for this assumption. Violations of the assumption may show up as curvature in this plot. The plot of the residuals versus the estimated values for the SHOES data set is shown in Figures 5.12 and 5.14, the latter showing only the data with estimated values between 6 and 12 pairs of shoes. Neither of these plots show clear curvature: The U shape in Figure 5.14 is mainly due to a few data points.

5.4.4.2 Plotting Residuals Versus Predictors

If you suspect that a particular predictor variable may have a nonlinear relationship with Y, you can examine the plot of the residuals versus that predictor variable for curvature or other patterns. In simple regression there is only one predictor, and the plot of the residuals versus this predictor for the SHOES data set has already been shown in Figure 5.13. This figure does not show any clear signs of curvature.

5.4.4.3 Remedies for Nonlinearity

When the linearity assumption is violated, the coefficients and estimated Y values from the regression are not unbiased estimates of the population values. There are three general types of remedy for violations of this assumption: (1) if you have left out an important predictor variable, including this variable may correct the problem; (2) transforming (or scaling) the Y and/or X variables may improve the linear relationship between those variables; (3) adding polynomials or interactions among the predictors may improve the linear relationship. Adding such terms has the disadvantage of complicating the model and adding new parameters to be fitted. Transforming data, scaling, and adding polynomials are covered in Chapter 7.

5.5 Confidence Intervals

SYSTAT will compute some types of confidence intervals for display in graphs. However, if you wish to report the actual values of the confidence intervals, you may compute them easily using SYSTAT's cumulative distribution functions and a hand-calculator. The formulas for a variety of confidence intervals and regions are provided below with examples from the regression on the SHOES data set.

5.5.1 Intercept

The bounds for the confidence intervals on the intercept from the regression can be found as follows:

$$\hat{\beta}_0 \pm t_{(\alpha/2,\, N-p-1)} \times SE_{\hat{\beta}_0} \qquad (5.7)$$

and the confidence interval itself can be expressed as

$$\hat{\beta}_0 - t_{(\alpha/2,\, N-p-1)} SE_{\hat{\beta}_0} \leq \beta_0 \leq \hat{\beta}_0 + t_{(\alpha/2,\, N-p-1)} SE_{\hat{\beta}_0} \qquad (5.8)$$

where β_0 is the population value of the intercept coefficient, $\hat{\beta}_0$ is the value of the estimated intercept coefficient from the regression output, $SE_{\hat{\beta}_0}$ is the standard error of the estimate, and $t_{(\alpha/2,\, N-p-1)}$ is the value of the t distribution with $N-p-1$ degrees of freedom that lies beyond $(1-\alpha/2) \times 100$ percent of the distribution. That is, if you want the 95% confidence interval ($\alpha = .05$), you must find the magnitude of t corresponding to the upper and lower 2.5% of the t distribution. The value for a t with 98 degrees of freedom corresponding to the upper 2.5% of the distribution is 1.984 (see section 2.2.3.2.2).

Equation 5.7 can be re-expressed using the labels from SYSTAT's regression output as:

$$\text{COEFFICIENT}_{\text{CONSTANT}} \pm t_{(\alpha/2,\, N-p-1)} \times \text{STD ERROR}_{\text{CONSTANT}} \qquad (5.9)$$

Plugging in the value of t and the values of the coefficient and standard error from the regression output in Table 5.4, you get

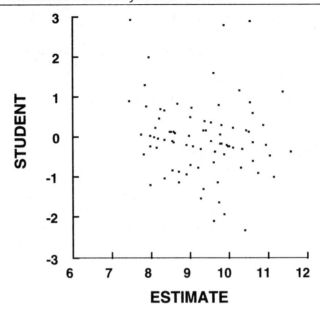

Figure 5.14 Plot of the studentized residuals versus the estimated values between 6 and 12 from the regression of shoes on income in the SHOES data.

$$7.249 \pm 1.984 \times 0.649 \tag{5.10}$$

or 7.249 ± 1.287. The confidence interval, therefore, is

$$5.962 \leq \beta_0 \leq 8.536 \tag{5.11}$$

When the assumptions underlying the regression are met, 95% of confidence intervals generated in this manner will contain the population value of the intercept.

5.5.2 Slope

The bounds for the confidence intervals on the slope from the regression are found in the same manner as those for the intercept:

$$\hat{\beta}_1 \pm t_{(\alpha/2, N-p-1)} \times SE_{\beta_1} \tag{5.12}$$

and, similarly, the confidence interval can be expressed as

$$\hat{\beta}_1 - t_{(\alpha/2, N-p-1)} SE_{\beta_1} \leq \beta_1 \leq \hat{\beta}_1 + t_{(\alpha/2, N-p-1)} SE_{\beta_1} \tag{5.13}$$

where β_1 is the population value of the slope, $\hat{\beta}_1$ is the value of the estimated slope from the regression output, SE_{β_1} is the standard error of the slope estimate, and $t_{(\alpha/2, N-p-1)}$ is the value of the t distribution with $N-p-1$ degrees of freedom that lies beyond $(1-\alpha/2) \times 100$ percent of the distribution.

Re-expressing Equation 5.12 using the labels from SYSTAT's regression output, you get the following:

$$\text{COEFFICIENT}_X \pm t_{(\alpha/2, N-p-1)} \times \text{STD ERROR}_X \qquad (5.14)$$

Plugging in the value of $t = 1.984$ (see section 2.2.3.2.2) and the values of the coefficient and standard error from the regression output in Table 5.4, you get

$$0.00014 \pm 1.984 \times 0.000024 \qquad (5.15)$$

or 0.00014 ± 0.000048. The resulting confidence interval is

$$0.000092 \leq \beta_0 \leq 0.000190 \qquad (5.16)$$

When the assumptions underlying the regression are met, 95% of confidence intervals generated in this manner will contain the population value of the slope.

5.5.3 Prediction Interval

You can also put confidence intervals on a new Y value predicted from the regression equation given some value of X (see, e.g., Weisberg, 1985, pp. 21–22). The predicted value of Y can be found using Equation 5.1, which is rewritten here:

$$\hat{Y}_* = \hat{\beta}_0 + \hat{\beta}_1 X_* \qquad (5.17)$$

where X_* is the particular value of the predictor variable being used, and $\hat{Y}_*$ is the corresponding prediction at that particular X value. This prediction will have three sources of variance: (1) the variance in the estimate of the intercept, (2) the variance in the estimate of the slope, and (3) the variance due to error. The calculation of the variance of the prediction given some particular value of X combines these three sources in the following formula:

$$\text{SE}\hat{Y}_* = \text{var}(\hat{Y}_* \mid X_*) = \hat{\sigma}_e \left[1 + \frac{1}{N} + \frac{(X_* - \overline{X})^2}{SS_X} \right]^{1/2} \qquad (5.18a)$$

where N is the number of cases, $\hat{\sigma}_e$ is the error variance, called *standard error of estimate* in SYSTAT regression output, SS_X is the sum of squares of the predictor variable, X is the mean of the predictor variable, and X_* is the particular X value from which the estimate of Y is being predicted. The mean and sum of squares for the predictor can be found from the simple summary statistics. For the SHOES data set this is done by typing

```
>STATS
>STATISTICS INCOME/MEAN VARIANCE
```

(see section 3.1). The output is shown in Table 5.11: The mean of the predictor, income, is 20,676.3, and its variance is 296,682,190.5. The sum of squares is equal to the variance multiplied by $N-1$:

$$SS_X = \text{var}(X) \times (N-1) \tag{5.19a}$$

$$= 296{,}682{,}190.5 \times 99 = 2{,}937{,}154{,}368.6 \tag{5.19b}$$

The standard error of estimate is shown in the regression output in Table 5.4 and is equal to 4.140.

Now suppose that you wish to find the confidence interval of the Y value predicted from an income of \$40,000. The variance of the prediction can be found by substituting the foregoing numbers into Equation 5.18a, which gives

$$SE_{\hat{Y}_*} = 4.140 \left[1 + \frac{1}{100} + \frac{(40{,}000 - 20{,}676.3)^2}{2{,}937{,}154{,}368.6} \right]^{1/2} = 4.415 \tag{5.18b}$$

To find the predicted value itself, substitute the coefficients from the regression output in Figure 5.4 into Equation 5.17, which gives the predicted Y value

$$\hat{Y}_* = (0.00014 \times 40{,}000) + 7.249 = 12.849 \tag{5.20}$$

The confidence interval around the prediction is found by the following expression:

$$\hat{Y}_* - t_{(\alpha/2,\, N-p-1)} SE_{\hat{Y}_*} \le Y_* \le \hat{Y}_* + t_{(\alpha/2,\, N-p-1)} SE_{\hat{Y}_*} \tag{5.21}$$

The value of $t_{(98)}$ that cuts off the top 97.5% of the variance is 1.984 (see section 2.2.3.2.2). Using this value, along with the prediction and standard error of the prediction, we find the bounds on the confidence interval are

$$\hat{Y}_* \pm t_{(\alpha/2,\, N-p-1)} SE_{\hat{Y}_*} = 12.849 \pm (1.984)(4.415)$$

$$= 12.849 \pm 8.759 \tag{5.22}$$

Finally, substituting these values into Equation 5.21, we find the 95% confidence interval:

$$4.089 \le Y_* \le 21.608 \tag{5.23}$$

Ninety-five percent of such confidence intervals would contain the true value of Y_*, which, in the example, means that

Table 5.11 Mean and variance for the income variable in the SHOES data set.

```
TOTAL OBSERVATIONS:      100

                       INCOME

N OF CASES               100
    MEAN             20676.3
    VARIANCE     296682190.5
```

there is about a 95% chance that a new family with an income of $40,000 will have between 4 and 22 pairs of shoes.

5.5.4 Fitted Values

Similar to the prediction interval, you can find the confidence interval for the *mean* value of Y given some value of X, that is, the confidence interval on the location of the fitted value of the regression line. This is usually only of interest when the regression model is known to be correct, but the parameters are estimated from the data (Weisberg, 1985, p. 22). The fitted value is computed in the same manner as the predicted value in Equations 5.17 and 5.20. However, this fitted value has a smaller variance than that for the prediction interval. The variance of the fitted value given some value of X is

$$\text{SE}_{\hat{Y}|X} = \text{var}(\hat{Y}|X) = \hat{\sigma}_e \left[\frac{1}{N} + \frac{(X - \bar{X})^2}{SS_X} \right]^{1/2} \quad (5.24a)$$

where N is the number of cases, $\hat{\sigma}_e$ is the error variance, called *standard error of estimate* in SYSTAT regression output, SS_X is the sum of squares of the predictor variable, $\bar{X}$ is the mean of the predictor variable, and X is the particular X value from which the fitted value of Y is being predicted. The mean and sum of squares for the predictor were shown in Table 5.11. The sum of squares is equal to the variance multiplied by $N-1$ and is shown in Equation 5.19b. The standard error of estimate, 4.14020, is shown in the regression output in Table 5.4.

Suppose that you wish to find the confidence interval around the fitted Y value from an income of $40,000. The variance of the prediction can be found by substituting the foregoing numbers into Equation 5.23, which gives

$$SE\hat{Y}_{|X} = 4.140 \left[\frac{1}{100} + \frac{(40{,}000 - 20{,}676.3)^2}{2{,}937{,}154{,}368.6} \right]^{1/2} = 1.533 \quad (5.24b)$$

Note that this standard error is smaller than that for the prediction interval in Equation 5.18b. The fitted value of Y, 12.849, was computed in Equation 5.20. The confidence interval around the fitted value is found by the following expression:

$$\hat{Y} - t_{(\alpha/2, N-p-1)} SE\hat{Y}_{|X} \le Y \le \hat{Y} + t_{(\alpha/2, N-p-1)} SE\hat{Y}_{|X} \quad (5.24)$$

The the value of $t_{(98)}$ that cuts off the top 97.5% of the variance is 1.984 (see section 2.2.3.2.2). Substituting this value, along with the fitted value and its standard error, into Equation 5.21, we find that the bounds on the confidence interval are

$$\hat{Y} \pm t_{(\alpha/2, N-p-1)} SE\hat{Y}_{|X} = 12.849 \pm (1.984)(1.533)$$

$$= 12.849 \pm 3.042 \quad (5.25)$$

Finally, substituting these values into Equation 5.20, we find the 95% confidence interval:

$$9.807 \le Y \le 15.891 \quad (5.26)$$

95% of such confidence intervals would contain the true value of the regression line at X.

5.5.5 Confidence Bands on the Regression Line

Suppose you wanted to draw bands around the regression line to show regions that would cover the true regression line 95% of the time. The regression lines covered by this region will vary both in their slope and intercept, so we want to compute a simultaneous 95% confidence region around both β_0 and β_1. From within this region, imagine constructing lines using all possible slope-intercept pairs, one line per pair. If you then plotted all of these lines you would fill in a 95% confidence region for the true regression line. SYSTAT will draw the confidence bands around such a region, shown in Figure 5.15 for the SHOES data set, from the following commands

▶ >GRAPH
>PLOT SHOES * INCOME/SMOOTH=LINEAR CONFI=.95

MACINTOSH

① Choose the menu item **Graph/Plot/Plot**.
② Select SHOES from the *Y variables* list and INCOME from the *X variable* list.
③ Click on **Smooth**.
④ Check **Linear**, check **Confidence**, type ".95" into the box, and click OK.
⑤ Click OK.

MS-DOS

① Choose the menu item **Graph/Plot/Data/2-D/Variables/Y**.
② Select SHOES from the variable list.
③ Choose the menu item **Graph/Plot/Data/2-D/Variables/X**.
④ Select INCOME from the variable list.
⑤ Choose the menu item **Graph/Plot/Data/2-D/Options/Smooth/Method/Linear/Confi/** and then type ".95" <ret>.
⑥ Choose the menu item **Graph/Plot/Data/2-D/Go!**

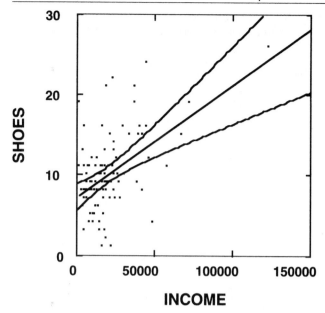

Figure 5.15 Regression line for shoes on income, showing the 95% confidence bands around the line.

5.6 The SHOES Data Set

The data set used in this chapter and Chapters 6 and 7 was created in SYSTAT using the math transformation functions. This data set can be recreated using those same functions by Macintosh users as shown in the next section. Unfortunately, DOS users will generate different random numbers, so their data set will look a little different. If you feel like typing, Table 5.12 shows the complete data set.

Because SYSTAT always uses the same random number seed when the program is started, Macintosh users can reconstruct this same data set using the commands below. This must be done before you use SYSTAT's random number generator after starting up the program.

In the SHOES data set I made an attempt to create a realistic relationship between variables, using distributions that approximated those seen in the real world. The first step in constructing the SHOES data set is to open a new file and fill the desired number of rows with missing entries. This is

MACINTOSH

① Choose the menu item **File/New**.

② Type in a new column label, such as "income", and hit <ret>.

③ Choose the menu item **Editor/Fill worksheet...**

④ Type the desired number of rows, "100", into the box and click **OK**.

⑤ Choose the menu item **Editor/Math**.

⑥ Set the variable INCOME equal to **int(xrn(5)*4000)**.

⑦ Set the variable TEMP equal to **4 − ((.0001) * income)^(1/3)**.

⑧ Set the variable FAMSIZE equal to **int(abs((temp − 1)*zrn − 1 + temp)) + 1**.

⑨ Set the variable SHOES equal to **int(abs(2 * famsize + .0005 * income + log(income/1000) * zrn)) + 1**.

accomplished with the *repeat* command. The data can then be generated with the following series of commands

```
> REPEAT 100
> LET INCOME = INT( XRN(5)*4000 )
> LET TEMP = 4 - ((.0001)*INCOME)^(1/3)
> LET FAMSIZE = INT(ABS((TEMP-1)*ZRN-1,
+TEMP)) + 1
> LET SHOES = INT(ABS(2*FAMSIZE+.0005*INCOME,
+LOG(INCOME/1000)*ZRN)) + 1
```

Table 5.12 The SHOES data set.

Case	Income	Fam Size	# of Shoes
1	18,244	1	9
2	21,883	1	7
3	15,439	4	11
4	16,915	6	16
5	42,742	1	15
6	18,953	1	8
7	12,895	5	12
8	5,293	1	3
9	67,397	1	15
10	37,658	3	21
11	22,174	1	9
12	10,491	1	5
13	6,513	2	7
14	9,358	1	8
15	19,712	2	9
16	9,724	3	9
17	39,203	1	9
18	25,972	5	12
19	29,599	1	16
20	4,299	6	13
21	1,826	9	19
22	12,666	2	6
23	5,142	6	16
24	23,684	4	22
25	4,736	5	11
26	46,765	3	12
27	10,371	4	12
28	40,082	3	15
29	17,688	2	5
30	7,989	3	8
31	7,967	4	11
32	24,203	4	8
33	14,569	2	3
34	21,712	3	15
35	6,715	2	8
36	9,387	2	5
37	37,133	1	19
38	7,300	4	11
39	122,855	1	26
40	19,257	2	9

Table 5.12 *(continued)*

Case	Income	Fam Size	# of Shoes
41	17,122	1	8
42	5,240	4	8
43	11,900	1	5
44	18,050	2	3
45	6,151	4	8
46	23,733	2	14
47	18,324	3	11
48	9,785	1	9
49	12,655	2	11
50	18,838	1	2
51	19,001	6	21
52	26,400	2	10
53	44,915	3	16
54	22,991	1	11
55	22,679	1	1
56	13,251	1	8
57	18,407	3	9
58	24,103	3	10
59	12,498	2	9
60	11,992	2	8
61	6,985	4	10
62	8,974	3	9
63	10,653	2	4
64	9,106	3	9
65	9,732	2	8
66	16,482	4	9
67	7,974	1	4
68	6,035	4	8
69	58,499	2	11
70	42,601	1	8
71	17,997	5	13
72	25,188	2	7
73	15,684	4	10
74	19,892	4	9
75	15,027	1	4
76	30,954	2	10
77	17,100	2	7
78	23,333	2	11
79	16,884	1	1
80	16,417	2	11

Table 5.12 *(continued)*

Case	Income	Fam Size	# of Shoes
81	27,878	1	7
82	20,533	3	11
83	14,693	4	8
84	44,805	2	24
85	13,011	4	12
86	34,280	1	15
87	40,602	1	16
88	20,620	1	9
89	4,251	3	6
90	1,533	5	11
91	18,158	2	10
92	6,372	4	9
93	27,122	1	9
94	15,022	3	10
95	14,183	1	6
96	48,822	1	4
97	24,233	4	13
98	5,355	2	7
99	72,463	1	19
100	3,632	3	8

6
Multiple Linear Regression

When you have multiple predictor variables that are all uncorrelated with each other, multiple regression can be treated essentially as a set of simultaneous simple regressions. However, in practice it will rarely be the case that the predictors are uncorrelated with one another. High correlations, called *collinearity*, can introduce a whole new set of difficulties in regression analysis and interpreting regression coefficients. It is the purpose of this chapter to pull together all the procedures for conducting a complete multiple regression analysis in SYSTAT, with particular emphasis on detecting and dealing with collinearity.

The analysis proceeds in four rough stages: (1) looking at the raw data, (2) estimating the regression coefficients, (3) diagnosing collinearity, and (4) checking the assumptions underlying the regression. Various confidence interval calculations are shown at the end of the chapter.

The SHOES data set was used in Chapter 5 to illustrate SYSTAT's simple regression procedures. In this chapter, in addition to family income and the number of pairs of shoes each family owns, we introduce a third measurement on the 100 families in the SHOES data set: the number of members in each family. The first five cases are shown in Table 6.1. (The complete data set is given in section 5.6, and can be reconstructed by Mac users using SYSTAT functions as described in that section.) Suppose that you now want to use both family income *and* family size to predict the number of pairs of shoes that a family owns. Following convention, I will call income (X_1) and family size (X_2) the *predictor* variables, and number of pairs of shoes (Y) the *outcome* variable. The goal of this chapter is to carry out a complete regression analysis of Y on X_1 and X_2 for this data set.

Table 6.1 The first five cases of the SHOES example data set in section 5.6. The names are fictitious; the numbers were generated in SYSTAT.

Case	Family	Income (X_1)	Fam. Size (X_2)	Prs. Shoes (Y)
1	Smith	18,244	1	9
2	Jones	21,883	1	7
3	Lincoln	15,439	4	11
4	Douglas	16,915	6	16
5	Gates	42,742	1	15
...				

6.1 Preliminary Diagnostics: Look at the Raw Data

The first step in data analysis almost always should be to look at plots of the data along with simple summary statistics. SYSTAT's graphics capabilities make this step quite easy. This section proceeds by looking at summary statistics, plotting variables separately to examine their distributions, plotting the joint (two-dimensional) distributions of the variables, and, finally, plotting the joint (three-dimensional) distribution of the variables using a *spin* plot.

6.1.1 Examining Single Variables

6.1.1.1 Summary Statistics

SYSTAT offers 13 summary statistics for single variables, all of which can be generated with a single command (see section 3.1). To generate all 13 statistics for the shoes (Y), income (X_1), and family size (X_2) variables, open the SHOES data set and type the following commands:

▶ >STATS
>STATISTICS SHOES INCOME FAMSIZE/ALL

MACINTOSH

① Choose the menu item **Stats/Stats/Statistics...**
② Select SHOES, INCOME, and FAMSIZE from the variable list.
③ Check the **All** option box.
④ Click **OK**.

MS-DOS

① Choose the menu item **Statistics/Stats/Statistics/Variables/**.
② Select SHOES, INCOME, and FAMSIZE from the variable list.
③ Choose the menu item **Statistics/Stats/Statistics/Options/All**.
④ Choose the menu item **Statistics/Stats/Statistics/Go!**

The "all" option at the end of the second command line tells SYSTAT to compute all 13 statistics. The output generated by these commands is shown in Table 6.2.

The statistics in Table 6.2 are described in section 3.1, and the shoes and income columns were discussed in section 5.1.1. The family size column in Table 6.2, however, is new. The mean family size is 2.56 members, with variance equal to 2.451, and minimum and maximum equal to 1 and 9, respectively.

We know that family size cannot be normally distributed: It is a discrete variable with only nine values in the SHOES data set, and it can't be less than one. Skewness for the family size variable is 1.135, with a standard deviation of $\sqrt{6/100} = 0.245$. Dividing the former by the latter to obtain an approximate value of Z gives 4.63, which is very unlikely due to chance. Thus, the population distributions of family size appears to be positively skewed. The value of kurtosis for family size is 1.625, with a standard deviation of $\sqrt{24/100} = 0.490$. Dividing kurtosis by its standard deviation gives $Z = 3.32$, which is unlikely due to chance, suggesting that the family size distribution is thick-tailed (Snedecor & Cochran, 1980). The plots in the next section will help us see just what the distributions actually look like.

Table 6.2 Summary statistics of the variables in the SHOES data set.

TOTAL OBSERVATIONS: 100

	SHOES	INCOME	FAMSIZE
N OF CASES	100	100	100
MINIMUM	1.000	1533.000	1.000
MAXIMUM	26.000	122855.000	9.000
RANGE	25.000	121322.000	8.000
MEAN	10.140	20676.310	2.560
VARIANCE	22.768	.296682E+09	2.451
STANDARD DEV	4.772	17224.465	1.566
STD. ERROR	0.477	1722.446	0.157
SKEWNESS(G1)	0.890	2.775	1.135
KURTOSIS(G2)	1.170	11.746	1.625
SUM	1014.000	2067631.000	256.000
C.V.	0.471	0.833	0.612
MEDIAN	9.000	17111.000	2.000

6.1.1.2 Histograms

A histogram shows the distribution of data points across a single variable (see section 3.2.1.1). Histograms for the shoes and income variables were shown in Figure 5.1. Figure 6.1 shows the histogram for the family size variable, which was created by typing

```
>GRAPH
>DENSITY FAMSIZE
```

From the heights of the bars you can read off the number of cases corresponding to each family size. For example, Figure 6.1 indicates that 32 "families" were, in fact, single individuals. At the opposite extreme, one family had nine members.

For our purposes, the most interesting feature of Figure 6.1 is that the distribution of family has a large positive skew. Another interesting feature is that family with nine members: We will want to keep an eye on this case to see if it has an undue influence on the regression analysis.

6.1.1.3 Box Plots

Box plots give a different visual impression than do histograms, and they contain more explicit information about the distribution of cases (see section 3.2.1.2). Box plots for the

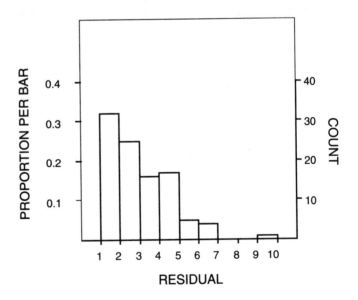

Figure 6.1 Histogram of the family size variable from the SHOES example data set.

shoes and income variable were shown in Figure 5.2. Such a plot for the family size variable is shown in Figure 6.2. This plot was created with the commands

```
>GRAPH
>BOX FAMSIZE
```

As in the histogram for family size in Figure 6.1, the box plot indicates positive skew. Furthermore, the box plot indicates that the largest family, the family with nine members, is an outlier. Again, we will want to keep an eye on this case.

6.1.1.4 Stem and Leaf Displays

Stem and leaf displays of the shoes and income variables were shown in Figure 5.3 (see also section 3.2.1.3). The stem and leaf display of the family size variable is shown in Figure 6.3. It was created with the commands

```
>GRAPH
>STEM FAMSIZE
```

Because the family size variable is discrete, with no decimal places, this plot looks essentially like a 'dit' plot (see section 3.2.1.1), with one circle (zero) for every case in the data set. This plot combines the information from the histogram and the box plot, but the only new information are the numerical values of the upper and lower hinges. In the plot, the family size that contains the median is marked with the letter *M*, and lines containing the upper and lower hinges (75th and 25th centiles) are marked with the letter *H*. The outlier that we found in the box plot is the case below the line "***outside values***".

6.1.2 Identifying Individual Cases

The graphical procedures described above help locate outliers. For example, we know that the family size variable in the

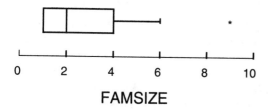

Figure 6.2 Box plot of the family size variable in the SHOES example data set.

```
STEM AND LEAF PLOT OF VARIABLE:
    FAMSIZE, N = 100

MINIMUM IS:              1.000
LOWER HINGE IS:            1.000
MEDIAN IS:               2.000
UPPER HINGE IS:            4.000
MAXIMUM IS:              9.000


    1   H0000000000000000000000000000000
    2   M000000000000000000000000
    3     000000000000000
    4   H00000000000000000
    5     00000
    6     0000
          ***OUTSIDE VALUES***
    9   0
```

Figure 6.3 Stem and leaf display of the family size variable in the SHOES data set.

SHOES data set has one family of size nine, and that this value is an outlier. To identify this case in the data set, you can simply type

>FIND FAMSIZE=9

MACINTOSH

① Choose the menu item **Data/Find Case...**
② Select FAMSIZE into the selection box on the left.
③ Choose the "=" operator and type "9" into the box on the right.
④ Click **OK**.

and SYSTAT will go to that case in the data set and highlight the value "9." (See section 3.2.1.5.) This is case 21, and it is shown in Table 6.3. Notice that although this family has a lot of members and a lot of shoes, they have a very small income. Could this be a typo? Because this case might substantially influence the regression, one would want to examine it very closely.

6.1.3 Plotting More Than One Variable

6.1.3.1 Plots of the Outcome Versus the Predictors

As in simple regression, in multiple regression to further help identify cases that may cause trouble, you can look at scatterplots of the outcome variable Y on the predictor variables. A scatterplot of shoes versus income for the SHOES data set was shown in Figure 5.4 (see section 3.2.2.1). A scatterplot of shoes versus family size is shown in Figure 6.4. This plot shows that the distribution has a megaphone shape opening to the left, which may suggest decreasing variance as family size in-

Table 6.3 The case with the largest family size in the SHOES data set.

Case	INCOME	FAMSIZE	SHOES
21	1,826.000	9.000	19.000

creases. We will want to explore this possibility in section 6.5.2. You can see also in this plot that the one outlier case, the family of size nine, is not particularly extreme on the shoe dimension.

There are a number of other scatterplots that you may want to examine, most of which were covered in section 5.1. Any of these plots may add to your understanding of the data. I will just show two more, the kernel density plots of the shoes and family size variables (see sections 3.2.2.2 and 5.1.7 for kernel density plots of shoes against the income variable). These plots are like histograms in 3-D: The height of the plot shows the proportion of cases falling into each two-dimensional interval defined by combinations of the Y and X variables. Unlike histograms, the family size variable is treated as continuous, which we know is not quite right. However, the 3-D surfaces in Figure 6.5 suggest that the distribution might be bimodal.

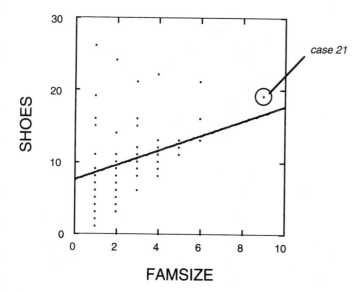

Figure 6.4 Scatterplot of shoes versus family size in the SHOES data set.

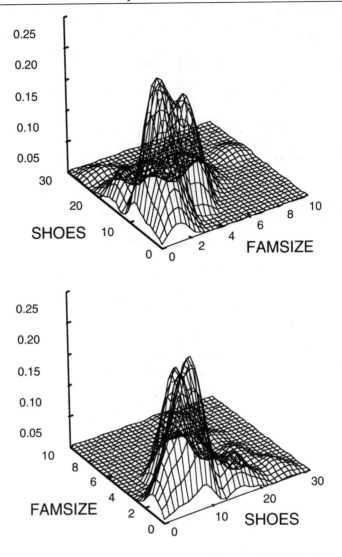

Figure 6.5 Kernel density plots of shoes against family size in the SHOES data set.

This is probably just a chance blip in our sample, but you might want to see whether the points in the two humps (roughly, above and below the line in Figure 6.4) might be from different populations. For example, did one subsample come from a cold climate where people tend to own more shoes, regardless of family size? (Because this is a hypothetical data set I will not pursue this any further, but you might want to in a real data set.)

6.1.3.2 Plots of Predictors Versus Predictors

One type of scatterplot, which can be useful for detecting and assessing collinearity, are scatterplots of each of the predictor variables against each of the other predictor variables. For the shoes data set, this plot can be created by typing (see section 3.2.1.1.1)

```
>GRAPH
>PLOT INCOME * FAMSIZE
```

This scatterplot is shown in Figure 6.6. From this plot it appears that family size and income are negatively correlated. The strength of this correlation may be heavily influenced by a few wealthy individuals, however, such as the person with an income of around $125,000 in Figure 6.6. The one family with nine members may also heavily influence this correlation. How these cases will affect the regression depends on their values on the shoes variable, but they certainly contribute to the collinearity of the two predictors.

6.1.4 Scatterplot Matrices: SPLOM

As the number of predictors increases, the number of two-dimensional scatterplots that you want to look at explodes. In

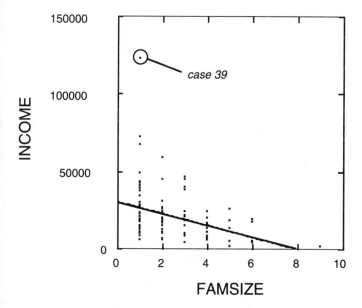

Figure 6.6 Scatterplot of the two predictors, income and family size, in the SHOES data set.

simple regression, we were only interested in plots of Y versus X. By adding a second predictor, we now want to look at plots of Y on X_1, Y on X_2, and X_1 on X_2. A third predictor would add three more plots to this list. Fortunately, SYSTAT provides a very useful procedure for plotting all of the variables against each other in a matrix of scatterplots. SYSTAT calls this type of graph a "SPLOM," for Scatter*PLOt* Matrix. To create a SPLOM, just type

```
>GRAPH
>SPLOM SHOES INCOME FAMSIZE
```

The resulting matrix is shown in Figure 6.7. This matrix shows all of the pairwise scatterplots between the variables that you select. The middle cell in the upper row in Figure 6.7 contains the scatterplot of shoes versus income from Figure 5.4. The plot in the upper right contains the scatterplot of shoes versus famsize, as shown in Figure 6.4. The plot in the middle row on the right is just the scatterplot of income versus family size from Figure 6.6. The plots in the lower half of the matrix are just the plots from the upper half with the axes transposed. This gives you a second view of each plot from a slightly different orientation, which can sometimes allow you to see patterns that did not catch your eye at first.

The SPLOM display offers two particularly useful options: graphing the best-fitting line and displaying the histograms

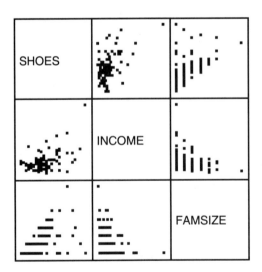

Figure 6.7 SPLOM of the shoes, income, and family size variables in the SHOES data set.

for each variable. These options are used in the following command:

```
>SPLOM SHOES INCOME FAMSIZE/SMOOTH=LINEAR,
DENSITY=HIST
```

The resulting SPLOM display is shown in Figure 6.8. In these plots, the angles of the lines help you quickly pick out correlated variables, the shapes of the point clouds help detect nonlinear relationships, and the points themselves help detect outliers and influential cases. The histograms of each variable are miniature versions of the histograms in Figure 5.1 and 6.1. These histograms show the distributions of the variable along the X-axis in its corresponding column. For example, the shoes histogram in the upper left of Figure 6.8 gives the distribution of the shoes variable on the X-axis in the two scatterplots in the leftmost column.

As you can see, a SPLOM display can replace a large number of individual graphs. Sometimes you may want to plot variables individually, to get a more detailed view or for publication purposes, but you should probably always begin with the SPLOM.

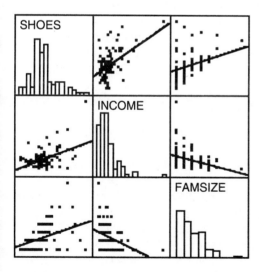

Figure 6.8 SPLOM of the shoes, income, and family size variables in the SHOES data set, with best-fitting lines and histograms for each variable.

6.2 Multiple Regression

To compute a multiple regression analysis, you first have to specify a linear model. This model tells SYSTAT which variable is to be treated as the outcome, and which variables are the predictors whose regression coefficients you wish to estimate. In multiple regression this equation takes the form

$$Y = \beta_0 X_0 + \beta_1 X_1 + + \beta_2 X_2 + \ldots + \beta_p X_p + e \qquad (6.1)$$

where Y is the outcome variable, X_0 is a constant equal to 1 (or 0 when the model is to contain no constant), and the β_j are the regression coefficients for the predictor variables X_j. In linear regression, values of β_j are chosen that minimize the squared deviations of the actual data points from predicted values, and if the data are multivariate normal, the expected values of these parameters are the "right answers" in the actual relationship between the variables. Of course, estimates are not perfect because there will be noise in real data, and we have probably omitted relevant variables. Any important variables that have been left out of the model, any error in recording data, or other noise in the system is lumped together into the term "residual variation" or "error variation." This is designated e and represents error in the system.

When you enter a model into SYSTAT, include the outcome, constant, and any predictor variables that you want, but leave off the term for the error component. Anything that you leave out of the model will be treated by SYSTAT as error variance. The commands for computing a multiple regression of shoes on income and family size in the SHOES data set is

```
>MGLH
>MODEL SHOES = CONSTANT + INCOME + FAMSIZE
>SAVE SHOES2.RES /MODEL
>PRINT = LONG
>ESTIMATE
```

The *model* command tells SYSTAT which variables to include in the regression. The *save* command will save to a file the residuals and other diagnostic information that you can later use to assess the validity of the assumptions underlying regression (more on this in section 6.5). I called this file "shoes2.res" to distinguish it from the "shoes.res" file that was created by the simple regression in Chapter 5. The *model* option after the slash in the *save* command asks SYSTAT to include the original variables in the new file with the residuals. The command *print=long* (the "extended results" option using Mac menus or

MACINTOSH

① Choose the menu item **Edit/Preferences...** and select the **Statistics/Extended** option.

② Choose the menu item **Stats/MGLH/Regression...**

③ With the cursor in the *Dependent* box, select SHOES from the variable list.

④ With the cursor in the *Independent* box after CONSTANT, select INCOME and FAMSIZE from the variable list.

⑤ Check **Save residuals**.

⑥ Click **OK**.

⑦ Type a filename, e.g., "shoes2.res" into the dialog box, and click **Save**.

⑧ Check **Save model and diagnostics**.

⑨ Click **OK**.

MS-DOS

① Choose the menu item **Utilities/Output/Results/Long**.

② Choose the menu item **Statistics/MGLH/Regression/Model/Variables/**.

③ Select SHOES from the variable list.

④ Select INCOME and FAMSIZE from the variable list.

⑤ Choose the menu item **Statistics/MGLH/Regression/Estimate/Save/**, hit <esc>, type, e.g., "shoes2.res", and then choose /**Model**.

⑥ Choose the menu item **Statistics/MGLH/Regression/Estimate/Go!**

"long results" using DOS menus) causes SYSTAT to give additional diagnostic information useful for detecting collinearity. The output of the regression analysis is shown in Table 6.4.

In this table I have inserted labels for various sections so that I can easily refer to them separately. Part B contains the standard information about the regression coefficients. The first two lines of part B give information about the overall fit

Table 6.4 Extended multiple regression output from shoes on income and family size in the SHOES data set.

A. Collinearity Diagnostics

```
EIGENVALUES OF UNIT SCALED X'X
                          1            2            3
                      2.452        0.463        0.085

CONDITION INDICES
                          1            2            3
                      1.000        2.302        5.360

VARIANCE PROPORTIONS
                          1            2            3
     CONSTANT         0.022        0.002        0.976
       INCOME         0.043        0.455        0.502
      FAMSIZE         0.031        0.208        0.761
```

B. Coefficient Information

```
DEP VAR: SHOES      N: 100    MULTIPLE R: 0.742    SQUARED MULTIPLE R: 0.551
ADJUSTED SQUARED MULTIPLE R: 0.542        STANDARD ERROR OF ESTIMATE: 3.230

VARIABLE   COEFFICIENT    STD ERROR    STD COEF   TOLERANCE      T       P(2 TAIL)

CONSTANT         1.567        0.872       0.000        .         1.797     0.075
INCOME     .195591E-03  .200921E-04       0.706      0.880       9.735    .10E-14
FAMSIZE          1.769        0.221       0.580      0.880       8.003    .26E-11


CORRELATION MATRIX OF REGRESSION COEFFICIENTS
                 CONSTANT       INCOME      FAMSIZE
   CONSTANT         1.000
     INCOME        -0.702        1.000
    FAMSIZE        -0.814        0.347        1.000
```

Table 6.4 *(continued)*

C. ANOVA

```
                    ANALYSIS OF VARIANCE

SOURCE          SUM-OF-SQUARES    DF    MEAN-SQUARE    F-RATIO         P

REGRESSION         1242.267        2       621.134     59.549  .999201E-15
RESIDUAL           1011.773       97        10.431
```

D. Case Diagnostics

```
WARNING: CASE 21 HAS LARGE LEVERAGE              (LEVERAGE             =    0.182)
WARNING: CASE 24 IS AN OUTLIER      (STUDENTIZED RESIDUAL   =    2.827)
WARNING: CASE 37 IS AN OUTLIER      (STUDENTIZED RESIDUAL   =    2.719)
WARNING: CASE 39 HAS LARGE LEVERAGE              (LEVERAGE             =    0.378)
WARNING: CASE 84 IS AN OUTLIER      (STUDENTIZED RESIDUAL   =    3.349)
WARNING: CASE 96 IS AN OUTLIER      (STUDENTIZED RESIDUAL   =   -2.913)
WARNING: CASE 99 HAS LARGE LEVERAGE              (LEVERAGE             =    0.101)

DURBIN-WATSON D STATISTIC      2.041
FIRST ORDER AUTOCORRELATION   -0.023

RESIDUALS HAVE BEEN SAVED
```

of the model. The multiple correlation between Y and the predictors is given as *multiple R* (part B, line 1). The square of this correlation, *squared multiple R*, gives the proportion of the variance in Y that is predictable by the X's in combination. This is a substantial .551, which means that 55.1% of the variance in shoes is predictable from income and family size, as shown in line 1 of part B. In line 2, the *adjusted squared multiple R* is an estimate of what the proportion of variance accounted for would be in a new sample of data from the same population, and it always will be somewhat lower than the unadjusted value. The *standard error of estimate*, 3.230, is equal to the square root of the *mean square* for the residual, 10.431, from part C.

The fourth, fifth, and sixth lines of part B give the best-fitting estimates of the coefficients in the linear equation (Equation 6.1) relating the X's to Y, along with their standard errors (*std error*) and significance tests. In Table 6.5 (page 181), the coefficient for the constant (the Y-intercept) is 1.567. The coefficient (slope) for the income variable is 0.000196: Holding family size constant, an increase in income of $10,000 is associated with an increase of about two pairs of shoes. The coefficient (slope) for family size is 1.769: Holding income constant, an increase of one family member is associated with an increase of about 1.769 pairs of shoes. Using these values, you can show the function relating shoes to income and family size and compute predicted values of the shoes variable by plugging the coefficient estimates into Equation 6.1:

$$\text{shoes} = (0.000196 \times \text{income}) + (1.769 \times \text{famsize}) + 1.567 \tag{6.2}$$

For a family of four with an income of $25,000, this regression equation predicts that they will own 13.5, or approximately 14, pairs of shoes.

Dividing the coefficients by their standard errors, you obtain the t's shown in lines 4, 5, and 6 of part B of Table 6.4. Both of the predictor variables are highly significant by any standard. The *standard coefficients* (std coef) are the partial correlations (strictly, "semi-partial" correlations; Howell, 1987) between each predictor and the residuals from the regression of Y on all the other predictors. Interestingly, in this data set these partial correlations are higher than their respective first order correlations in Table 6.5. For example, the correlation between income and shoes in Table 6.5 is .505; in Table 6.4 the partial correlation between shoes and income after family size is taken into account is .706. This increase is due to the negative correlation between income and family size: Because family size is positively correlated with shoes, it suppresses the first order relationship between income and shoes.

The analysis of variance table in part C provides a test of the overall fit of the regression. The fit of the regression in the SHOES data set is highly reliable, $F_{(2,97)} \approx 59.5$, $p = 1 \times 10^{-15}$. Dividing the sum-of-squares for the regression by the total sum-of-squares gives the *squared multiple R* in line 1 of the output.

The remainder of the output in Table 6.4, part A, tolerance, and part D, is discussed below.

6.3 Collinearity Diagnostics

Collinearity (or "multicollinearity," or "ill-conditioning," or "perfect dependency") occurs when a predictor variable is perfectly correlated with another predictor or combination of predictors. In such a case the regression coefficients for the set of predictors are indeterminate, and the collinear predictor adds no new information to the regression. In the case of near collinearity, in which a predictor variable is very highly, but not perfectly, correlated with another predictor or combination of predictors, the regression coefficients can be estimated, but the variances of those estimates will be *degraded*: the magnitude of the estimated variance will be determined primarily by the collinearity relation. Such degraded variances will be greatly inflated, significance tests will lack power, confidence and prediction intervals will be uselessly large, and the coefficient estimates themselves will be highly unstable. (In the remainder of this section I will use "collinearity" and "dependency" interchangeably to mean "near collinearity.") Collinearity is usually not a problem in most social science data, but it frequently turns up in econometrics.

Not all degrading collinearity is harmful (Belsley, Kuh, & Welsch, 1980, pp. 115–117). It is possible to have one or more dependencies among your predictors and still have one or more predictors that are not involved in any of those dependencies: If you are fortunate enough to be only interested in the uninvolved predictors, then you have no problem. Furthermore, if you are only interested in testing whether a coefficient is significantly different from zero and you find that it is, then the lack of power of your test is irrelevant. As Belsley, Kuh, and Welsch state (1980, p. 116), "collinearity doesn't hurt so long as it doesn't bite."

This section presents three methods of diagnosing collinearity, in order of usefulness (least to most). Only the third method, employing condition indices, allows you to determine the number of dependencies and which predictors are involved in those dependencies. However, the first two methods are more widely known so they deserve mention.

6.3.1 Detecting Collinearity with Correlations

Graphs are indispensable for helping you see patterns in the data and which cases may be responsible for those patterns. However, graphs should always be supplemented with correlations to help detect collinearity. The first order correlation among the predictors, i.e., the pairwise Pearson product-moment correlations, can help locate pairs of redundant predictors. When two predictors are very highly correlated, say $r > .9$, one predictor does not add much information that was not already contained in the other, but it can produce more poorly estimated regression coefficients. In such a case you should consider dropping one of those predictors. The loss in variance accounted for will be small, but the remaining coefficients may be considerably better estimated.

To look at all pairwise correlations among p predictors, simply type the commands

```
>CORR
>PEARSON PREDICTOR1 PREDICTOR2 ... PREDICTORP
```

For example, to find all pairwise correlations between the three variables in the SHOES data set type

```
>CORR
>PEARSON SHOES INCOME FAMSIZE
```

These commands produce a matrix of pairwise Pearson product-moment correlations between the variables given in the command. This matrix is shown in Table 6.5.

The resulting correlation between the income and family size variables is −.347. This correlation may cause interpretive difficulties, but in itself it is not large enough to cause collinearity problems.

> **MACINTOSH**
> ① Choose the menu item **Stats/Corr/Pearson...**
> ② Select SHOES, INCOME, and FAMSIZE from the variable list.
> ③ Click **OK**.
>
> **MS-DOS**
> ① Choose the menu item **Statistics/Corr/Pearson/Variables/** and select SHOES, INCOME, and FAMSIZE from the variable list.
> ② Choose the menu item **Statistics/Corr/Pearson/Go!**

Table 6.5 Matrix of all pairwise Pearson product-moment correlations between the shoes, income, and family size variable in the SHOES data set.

PEARSON CORRELATION MATRIX

	SHOES	INCOME	FAMSIZE
SHOES	1.000		
INCOME	0.505	1.000	
FAMSIZE	0.336	-0.347	1.000

NUMBER OF OBSERVATIONS: 100

6.3.2 Tolerance

Tolerance is a measure of collinearity in multiple regression that is automatically computed by SYSTAT whenever you compute a regression. For the SHOES data set, the values of tolerance are shown in lines 5 and 6 of part B in Table 6.4. Tolerance is equal to 1 minus the proportion of variance in each predictor accounted for by all other predictors, that is, the proportion of variance in X_j *not* accounted for by the other X's. When there are only two predictors, as in the present example, this is just 1 minus the square of the correlation between the two predictors. The correlation between income and family size from Table 6.5 was −.347, so 1−(−.347)² = .880, the value of tolerance given for each of the predictors in Table 6.4. When there are more than two predictors, tolerance is 1 minus the R^2 from the regression of each X_j on all the other X's. When the value of tolerance gets very close to zero, that predictor is close to being a linear combination of the other predictors. Such a predictor does not add much new information to the regression, but can degrade the estimates of coefficients. One difficulty with tolerance as a diagnostic is that we have no really good rules for how small is "small." Values less than 0.1 may be trouble, and values less than 0.01 almost certainly are, but these are very rough heuristics. A second problem is that when multiple predictors have low tolerance, it is difficult to determine the precise source of the dependency or even how many dependencies there are. Thus, tolerance is a better measure of when you don't have a problem than when you do: When all of the values of tolerance are near 1, collinearity is not a problem; when one or more values of tolerance are low, you should proceed to the procedures in the next section to find out why.

6.3.3 Condition Indices

The number of collinearities, or *dependencies*, among a set of predictors is equal to the number of large *condition indices* (CIs) corresponding to the eigenvalues of those predictors. The condition indices (κ_j) are equal to the square-roots of the ratios of the largest eigenvalue ($\lambda_{largest}$) to each individual eigenvalue (λ_j). Thus,

$$k_j = \sqrt{\lambda_{largest} / \lambda_j}$$

The largest condition index is called the condition *number*.

The particular predictors involved in each dependency can be determined by finding two or more predictors that have most of their variance (say, > 50%) determined by that dependency. These considerations led to the double condition for diagnosing a degrading collinearity (Belsley, Kuh, & Welsch, 1980, p. 112): A degrading collinearity has a high condition index (CI) and is associated with high variance proportions for two or more estimated regression coefficient variances.

6.3.3.1 Procedure

To apply these diagnostic criteria, we need some heuristic for what counts as a "high CI" and what counts as a "high variance proportion." Belsley, Kuh, and Welsch (1980, pp. 117–159) provide such heuristics based on empirical trials. A CI between roughly 10 and 30 usually indicates moderate collinearity, and above 30 usually indicates serious collinearity. A "high" variance proportion is 0.5 or greater, which means that the coefficient has half or more of its variance determined by the collinearity relation. The following diagnostic procedure is adapted from their recommendations (pp. 156–159).

Step 1: This is the hardest step, but fortunately SYSTAT does this step for you when you choose extended output for the regression analysis, as shown in part A of Table 6.4: Scale the matrix X'X to have 1's in the diagonal, and then compute the eigenvalues of the principal components of this matrix, the condition indices (CIs), and the variance proportions.

Step 2: Find the number of potential collinearities by counting the CIs over some chosen threshold—over 30, for example.

Step 3: When one or more potential collinearities are found, examine the variance proportions to locate predictors whose coefficients have more than 50% of their variance associated with a single high CI. Two cases each have their own special considerations:

Case 1: Only one large CI was found in step 2. Look for *two or more* predictors that have variance proportions greater than about .5 for the the high CI.

Case 2: Two or more large CIs were found in step 2.

a. *Competing dependencies*. When two or more CIs are of the same order of magnitude, for example, CIs of 30 and 40, they are said to be *competing*, and any predictors that are involved in competing dependencies can have their variance arbitrarily distributed among them; that is, dependencies of roughly equal magnitude can be confounded. So, to determine whether a predictor is involved in a set of competing dependencies, add the variance proportions for the predictor across the competing CIs: Any predictor that sums to more than about .5 is involved in at least one of the competing dependencies; you just can't tell which.

b. *Dominating dependencies*. When two or more CIs are of different orders of magnitude, for example, CIs of 30 and 300, the larger is said to *dominate* the smaller. As with competing dependencies, any predictors that are involved in both dominated and dominating dependencies can have their variance distributed among those dependencies, so any predictor that sums to more than about .5 across those dependencies is involved in at least one of the dependencies. Furthermore, when a predictor is involved in the dominating (larger) dependency, its simultaneous involvement in a dominated (smaller) dependency can be masked; that is, any predictors that have variance proportions greater than about .5 for the dominating (larger) CI may also be involved in the dominated (smaller) CI, even though they do not appear to be involved in the smaller one.

Step 4: To further analyze competing and/or dominated (smaller) CIs to see which predictors are involved, you can perform auxiliary regressions. More on this below.

Step 5: You may conclude that there is no evidence of degraded coefficient estimates for a predictor when (1) the predictor is not associated with any high CI, or (2) the predictor is the only one associated with a noncompeting, nondominated CI, or (3) the predictor is the only one associated with a dominated CI and is not a linear combination of the predictors involved in the CI(s) that dominate its own, which can be determined through auxiliary regressions.

In principle, once you have discovered the number of dependencies in step 2, you could compute all possible regressions among a set of predictors in step 4 to help determine which predictors are involved in each dependency. However, Belsley, Kuh, and Welsch (1980; see pp. 113, 121, 142–143, 155–156, 159–160) offer some advice on choosing and interpreting one auxiliary regression for each confounded (competing, dominated, or dominating) dependency. First, choose the one predictor most clearly associated with each confounded dependency. When two or more predictors are roughly equally associated with a single dependency, choose the one that is least associated with the other dependencies. When these two or more predictors are also associated roughly equally with other dependencies, choose the one whose association is with the most removed dependency, i.e., skips the most CIs. Second, once you have chosen one predictor for each confounded dependency, regress each of those predictors on the unchosen predictors involved in the confounded dependencies. The *t* tests of the predictors in these regressions pick out which predictors are "significantly" involved in each dependency.

For example, if X_1 is the only predictor involved with a CI of 30, X_2 is the only predictor involved with a CI of 40, and X_3 and X_4 are involved with a CI of 300, it may or may not be the case that X_2 is a linear combination of X_3 or X_4, and X_1 is a linear combination of X_2 and X_3 or X_4. To sort out these dependencies, choose the highest variance proportions from each of the smaller dependencies, in this case X_1 and X_2, respectively, and regress them both on X_3 or X_4.

Scaling problems can both produce dependencies and mask them in the CIs. When a predictor and the constant are both associated with the same CI, it suggests that at least that dependency might be reduced by *centering* that predictor, i.e., subtracting the mean of the predictor from each case.

6.3.3.2 Example

Unfortunately (or fortunately), the SHOES data set does not provide a very illustrative example of the above procedures because none of the condition indices are large. In part A of Table 6.4 you can see that the largest condition index is 5.36. There are dependencies among the predictors, but they are not large enough to degrade the coefficient estimates.

So, to better illustrate the procedures, I added two new variables to the SHOES data set: an individual income variable, which is just the family income divided by the number of family members, and a variable containing last year's income, which I contrived with the following command:

```
>LET LASTYEAR = INCOME - (INCOME/20*ZRN)
```

Thus, both of these new variables have dependencies with the income and family size variables. The first order correlation matrix for the predictors is shown in Table 6.6. As you can see, some of the correlations are very high.

To further analyze these dependencies, perform the regression of shoes on these four predictors using extended output, and look at the condition indices. These are shown in Table 6.7. The largest condition index, 64.131, exceeds our heuristic threshold of 30 and dominates all the others. The *income* and *lastyear* variables have more than 99% of their variance associated with this CI. The next two largest CIs, 8.848 and 6.534, do not exceed the threshold and are really too small to worry about, but they may serve as examples of competing dependencies. The *constant* is clearly associated with the third CI and *indincom* is clearly associated with the fourth, but the *family*

Table 6.6 Correlation matrix of the income and family size variables from the SHOES data set, along with two new hypothetical variables, income per household member (INDINCOM) and last year's income (LASTYEAR).

PEARSON CORRELATION MATRIX

	INCOME	FAMSIZE	INDINCOM	LASTYEAR
INCOME	1.000			
FAMSIZE	-0.347	1.000		
INDINCOM	0.904	-0.514	1.000	
LASTYEAR	0.998	-0.340	0.899	1.000

NUMBER OF OBSERVATIONS: 100

Table 6.7 Eigenvalues, condition indices, and variance proportions from the regression of shoes on the four predictor variables shown in Table 6.6.

EIGENVALUES OF UNIT SCALED X'X

1	2	3	4	5
3.945	0.912	0.092	0.050	0.001

CONDITION INDICES

1	2	3	4	5
1.000	2.080	6.534	8.848	64.131

VARIANCE PROPORTIONS

	1	2	3	4	5
CONSTANT	0.007	0.023	0.887	0.084	0.000
INCOME	0.000	0.000	0.001	0.005	0.993
FAMSIZE	0.006	0.100	0.453	0.441	0.000
INDINCOM	0.004	0.021	0.023	0.928	0.023
LASTYEAR	0.000	0.000	0.001	0.006	0.992

size variable appears to be distributed between the two. When we sum family size across the two competing CIs, it does exceed .5. Furthermore, because they are involved with the dominating CI, the *income* and *lastyear* variables' involvement with the smaller CIs may be masked.

Thus, we know that income and lastyear are highly dependent and have seriously degraded coefficients, and we know

Table 6.8 The top panel shows part of the output from the regression of income on family size and last year's income, and the bottom panel shows part of the output from the regression of indincom on family size and last year's income.

```
DEP VAR: INDINCOM   N: 100   MULTIPLE R: 0.929   SQUARED MULTIPLE R: 0.863
ADJUSTED SQUARED MULTIPLE R: 0.861    STANDARD ERROR OF ESTIMATE: 6544.292

VARIABLE      COEFFICIENT    STD ERROR    STD COEF   TOLERANCE      T      P(2 TAIL)

CONSTANT        2689.368      1766.763      0.000        .         1.522     0.131
INCOME             0.840         0.041      0.825      0.880      20.631     0.000
FAMSIZE        -2549.136       447.934     -0.228      0.880      -5.691     0.000

DEP VAR: FAMSIZE    N: 100   MULTIPLE R: 0.347   SQUARED MULTIPLE R: 0.120
ADJUSTED SQUARED MULTIPLE R: 0.111    STANDARD ERROR OF ESTIMATE: 1.476

VARIABLE      COEFFICIENT    STD ERROR    STD COEF   TOLERANCE      T      P(2 TAIL)

CONSTANT           3.212         0.231      0.000        .        13.888     0.000
INCOME            -0.000         0.000     -0.347      1.000      -3.661     0.000
```

that family size, indincom, and the constant may all enter into milder dependencies that do not seriously degrade their coefficients. We do not know whether income and lastyear enter into the smaller dependencies because the dominating CI may mask such relations. To uncover more about these dependencies, you can perform auxiliary regressions. To determine whether income, lastyear, and family size enter into the second largest dependency, regress indincom on income and family size (you should omit lastyear because of its known collinearity with income). This regression is shown in the top panel of Table 6.8. This significant *t*'s in this regression indicate that indincom is a linear combination of both income (and hence lastyear) and family size.

To see whether income and lastyear enter into the third largest dependency, regress family size on income (again, omitting lastyear). This regression is shown in the bottom panel of Table 6.8. Income does appear to be associated with the third dependency. Also note that the constant is significant in this regression.

Do these dependencies make sense given the way this data set was constructed? Yes. The largest CI properly picked out the near dependency between income and lastyear, which were related by the equation *lastyear = income − income/(20*zrn)*. In the second dependency, indincom was related to all four of the other predictors through the equation *indincom = income/famsize*. Finally, family size was related to income (and hence lastyear) through the equations given in section 5.6.

It should be emphasized that the analyses of the third and fourth CIs were performed for illustrative purposes only—both of those CIs are too small to cause concern.

6.4 Case Diagnostics

6.4.1 Studentized Residuals

When you ask SYSTAT to save the residuals from a regression analysis, it saves the studentized residuals in a column labelled STUDENT in the file that you specify. To examine those residuals, open that file in the data editor (see section 1.1.4 for opening files). The studentized residuals are "externally studentized," which means they are approximately distributed as t with $N-p-1$ degrees of freedom. Part D of the regression output in Table 6.4 indicates that four cases have unusually large residuals: cases 21, 24, 51, and 84. All four of these cases have absolute values greater than 2.627, which we would only expect to find about 1 time in 100 due to chance in a t distribution with 97 degrees of freedom (see section 2.2.3.2.2 for how to find critical values of t). This many large residuals suggest that the residuals may not be normally distributed.

However, you should not stop with the warnings provided by SYSTAT. A good way to display studentized residuals is with a stem and leaf display and box plots, as shown in Figure 6.9. The stem and leaf display and box plot in Figure 6.9 were created by typing the following commands (see sections 3.2.1.2 and 3.2.1.3):

```
>GRAPH
>STEM STUDENT
>BOX STUDENT
```

Three of the large studentized residuals listed in the regression output in Table 6.4 and a fourth large residual are shown to the right of the upper fence in the box plot in Figure 6.9. Two large negative residuals, one of which is listed in Table 6.4, are shown to the right of the lower fence in the same box plot.

Perhaps the easiest way to identify the individual cases corresponding to large and small residuals is to sort the file by STUDENT and then reopen the newly sorted file (see section 2.1.1). Table 6.9 shows the first six and last six cases in this

```
STEM AND LEAF PLOT OF VARIABLE: STUDENT, N =
100

MINIMUM IS:         -2.913
LOWER HINGE IS:     -0.532
MEDIAN IS:           0.023
UPPER HINGE IS:      0.413
MAXIMUM IS:          3.349
```

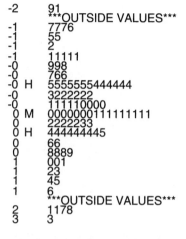

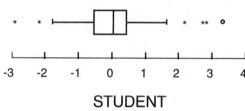

Figure 6.9 Stem and leaf display and box plot of studentized residuals from the regression of shoes on income and family size in the SHOES data set.

newly sorted file, containing the six largest negative and positive studentized residuals. Seven of these studentized residuals have absolute values greater than 2.0, which is not far from the five we would (roughly) expect due to chance in a t distribution with 97 degrees of freedom. You should examine such cases individually to determine whether they are unusual in some other respect and whether the data should be transformed in some manner, which I will return to in Chapter 7.

Table 6.9 The six largest negative and positive studentized residuals from the regression of shoes on income and family size in the SHOES data set.

Case	STUDENT
1	−2.913
2	−2.158
3	−1.792
4	−1.787
5	−1.775
6	−1.700
.	
.	
.	
95	1.647
96	2.159
97	2.192
98	2.719
99	2.827
100	3.349

6.4.2 Leverage

Part D of the regression output in Table 6.4 also lists three cases, 21, 39, and 99, as having large values of leverage (see also section 5.3.2), usually denoted symbolically as h_i or h_{ii}. Dividing the number of coefficients, $p+1$, by N gives the mean value of h_i, in this case 3/100= 0.03. A rule of thumb is to take any h_i greater than twice the mean as "large", that is, when

$$h_i > 2(p+1)/N \tag{6.3}$$

For a data set with 3 coefficients and 100 cases, this value is 0.06. This is the criterion used by SYSTAT to bring the cases 21, 39, and 99 to our attention in Table 6.4. Such cases should be looked at more closely to see if they are unusual in any other respect. For small values of p this criterion tends to bring too many cases to our attention. A more accurate method is provided by Equation 5.6b: The $F_{(1,97)}$ value corresponding to $\alpha=.05$ is 3.94 (see section 2.2.3.3.2). Solving backwards this gives $h_i = 0.084$ as the criterion value, which means that you should only expect about five cases in 100 to have $h_i > .084$ due to chance in such a data set. The three cases picked out by SYSTAT's heuristic all meet this more stringent criterion. Summary statistics for the leverages in the example data set are shown in Table 6.10.

As noted in the previous chapter, it is not clear what "due to chance" means in this context. In regression we assume that the X values are fixed, not randomly sampled. And high leverage is high leverage, whether the problem case was drawn from the same distribution as the rest of the X values or not. In general, leverage should only be used as a heuristic for locating potential problem cases and for understanding the cause of large Cook's Ds (discussed in the next section)—not as a basis for deleting cases.

6.4.3 Cook's D

Cook's D provides a measure that takes into account both leverage and the residuals and indicates how much actual influence each case has on the slope of the regression line (see section 5.3.3). Values of Cook's D are saved as "COOK" in the file along with leverage and the studentized residuals. Summary statistics (see section 3.1) are shown in Table 6.10. As a heuristic, you should examine cases with Cook's D ≥ 1.0 (see section 5.3.3). However, in the regression of shoes on income and family size in the SHOES data set, the largest value of Cook's D is .107 (shown in Table 6.10), which is well within the acceptable range. Thus, some cases in this data set have high leverage and others have large residuals, but no cases have high influence on the regression coefficients. If you compute the regression excluding the case with the largest value of Cook's D, you will find that the coefficients will change only slightly with such cases deleted.

Do not place too much confidence in Cook's D alone for finding influential cases. Cook's D only considers changes in

Table 6.10 Summary statistics for leverage and Cook's D for the regression of shoes on income and family size in the SHOES data set.

```
TOTAL OBSERVATIONS:     100

                      LEVERAGE            COOK

   N OF CASES             100             100
   MINIMUM             0.0108      .658106E-07
   MAXIMUM             0.3784           0.1068
   MEAN                0.0300           0.0099
   SUM                 3.0000           0.9929
```

the coefficients resulting from deletion of single cases; if two highly influential cases lie near one another, they may mask each other's effects in the calculation of Cook's D because deleting either of them alone will have little impact on the regression. Graphs are indispensable for diagnosing such situations.

6.5 Assumption Diagnostics

Linear regression makes four assumptions about the distribution of errors: They are normally distributed, independent, have constant variance, and are unbiased. Each of these assumptions is examined below, and SYSTAT procedures are illustrated for assessing their validity. The importance of checking these assumptions cannot be overemphasized: No regression results should be taken seriously until the validity of the assumptions has been assessed. For more detailed descriptions of these assumptions, see section 5.4.

6.5.1 Normality

The normality assumption states that, for any combination of X's, the errors have a normal distribution. The values of the regression coefficients do not depend on this assumption, but their F and t tests, and corresponding confidence intervals, do. Fortunately, F's and t's are fairly robust to violations of the normality assumption, but serious violations may call for a transformation of some sort to make the errors look more normal. It turns out that in real data, transformations that make the errors more normal also often make them have more constant variance (see section 6.5.2).

6.5.1.1 Stem and Leaf Displays and Box Plots

We do not get to see the real errors in our data, but must instead use the residuals from our regression analysis. It turns out that in looking for cases with large studentized residuals in the previous section, I have already discussed two of the graphical methods used in examining the normality assumption: stem and leaf displays and box plots. These plots were shown in Figure 6.9. The stem and leaf display shows seven outliers in the 100 cases, which is more than we would expect in a normal distribution (see section 3.2.1.2). (Under the normality assumption, the studentized residuals are actually distributed as t. But for N greater than about 30 they will closely approximate the normal.) Both plots look fairly symmetric, suggesting that any deviation from normality is due to thick tails, rather than skewness.

6.5.1.2 Skewness and Kurtosis

If you do not trust your interpretation of the stem and leaf or box plots, skewness and kurtosis can be computed on the studentized residuals to help assess their normality. These statistics are shown in Table 6.11 along with a few other statistics that were generated by the following commands (see section 3.1):

```
>STATS
>STATISTICS STUDENT/N MINIMUM MAXIMUM MEAN,
SKEWNESS KURTOSIS
```

The standard deviation for skewness for 100 cases (see section 3.1) is 0.245, so the value of skewness in Table 6.11 divided by its standard deviation gives $Z = 1.78$, which is not significant (see section 2.2.3.1.1).

The value of kurtosis in Table 6.11 is 1.48, which indicates that the studentized residuals appear to have somewhat thicker tails than a normal, but this value is not significant (Snedecor & Cochran, 1980, p. 492).

Table 6.11 Summary statistics for the studentized residuals from the regression of shoes on income and family size in the SHOES data set.

```
TOTAL OBSERVATIONS:      100

                     STUDENT
   N OF CASES            100
   MINIMUM           -2.9128
   MAXIMUM            3.3486
   MEAN               0.0016
   SKEWNESS(G1)       0.4357
   KURTOSIS(G2)       1.4801
```

6.5.1.3 Normal Probability Plots

One of the clearest methods of determining the normality of the studentized residuals is by displaying them in a *normal probability plot*, also called a *rankit* plot (see section 3.2.1.4). These plots can be generated easily in SYSTAT using the following commands:

```
>GRAPH
>PPLOT RESIDUAL/NORMAL
```

In this plot, if the residuals are drawn from a normal distribution they should fall on a straight line. The slight f shape in Figure 6.10 agrees with the value of kurtosis found in the previous section: The residuals tend to have thick tails. Skewness is too slight to detect in this plot.

6.5.1.4 Summary

The residuals from the regression of shoes on income and family size show some positive kurtosis and a very small positive skew. Because F and t tests are robust to even moderately large departures from normality, the residuals in this analysis are close enough to normal not to cause any concern. See section 5.4.1 for more information on this assumption, and section 7.1.1.1 for transformations that may help make the residuals more normal.

6.5.2 Equal Variance

The second assumption underlying regression is that the residuals have the same variance at every combination of X's.

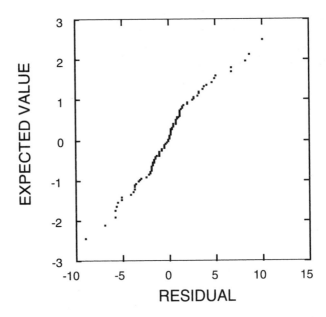

Figure 6.10 Normal probability plot of the residuals from the regression of shoes on income and family size using the SHOES data set.

See section 5.4.2 for more information. Violations of this assumption can often be seen in a plot of the studentized residuals versus the predicted values, and formal tests, such as the *score test* discussed below, are also available.

6.5.2.1 Plotting Studentized Residuals Versus Predicted Values

To assess the equal variance assumption graphically in multiple regression, plot the studentized residuals against the estimated (also "predicted" or "fitted") Y values from the regression line. This can be accomplished in SYSTAT by opening the file containing the diagnostics saved from the regression and using the following commands (see section 3.2.2.1):

```
>GRAPH
>PLOT STUDENT*ESTIMATE
```

This scatterplot is shown in the left panel of Figure 6.11. Although there are a few more extreme cases near the middle on the range of estimated values, there is no clear evidence that the variance changes systematically across the X-axis. If we delete case 39, however, unequal variance appears more plausible, as shown by the rough envelope formed by the lines in the right panel of Figure 6.11. Beware of using the envelope as

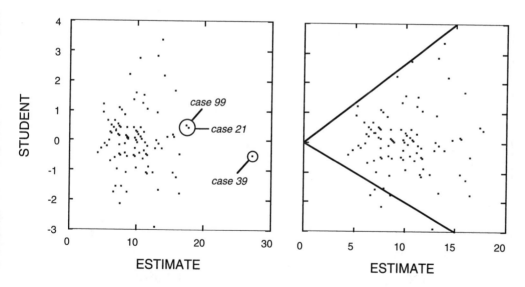

Figure 6.11 On the left is the plot of the studentized residuals versus the estimated values from the regression of shoes on income and family size in the SHOES data set. On the right is the same plot excluding case 39.

a sign of unequal variance, however, especially when the number of cases is decreasing as the envelope opens up. In the right panel of Figure 6.11 the residuals do appear to have an "opening megaphone" shape.

6.5.2.2 Plotting Studentized Residuals Versus Predictor Variables

A second type of plot that can help locate deviations from equal variance are plots of the studentized residuals versus the predictor (X) variables. The plots in Figure 6.12 were created using the command

```
>PLOT STUDENT*INCOME
>PLOT STUDENT*FAMSIZE
```

Again, neither of these plots show clear signs of unequal variance. Both plots have less spread towards the right, but this may be accounted for in large part by the fewer numbers of cases to the right.

6.5.2.3 Score Test for Nonconstant Variance

A formal test of the assumption of equal variance is provided by the "score test," which was described in section 5.4.2.3. The commands for computing each step of the score test in SYSTAT for the SHOES data set are as follows:

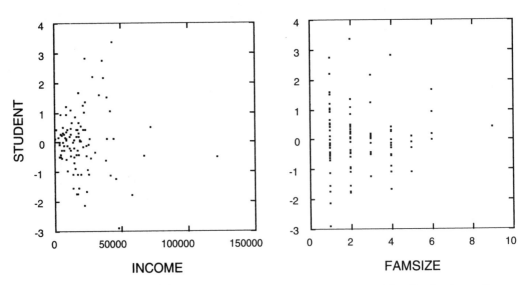

Figure 6.12 Plots of the studentized residuals versus income and family size from the regression for the SHOES data set.

I. Regress Y on all X's (the output from this regression is shown in Table 6.4):

```
>MGLH
>MODEL SHOES = CONSTANT+INCOME+FAMSIZE
>SAVE   "SHOES2.RES"/ MODEL
>ESTIMATE
```

II. The SS$_{residual}$ from the regression output in Table 6.4 is 1011.773, which, divided by $N = 100$, equals 10.11773.

III. To create the column of scaled squared residuals, type

```
>EDIT SHOES.RES
>LET U = (RESIDUAL^2)/10.11773
```

IV. Regress u on the estimated values of Y using, for example, the commands

```
>MGLH
>MODEL U = CONSTANT + ESTIMATE
>SAVE   U2.RES/MODEL
>ESTIMATE
```

(The output for this regression is shown in Table 6.12.)

V. The SS$_{regression}$ from the output in Table 6.12 is 9.1791. The score test, then, is equal to

$\chi^2_{(2)} = 9.179 / 2 = 4.590$, $p = .10$

Therefore, the score test does not indicate a statistically significant departure from equal variance, even though the assumption appeared to be violated in the right panel of Figure 6.11.

6.5.2.4 Remedies

As a rough rule, if you can't see unequal variance in the plot of the studentized residuals versus the estimated values (e.g., Figure 6.11), you don't have a serious problem. If you do not trust your eyes, try the score test. F's and t's are fairly robust to moderate departures from equal variance, but when the departure from equality is substantial, the most common remedies are (1) to perform a *variance stabilizing transformation* on the outcome (Y) variable (see section 7.1.1.2), or (2) do weighted least squares regression, which is not covered in this book but is discussed in the SYSTAT *Statistics* manual. Transforming variables is covered in Chapter 7, but, as mentioned above, you should be aware that transforming the Y variable to meet the equal variance assumption may produce violations of the

Table 6.12 Regression of the scales squared residuals u_i on the estimated values from the regression of shoes on income and family size.

```
DEP VAR: U      N: 100   MULTIPLE R: 0.168      SQUARED MULTIPLE R: 0.028
ADJUSTED SQUARED MULTIPLE R: 0.018    STANDARD ERROR OF ESTIMATE: 1.7985

VARIABLE     COEFFICIENT    STD ERROR    STD COEF   TOLERANCE      T      P(2 TAIL)

CONSTANT        0.1284        0.5478      0.0000        .        0.2343    0.8152
ESTIMATE        0.0860        0.0510      0.1678     1.0000      1.6845    0.0953

                         ANALYSIS OF VARIANCE

SOURCE       SUM-OF-SQUARES    DF    MEAN-SQUARE      F-RATIO        P

REGRESSION         9.1791       1       9.1791         2.8376      0.0953
RESIDUAL         317.0075      98       3.2348
```

linearity assumption. To recover linearity, you may have to transform one or more of the X variables.

6.5.3 Independence

The third assumption underlying regression is that the errors are uncorrelated with one another. When the normality assumption is met, uncorrelated errors are also *independent*, so this assumption is often called the independence assumption. In practice, violations of this assumption are difficult to distinguish graphically from violations of linearity, and whether you should attribute the violation to independence or linearity largely depends on the design of the study and the way the data were collected.

6.5.3.1 Examine the Design

Is there any reason that one observation in your data set might be related to, or be predictable from, another? The most obvious way that errors might be correlated is in time: Perhaps you made more positive data-recording errors later in the day, or perhaps you collected multiple data points from the same subject over time. If so, you should plot the residuals versus time and look for "snakiness" in the plot—positive errors tending to follow positive errors and negative errors tending to follow negative errors. If more than one of your data points were collected from the same person or group, their errors may be correlated, and you should adopt some type of *repeated measures* design (see Chapter 11), including a variable for person, family, and so on. If there is any reason that errors could be correlated with each other across one or more of the predictor variables, then you should plot the residuals against those predictors and examine those plots for snakiness.

If, after you have considered in full the design of your study, you can find no reason that the errors would be correlated, then they probably aren't. Patterns in the residual plots are most likely due to violations of linearity, which often can be corrected with transformations (see below). For example, in the SHOES data set we have supposed that we have data on the number of pairs of shoes that each family owns and each family's income. If the families in the study do not influence each other, and are not influenced by the study itself (for example, if wealthier families heard about the study and rushed out to buy more shoes), then we are probably safe to assume that the errors are uncorrelated.

6.5.3.2 Durbin-Watson

The Durbin-Watson statistic is a measure of the correlation among the errors when the cases are equally spaced in time. A value less than about 1.4 or greater than about 2.6 indicates a possible violation of the independence assumption. In the SHOES data set the cases are not spaced in time at all as far as we know, so the Durbin-Watson statistic is not of much interest. The value of 2.041 in Table 6.4 is very near the middle of this range, which is what we would expect.

6.5.3.3 Autocorrelation and the ACF Plot

The first order autocorrelation shown in Table 6.4 is the correlation between the column of residuals for case 1 through $N–1$ with the column of residuals for case 2 through N. In other words, it's the correlation of the residual for each case with the residual for the next case. The *order* of the autocorrelation is determined by the size of the "lag": How many cases down you would shift the second column before computing the correlation. If you suspect that your errors may be correlated over some variable other than case, such as time or estimated Y value, you can sort the data file by that variable and compute the autocorrelation using the *ACF* plot. In the example data set we have no reason to suspect that cases are correlated with one another. See section 5.4.3.3 for more on this topic.

6.5.3.4 Plot of Residuals Versus Estimated Values

Correlated errors may show up as patterns in scatterplots of the residuals versus the estimated Y values. This plot was shown in Figure 6.11. When errors are positively correlated, you would expect to see a "snakey" pattern, with positive residuals tending to follow positive residuals and negative residuals tending to follow negative residuals. No snakiness is apparent in that figure.

6.5.3.5 Plotting Residuals Versus Other Variables

In general, you should plot the residuals versus the variable or variables over which you think they may be correlated. If you think they are correlated in time, then plot the residuals versus time. If you think they are correlated in one or more of the predictors, then plot the residuals versus those predictors. Plots of the studentized residual versus the predictors in the SHOES data set were shown in Figure 6.12. The left panel shows some signs of snakiness as it trails off to the right, but this is due to a very small number of data points.

6.5.3.6 Remedies for Correlated Errors

Correlated errors are a result of the "system" that produced the data, so the correlation either exists or it doesn't. Transformations of the data won't make it go away, but there are ways of modelling the correlation within the regression framework that can be found in very advanced textbooks. Fortunately, even when the independence assumption is violated, the coefficients in the regression are still unbiased estimates of the population values. Only the standard errors, significance tests, and confidence intervals are wrong. When the errors are positively correlated, p-values and confidence intervals will be too small, and readers should be warned of this.

When the residuals are correlated over time, a time series analysis will often be appropriate. Such procedures are available with SYSTAT's *Series* command, but they are beyond the scope of this book.

6.5.4 Linearity

The fourth assumption underlying regression is that the mean value of Y at each value of X is a straight line function of X. In other words, all the members of the population are described by the same linear model. Linearity is often described as an assumption that the errors are *unbiased*, with an expected value of zero, which would not be true, for example, if you have left out an important X variable. In reality, the linear model is almost always only an approximation of the true relationship between variables; but it is often close enough not to be misleading, at least within the restricted range of the data. When this assumption is violated, the least squares estimates do not give unbiased estimates of coefficients or estimated values. The diagnostic procedures below are graphical, but formal tests do exist (e.g., Weisberg, 1985, pp. 89–95).

6.5.4.1 Plotting Residuals Versus Estimated Values

Plots of the residuals versus the estimated Y values from the regression are often used as a diagnostic for this assumption. Violation of the assumption may show up as curvature in this plot. The plot of the residuals versus the estimated values for the SHOES data set was shown in Figure 6.11. This plot showed no signs of curvature.

6.5.4.2 Added Variable Plots

Another type of plot that is very useful for detecting non-linearity are *added variable plots*, also called *partial regression leverage plots*. These plots show the part of the outcome variable that is unexplained by all of the X's, excluding X_j, as a function of the part of X_j that is unexplained by all of the other X's. That is, it's a plot of the residuals from the regression of Y on $X_1 \ldots X_p$, excluding X_j, versus the residuals from the regression of X_j on $X_1 \ldots X_p$, excluding X_j. These plots require a lot of labor to create.

To create these plots, select a predictor of interest and regress Y on all of the predictors *except* that one, saving the residuals. Then regress the chosen predictor on the other predictors, again saving the residuals. For example, to create an added variable plot for the family size variable in the SHOES data set, first regress shoes on all the predictor variables except family size, which in this case is just the variable income (this was the same regression that we computed in Chapter 5):

```
>MGLH
>MODEL SHOES = CONSTANT + INCOME
>SAVE "SHO_ON_I.RES" /RESID
>EST
```

Second, regress the family size variable on all of the other predictors, which, again, is just the variable income:

```
>MGLH
```

```
>MODEL FAMSIZE = CONSTANT + INCOME
>SAVE "FAM_ON_I.RES" /RESID
>EST
```

Now you need to get the residuals into the same file. You can do this by abutting the two files with the commands

```
>DATA
>USE "SHO_ON_I" "FAM_ON_I"
>SAVE "FAM.ADD"
>RUN
```

Open this new file "fam.add" with the data editor. The first column labelled "residual" in this file contains the residuals from the first regression, so you should relabel them something like "SONI" (for Shoes ON Income). The second column labelled "residual" in this file contains the residuals from the second regression, so you should relabel them something like "FONI" (for Famsize ON Income). Finally, you can create the added variable plot for family size shown in Figure 6.13 with the commands

```
>GRAPH
>PLOT SONI*FONI
```

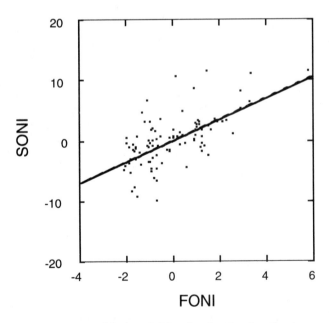

Figure 6.13 Added variable plot for the family size variable in the SHOES data set.

This plot does not show any clear indication of curvature, but it does show one case to the right that has high leverage on this line and, hence, on the regression coefficient for family size. In fact, this is case 21, which SYSTAT brought to our attention in Table 6.4 as having high leverage. Fortunately, this case has a small value of Cook's D, indicating that dropping it from the regression would have little impact on the coefficients.

Figure 6.14 shows the added variable plot for the income variable, which was generated in a manner analogous to the added variable plot for family size. This plot shows the residuals from the regression of shoes on family size (SONF) versus the residuals from the regression of income of family size (IONF). Setting aside the furthest point to the right, the envelope of this plot shows some indication of upward curvature—like a bow shooting an arrow towards the ground to the right. However, this envelope is based on a small number of cases towards the right, and the furthest point to the right seems to bend the curve the other way. This point is case 39, which SYSTAT indicated had high leverage in Table 6.4, but, again, it does not have high influence on the regression coefficients. In Chapter 7 we will examine the effects of some transformations of variables on this plot.

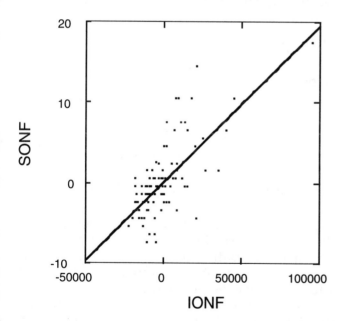

Figure 6.14 Added variable plot for the income variable in the SHOES data set.

6.5.4.3 Remedies for Nonlinearity

When the linearity assumption is violated, the coefficients and estimated Y values from the regression are not unbiased estimates of the population values. There are three general types of remedy for violations of this assumption: (1) If you have left out an important predictor variable, including this variable may correct the problem; (2) Transforming (or scaling) the Y and/or X variables may improve the linear relationship between those variables; (3) Adding polynomials or interactions among the predictors may improve the linear relationship. Adding such terms has the disadvantage of complicating the model and adding new parameters to be fitted. Transformations, scaling, and adding polynomials are covered in Chapter 7.

6.6 Confidence Intervals

SYSTAT will compute some types of confidence intervals for display in graphs. However, if you wish to report the actual values of the confidence intervals, you may compute them easily using SYSTAT's cumulative distribution functions and a hand calculator. The formulas for a variety of confidence intervals are provided below with examples from the example data regression.

6.6.1 Intercept

The bounds for the confidence intervals on the intercept from the regression can be found as follows:

$$\hat{\beta}_0 \pm t_{(\alpha/2,\ N-p-1)} \times SE_{\beta_0} \tag{6.4}$$

and the confidence interval itself can be expressed as

$$\hat{\beta}_0 - t_{(\alpha/2,\ N-p-1)} SE_{\beta_0} \leq \beta_0 \leq \hat{\beta}_0 + t_{(\alpha/2,\ N-p-1)} SE_{\beta_0} \tag{6.5}$$

where β_0 is the population value of the intercept coefficient, $\hat{\beta}_0$ is the value of the estimated intercept coefficient from the regression output, SE_{β_0} is the standard error of the estimate, and $t_{(\alpha/2,\ N-p-1)}$ is the value of the t distribution with $N-p-1$ degrees of freedom that lie beyond $(1-\alpha/2) \times 100$ percent of the distribution. In other words, if you want the 95% confidence interval ($\alpha = .05$), you must find the magnitude of t corresponding to the upper and lower 2.5% of the *t* distribution. The value for a *t* with 97 degrees of freedom corresponding to the upper 2.5% of the distribution is 1.985 (see section 2.2.3.2.2).

Equation 6.5 can be re-expressed using the labels from SYSTAT's regression output as:

$$\text{COEFFICIENT}_{\text{CONSTANT}} \pm t_{(\alpha/2,\, N-p-1)} \times \text{STD ERROR}_{\text{CONSTANT}} \quad (6.6)$$

Plugging in the value of *t* and the values of the coefficient and standard error from the regression output in Table 6.4, you get

$$1.567 \pm 1.985 \times 0.872 \quad (6.7)$$

or 1.567 ± 1.731. The confidence interval, therefore, is

$$-0.164 \le \beta_0 \le 3.298 \quad (6.8)$$

When the assumptions underlying the regression are met, 95% of confidence intervals generated in this manner will contain the population value of the intercept.

6.6.2 Slopes

The bounds for the confidence intervals on the slopes from the regression are found in the same manner as those for the intercept:

$$\hat{\beta}_j \pm t_{(\alpha/2,\, N-p-1)} \times \text{SE}_{\beta_j} \quad (6.9)$$

and, likewise, the confidence interval itself can be expressed as

$$\hat{\beta}_j - t_{(\alpha/2,\, N-p-1)} \text{SE}_{\beta_j} \le \beta_j \le \hat{\beta}_j + t_{(\alpha/2,\, N-p-1)} \text{SE}_{\beta_j} \quad (6.10)$$

where β_j is the population value of the slope, $\hat{\beta}_j$ is the value of the estimated slope from the regression output, SE_{β_j} is the standard error of the slope estimate, and $t_{(\alpha/2,\, N-p-1)}$ is the value of the *t* distribution with $N-p-1$ degrees of freedom that lie beyond $(1-\alpha/2) \times 100$ percent of the distribution.

Re-expressing Equation 6.10 using the labels from SYSTAT's regression output, you get the following:

$$\text{COEFFICIENT}_{X1} \pm t_{(\alpha/2,\, N-p-1)} \times \text{STD ERROR}_{X1} \quad (6.11)$$

Plugging in the value of $t = 1.985$ (see section 2.2.3.2.2) and the values of the coefficient and standard error from the regression output in Table 6.4, for the income variable you find:

$$0.000196 \pm 1.985 \times 0.000020 \quad (6.12)$$

or 0.000196 ± 0.000040, with the resulting confidence interval:

$$0.000156 \leq \beta_1 \leq 0.000236 \qquad (6.13)$$

Similarly, for the famsize variable you find:

$$1.769 \pm 1.985 \times 0.221 \qquad (6.14)$$

or 1.769 ± 0.439, with the resulting confidence interval:

$$1.330 \leq \beta_1 \leq 2.208 \qquad (6.15)$$

When the assumptions underlying the regression are met, 95% of confidence intervals generated in this manner will contain the population value of their respective coefficients.

7

Transformations

There are three basic reasons to transform variables, corresponding to three of the assumptions underlying regression: to normalize distributions, to equalize variances, and to linearize relationships. You can transform the outcome variable and/or one or more of the predictor variables in an attempt to achieve these goals. However, you should realize that a transformation that causes the data to meet one assumption may also cause the data to violate another assumption that was previously met. Similarly, in multiple regression, a transformation that linearizes the relationship between the outcome and one predictor may add curvature to the relationship between the outcome and another predictor. This can make appropriate transformations difficult to discover by trial and error.

In addition to improving the fit of the model, transformations should also *make sense*. This latter criterion can help you avoid a lot of trial and error: Often you will want to transform a variable for theoretical reasons, because the transformed scale makes sense, before you even look at the data. Other times you may use the data to suggest transformations, but you should try not to settle on a transformation unless the variable has some meaningful interpretation on that scale. This stricture may be easier to state than to live up to in all cases, but at least it gives us something to strive for. The fact is that theories based on nonmeaningful scales do not tend to survive.

The physical act of transforming variables in SYSTAT is trivial and simply involves creating a new column containing the transformed variable using SYSTAT's math functions. This chapter attempts to pull together some guidelines for choosing transformations, first for the outcome variable, and then for the predictor variables. For more information on transformations I particularly recommend Mosteller and Tukey (1977,

Chapters 4, 5, and 6), Snedecor and Cochran (1980, pp. 287–292), and Weisberg (1985, pp. 133–135, 140–159).

7.1 Transforming the Outcome Variable

7.1.1 The Ladder of Transformations

We can think of transformations of the outcome variable as being ordered by the power p to which Y is raised, for example, ... Y^{-2}, Y^{-1}, $Y^{-.5}$, Y^0, $Y^{.5}$, Y^1, Y^2, ...; letting Y^0 be $\log(Y)$, Mosteller and Tukey (1977) call this ordering the *ladder of transformations*. In principle, Y could be raised to any power from negative to positive infinity, but you very rarely will see transformations outside the interval −3 to +3 in real data, and in fact, most researchers would be reluctant to even go beyond −2 or +2. Moving towards large positive powers is called moving up the ladder, whereas moving towards large negative powers is called moving down the ladder. The reason this ladder is useful is that we know which direction to move to improve the fit of models under various conditions with certain types of data.

7.1.1.1 Normalizing Transformations

When your outcome variable is not normally distributed and you would like it to be, or when your residuals are not normally distributed and you would like them to be, the shape of the non-normality can suggest transformations of the outcome variable up or down the ladder that might recover normality. Transformations up the ladder tend to pull in small values and spread out large values. This can help remove negative skew. Transformations down the ladder tend to spread out small values and pull in large values, which can help remove positive skew. (These transformations only have the described effect when the outcome variable is strictly positive.)

Some type of data tend to be skewed one way or the other, but positive skew is much more common. For example, counts and latencies both tend to be positively skewed. Depending on the severity of the skew, you can consider transformations further and further down the ladder. Table 7.1 summarizes the normalizing effects of various transformations and lists some types of data for which they are often appropriate.

To illustrate, consider the shoes variable in the SHOES data set (used in Chapters 5 and 6). This variable consists of counts of numbers of shoes, which we just said often tend to be positively skewed. The histogram of this variable is shown in the left panel of Figure 7.1. As expected, the distribution shows

Table 7.1 The ladder of transformations.

p	Form	In SYSTAT	Normalizes When	Equalizes Variance When	Linearizes Y on X When
2	Y^2	let Y=Y^2	residuals are negatively skewed and $Y_i \geq 0$	$\hat{\sigma}_e^2 \propto 1/\hat{Y}_i$	bowed up
1	Y	—	—	—	—
.5	$\sqrt{Y}$	let Y=sqr(Y)	Y_i are counts of rare events (Poisson distribution)	$\hat{\sigma}_e^2 \propto \hat{Y}_i$	bowed down
	$\sqrt{Y}+\sqrt{Y+1}$	+sqr(Y+1)	(add when some counts are zero or very small)		
0	log Y	let Y=log(Y)	residuals are pos. skewed and $Y_i > 0$ (try when Y are amounts or counts and $Y_{max}/Y_{min} > 10$)	$\hat{\sigma}_e \propto \hat{Y}_i$	bowed down
	log(Y+1)	let Y=log(Y+1)	(use when some values are zero)		
−.5	$-1/\sqrt{Y}$	let Y= -1/sqr(Y)	$Y_i > 0$; some values very large but most bunched near 0	—	bowed down
−1	$-1/Y$	let Y= -1/Y	$Y_i > 0$; some very large but most bunched near 0 (converts latencies to rates, etc.)	$\hat{\sigma}_e \propto \hat{Y}_i^2$	bowed down
	−1/(Y+1)	let Y = -1/(Y+1)	(use when some values are zero)		
−2	$-1/Y^2$	let Y = -1/Y^2	pulls in very large values and spreads out small values	—	bowed down
*	$\sin^{-1}(\sqrt{Y})$	let Y=asn(sqr(Y))	Y_i are rates or binomial proportions	$\hat{\sigma}_e^2 \propto (\hat{Y}_i)(1-\hat{Y}_i)$	—

clear positive skew. This skew suggests a transformation down the ladder, and Table 7.1 mentions that the square root transformation is often appropriate for counts. The histogram for the square root of the shoes variable is shown in the right panel of Figure 7.1 and, sure enough, this transformed distribution looks much more symmetric than the untransformed variable.

Does this result mean that we should be using the squareroot of shoes rather than shoes itself? Not necessarily. First, the regression does not assume that Y is normally distributed, only that the residuals are normally distributed. Second, even if this transformation made the residuals look more normal, we would not want to use it if the variable "square root of the number of shoes" was not meaningful: The significance tests in the regression are fairly robust against violations of normality, so you wouldn't want to give up meaningfulness just to gain slightly more accurate p-values. Third, the square root transformation did not improve the normality of the residuals: If anything, it made the distribution of residuals even thicker-tailed, as you can see in the normal probability plot in Figure 7.2 (see section 3.2.1.4 for more on interpreting normal probability plots).

Finally, transformations like this one that improve normality might help or harm the other assumptions: Although regression is also fairly robust against unequal variance, we

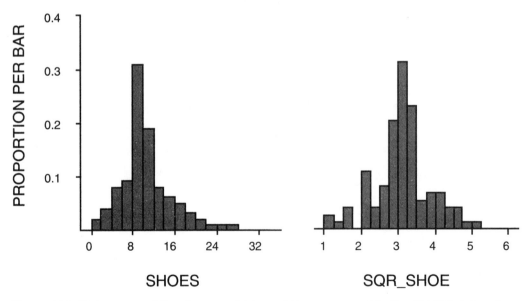

Figure 7.1 Histograms of the shoes variable and its square root in the SHOES data set.

would not want to improve normality at the expense of linearity. For any transformation we must assess the effects that the transformation has on the validity of *all* of the assumptions before deciding to stick with that transformation.

7.1.1.2 Variance Stabilizing Transformations

When the variance of the residuals is not constant across the predicted values of the outcome variable, i.e., the equal variance assumption is violated, it is often possible to find a transformation of the outcome variable that will make the residuals more equally distributed. Again, we can look to the ladder of transformations for guidance. Transformation powers greater than one tend to spread apart larger values, so they should be used when the the variance decreases as the predicted value of Y increases. For example, squaring the outcome variable can stabilize the variance in cases where the variance is inversely proportional to the predicted value of Y.

Transformation powers less than one tend to pull in larger values, so they can help equalize the variance when variance is an increasing function of the predicted value of Y. The log

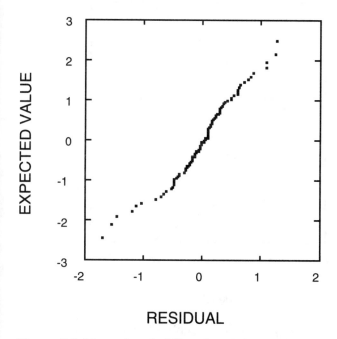

Figure 7.2 Normal probability plot of the residuals from the regression of the square root of shoes on income and family size in the SHOES data set.

transformation, for example, is probably the most used variance stabilizing transformation because it equalizes the variance when the standard deviation is proportional to the predicted value of Y, i.e., when the magnitude of error tends to be a percentage of Y. The next to last column of Table 7.1 gives conditions in which particular transformations are especially appropriate.

To illustrate, consider the plot of the studentized residuals versus the fitted values from the regression of shoes on income and family size. This plot was shown in the right panel of Figure 6.11 and is reproduced here in Figure 7.3 (this plot excludes case 39, which tended to make the rest of the plot bunch together; see Figure 6.11).

This plot shows signs of increasing variance with the predicted value of the outcome variable. (But beware of making too much of the shape of the envelope around the data, as shown by the two lines in the figure. The envelope can be overly influenced by a few data points, such as the three residuals above about 2.5 on the Y-axis in Figure 7.3.) Would

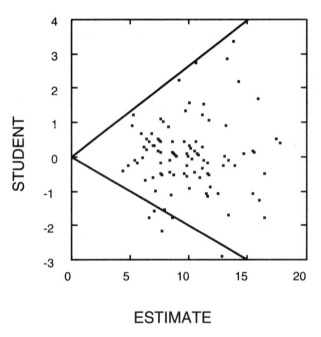

Figure 7.3 Scatterplot of the studentized residuals versus the predicted values from the regression of shoes on income and family size in the SHOES data set (this plot excludes case 39).

a transformation down the ladder make the variance in this plot look more constant? Figure 7.4 shows the plot of the studentized residuals versus the predicted values from the regression of the square root of shoes on income and family size in the SHOES data set. Indeed, this plot shows no hint of a megaphone shape opening to the right. In fact, if you squint your eyes, it almost appears that the core of the data has a megaphone shape pointing to the left.

Obviously, these plots are not easy to read. You can usually find anything you are looking for if you squint, stand on your head, or look at the plot after staring at the sun. Because the significance tests in regression are fairly robust against violations of the equal variance assumption, for most practical purposes you can follow the rule "if you can't see it, it's not a problem." And even if you think you can see it, your eyes may be misleading you: The score test in Chapter 6 (section 6.5.2.3) did not indicate statistically significant deviation from unequal variance in these residuals.

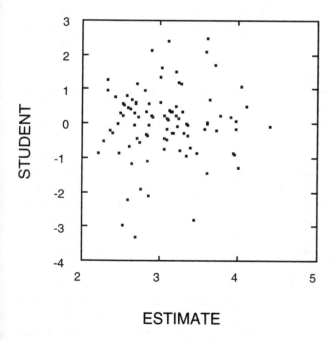

Figure 7.4 Scatterplot of the studentized residuals versus the predicted values from the regression of the square-root of shoes on income and family size in the SHOES data set (this plot excludes case 39).

Should we be using the square root of shoes? Both the normality of the shoes variable and the equal variance assumption appear to be improved by using this variable. However, the normality of the residuals is not (see Figure 7.2). Furthermore, we have yet to assess the effect of this transformation on the linearity assumption. I turn to that next.

7.1.1.3 Linearizing Transformations

Curvature in plots of the residuals versus estimated values or in added variable plots (see section 6.4.5) constitutes evidence that the linearity assumption is violated. We should distinguish two types of curvature: (1) the outcome variable is a monotonically increasing function of the predictors, but the rate of increase is not constant; and (2) the outcome variable reaches its maximum level within the range of the predictors; i.e., it rises and then starts back down. In the first case, some transformation of the outcome (or of one or more predictors) may linearize the function. In the second case, you may have to add polynomials of the predictors to linearize the function (see section 7.2.2 below).

When the former appears to be the case, the direction of the curvature in the plots of the outcome versus the predictors or in the added variable plots can guide us as to which transformations we should try. Figure 7.5 provides a mnemonic for these transformations, adapted from Mosteller and Tukey (1977, p. 84). If the curve were a bow shooting an arrow, the direction in which the arrow is pointed relative to the X- or Y-axis indicates whether the transformation of the X and Y variable should be up or down the ladder. Arrows being shot upwards point to higher values on the Y-axis, so transformations of Y up the ladder would tend to linearize the curve. Arrows being shot toward the right point to higher values on the X-axis, so transformations of X up the ladder would tend to linearize the curve. For example, when the curve is bowed up to the right, like the curve in the upper right quadrant of Figure 7.5, transformations of Y up the ladder and transformations of X up the ladder will tend to straighten the curve.

To illustrate, suppose that we did carry out the regression of the square root of shoes on family size and income, as discussed in the previous two sections. Figure 7.6 shows the added variable plot for the family size variable in this regression (see section 6.5.4.2). The Y-axis gives the values of the residuals from the regression of the square root of shoes on income (SQS_ON_I), and the X-axis gives the values of the residuals from the regression of family size on income (F_ON_I). Clearly,

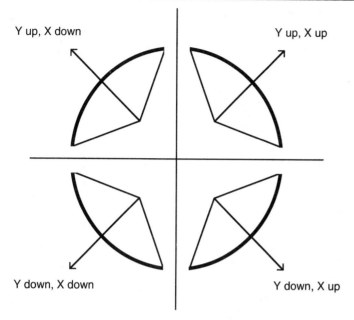

Figure 7.5 When a plot is curved in the shape of one of the "bows," the arrows point in the direction, relative to the X- or Y-axis, of the transformation up or down the ladder that will tend to linearize the curve.

the cloud of points is bowed up to the left, as the curve that I added is intended to highlight. According to Figure 7.5, a curve of this shape can be straightened by transforming Y up the ladder or X down the ladder. But remember that we got to this plot by transforming Y (shoes) down the ladder, by taking the square root. This plot indicates that the square root transformation was ill-advised. We would have been better off sticking with the untransformed shoes variable, i.e., moving back up the ladder to where we started. The added variable plot for the family size variable from the regression of the untransformed shoes variable on family size and income is shown in Figure 7.7 (reproduced from Figure 6.13). Although there may be some curvature left in this plot as well, if you ignore the case on the far right, this cloud of points is much less curved than Figure 7.6. Thus, moving up the ladder did in fact help straighten the curve.

This example shows that transformations aimed at improving one assumption can actually harm others. In the pecking order of assumptions, linearity is more important than normality or equal variance because when it is violated the

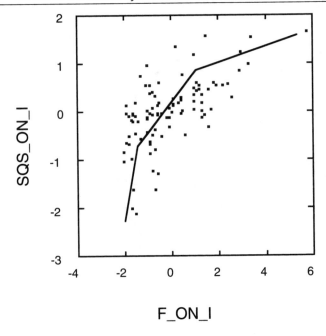

Figure 7.6 Added variable plot for the family size variable in the regression of the square root of shoes on family size and income in the SHOES data set.

coefficients from the regression are not unbiased estimates of the population coefficients. Thus, the square root transformation of the shoes variable is not a good choice.

7.1.2 Finding Transformations: Box-Cox

When your eyes and your theory are having a hard time deciding on a transformation, you might turn to a more formal method of finding a transformation. Box and Cox (1964) presented a method of choosing a transformation of the outcome variable to make the residuals look as much like a sample from a normal distribution as possible. This method involves systematically computing the log-likelihood of a number of transformations to determine for which transformation the residuals would most likely come from a normal distribution given the data. The transformations are expressed as powers of the outcome variable, just as in the ladder of transformations in section 7.1.1, and it is conventional to examine selected powers between –2 and +2, again letting 0 represent the natural log transformation. I have broken a variation of the Box-Cox method (see Weisberg, 1985, pp.

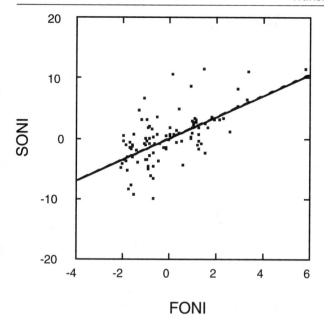

Figure 7.7 Added variable plot for the family size variable in the regression of the untransformed shoes variable on family size and income.

147–151) into five SYSTAT-appropriate steps, using the data from the SHOES data set to illustrate each.

Step 1: Compute the geometric mean of the outcome variable.

1a. Compute $\ln(Y)$. To compute the natural log of the shoes variable in the SHOES data set, type the following command (see also section 2.2.1):

```
>LET LN_SHOES = LOG(SHOES)
```

where ln_shoes is a new column containing the transformed outcome variable.

1b. Compute $\Sigma \ln(Y)$. Find the sum of the column of logged values that you just generated using the statistics command:

```
>STATS
>STATISTICS LN_SHOES / SUM N
```

The output from this command is shown in Table 7.2.

Table 7.2 Sum of the column of logged values of the shoes variable in the SHOES data set.

```
TOTAL OBSERVATIONS:      100

                      LN_SHOES

N OF CASES                 100
SUM                    218.874
```

1c. Compute the geometric mean: $GM = e^{\Sigma \ln(Y)/N}$. This step is easily performed on a hand calculator using the values from Table 7.2, but if you don't have one handy you can do it in SYSTAT:

```
>LET GM = EXP(218.874/100)
```

This command will fill the entire column labelled GM with the value 8.924, which is the geometric mean of the shoes variable.

Step 2: *Transform the outcome variable up and down the ladder, from –2 to +2.*

Create five additional columns, one for Y raised to each of the powers –2, –1, –.5, .5, and 2. (We are also interested in Y^0 and Y^1, but these two columns already exist: We use $\ln(Y)$ in place of Y^0 and, of course, Y^1 is just Y.) In the commands that follow, I use the underscore followed by the letter *m* to indicate "minus" in my column labels, and the letter *p* to indicate "point." Thus, shoes raised to the –.5 power is labelled SHOE_MP5. You may come up with another labelling system that you prefer. (Alternative commands for the same transformations are found in Table 7.1.)

```
>LET SHOE_M2  = -SHOES^(-2)
>LET SHOE_M1  = -SHOES^(-1)
>LET SHOE_MP5 = -SHOES^(-.5)
>LET SHOE_P5  =  SHOES^(.5)
>LET SHOE_2   =  SHOES^(2)
```

Step 3: *Regress each Y^p on all of the X variables.*

Compute seven regressions of Y^p on all of the X variables, one for Y raised to each of the powers –2,

Table 7.3 The first column shows powers of Y (shoes), and the second column the sums-of-squares from the regressions of Y raised to the powers on the income and family size variables in the shoes data set.

Powers of Y	$SS_{residual}$
−2	1.877
−1	1.706
−0.5	1.173
0	18.779
0.5	28.607
1	1011.773
2	652813.262

−1, −.5, 0, .5, 1, and 2 (remember, we use ln(Y) in place of Y^0). For each regression, write down the $SS_{residual}$. (This may seem like a lot of work, but if you use menus, use standard [short] output, and don't save residuals, it will only take a few minutes.)

```
>MGLH
>MODEL SHOE_M2=CONSTANT+INCOME+FAMSIZE
>ESTIMATE
>MODEL SHOE_M1=CONSTANT+INCOME+FAMSIZE
>ESTIMATE
>MODEL SHOE_MP5=CONSTANT+INCOME+FAMSIZE
>ESTIMATE
>MODEL LN_SHOES=CONSTANT+INCOME+FAMSIZE
>ESTIMATE
>MODEL SHOE_P5=CONSTANT+INCOME+FAMSIZE
>ESTIMATE
>MODEL SHOES=CONSTANT+INCOME+FAMSIZE
>ESTIMATE
>MODEL SHOE_2=CONSTANT+INCOME+FAMSIZE
>ESTIMATE
```

The powers of Y and their corresponding SS are shown in Table 7.3.

Step 4: *Compute the log-likelihood for each of the powers of Y.*

Open a new data editor worksheet and enter the two columns of data from Table 7.3. These two columns are the first two columns of Table 7.4, using 'P' as the label for the powers and 'SS' for the column of

SS_residual. You can then compute the log-likelihood L for each power of Y using the equation

$$L(P) = N \ln(|P|) + N(P-1)\ln(GM) - F(N,2) \ln(SS_{residual}) \quad (7.1)$$

where P is the power of transformation of Y, N is the number of cases, $SS_{residual}$ is from the corresponding regressions, and GM is the geometric mean of the Y variable. When $P=0$, you should use the equation

$$L(0) = - N \ln(GM) - F(N,2) \ln(SS_{residual}) \quad (7.2)$$

For the shoes data set, $N = 100$ and $GM = 9.399$. Thus, to evaluate these expressions for each of the row in the new data set, you can use the following commands:

```
>IF P<>0 THEN LET LOGLIKLI=
100*LOG(ABS(P))+100*(P-1)*LOG(9.399)
-50*LOG(SS)
>IF P=0 THEN LET LOGLIKLI=
-100*LOG(9.399)-50*LOG(SS)
```

The resulting log-likelihoods are shown in the third column of Table 7.4.

Step 5: *Plot the log-likelihoods against the powers of the transformations of Y.*

You can do this using the commands:

```
>GRAPH
>PLOT LOGLIKLI*P
```

Table 7.4 The SYSTAT data editor window. The column P shows powers of Y (shoes), the column SS shows the sums of squares from the regressions of Y raised to the powers on the income and family size variables in the shoes data set, and the column LOGLIKLI gives the log-likelihoods of those powers.

	P	SS	LOGLIKLI
1	-2.000	1.877	-634.350
2	-1.000	1.706	-474.828
3	-0.500	1.173	-413.383
4	0.000	18.779	-370.697
5	0.500	28.607	-349.027
6	1.000	1011.773	-345.973
7	2.000	652813.262	-376.077

This plot is shown in Figure 7.8. The maximum value of this function represents the optimal power transformation of Y, and, because these functions are unimodal, we can read the approximate maximum off of the plot. In Figure 7.8, the log-likelihood increases with powers of Y up to about 0.8 and then starts decreasing.

If you wanted to find the maximum with greater precision, you could repeat steps 2 through 5 above for values of P between .5 and 2. However, you would usually not want to transform a variable by a relatively meaningless power like 0.8, but, instead, use the nearest meaningful power. In the present example we are lucky: The maximum is around .8, which is very close to the power of 1.0, i.e., no transformation at all. This power is down the ladder from 1.0, consistent with the direction that the unequal variance in the scatterplot of the residuals versus the predicted values (Figure 7.3) suggested we should move; furthermore, it is up the ladder from .5, which is the direction the added variable plot in Figure 7.6 suggested we should move. However, the simplicity of interpretation gained by using 1.0 as the power far outweighs the very slight

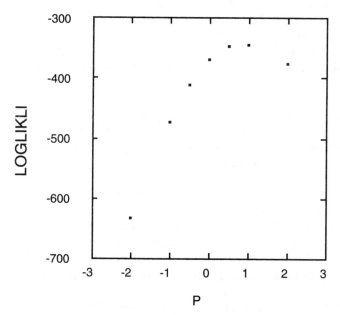

Figure 7.8 Log-likelihoods of the powers P against the powers themselves.

increase in log-likelihood that you would gain by using 0.8. Thus, the Box-Cox method is probably trying to tell us that the best transformation for the shoes variable is no transformation at all.

7.2 Transforming the Predictor Variables

As the examples in previous sections indicate, transformations that help remedy violations of one assumption can actually make other violations worse. Furthermore, transforming the outcome variable to straighten plots of Y against one of the predictors can make previously straight plots of Y against other predictors curve. Sometimes you can avoid this bind by transforming one or more of the predictors, instead of, or in addition to, the outcome variable.

Predictors variable can also be transformed using the ladder of transformations. When the outcome variable is a monotonic function of the predictors, but does not increase or decrease at a constant rate within the range of predictors, we would usually try to use powers within the range -2 to $+2$. These transformations can be thought of very much the same way as transformations of Y, and Figure 7.5 shows which directions of X up or down the ladder will tend to straighten different curves. When the outcome reaches its maximum (or minimum) within the range of predictor, i.e., it increases to a point and then starts back down, we might want to use powers including or more extreme than -2 to $+2$, and we might want to try interactions such as $X_1 X_2$. The qualitative difference between this type of transformation and those within the range -2 to $+2$ is that the transformed predictor is usually added to the model in addition to the untransformed predictor. So instead of *replacing* X_1 with X_1^2, you now have a model that contains $X_1 + X_1^2$. This sort of polynomial regression is discussed briefly in section 7.2.2.

7.2.1 The Ladder of Transformations

The effect of transformations up and down the ladder is much the same for predictor variables as it is for outcome variables. Moving up the ladder tends to spread apart large values and pull in small values, and moving down the ladder tends to pull in large values and spread apart small values (again, assuming all values are strictly positive). Although transforming predictors can affect the normality of the residuals and the equality of their variance, the effect is not as readily predictable as for transformations of Y, so I will focus instead on the effects on linearity. The last column of Table 7.1 describes the shape of the curve that tends to be linearized by a given transformation of Y; this also works for transformations of X, if you replace the word "up" with the word "right" and "down" with "left," consistent with Figure 7.5. For example, the square root transformation of X tends to straighten curves that are bowed left, such as the two curves on the left side of Figure 7.5: These curves point in the "down" direction on the X-axis, and .5 is a transformation down the ladder.

Consider the added variable plot for the family size variable in the regression of the square root of shoes on family size and income, shown in Figure 7.6. This plot is bowed up to the left, and Figure 7.5 indicates that such a curve would tend to be straightened by a transformation of X down the ladder. Suppose we now take the square root of family size and recompute the regression. Figure 7.9 shows the added variable plot for *square root of family size* in the regression of the square root of shoes on income and the square root of family size. Indeed, this plot is less obviously curved than the plot in Figure 7.6, indicating that the transformation of family size down the ladder helped.

So was our transformation a success? Figure 7.10 shows the normal probability plot of the residuals from this regression and the scatterplot of the studentized residual against the fitted values. The normal probability plot is reasonably straight, indicating that the normality assumption is not badly violated, and the studentized residuals do not show a clear increasing or decreasing trend over the fitted values. Thus, all three assumptions appear to be met reasonably well.

There is one loose end to tie up, however. We still have not examined the added variable plot for the income variable. This plot is shown in Figure 7.11. Unfortunately, if we take the point on the far right seriously, this plot is clearly bowed up to the left (usually you wouldn't want to let a single point influence your decision so strongly, but this is hypothetical data anyway). This suggests a transformation of shoes up the ladder, or a transformation of income down the ladder.

Does a transformation of income down the ladder make sense? Yes. Some readers might have been impatiently wondering when I was going to get around to taking the log of income. In general, it is often a good idea to try the log of a variable when the ratio of the maximum value to the minimum value is greater than ten or so. In fact, income has very natural meaning on a log scale: You can imagine that other variables might be associated with *percent change* in income, which the log scale provides. Furthermore, income is positively skewed, and the log transformation will tend to pull in the large values. For these reasons income is often the textbook case of a variable that should be logged.

Figure 7.12 shows the new added variable plots for log income and square root of family size in the regression of the square root of shoes on those two variables. Neither plot shows clear curvature, although the plot on the left may tend to bow down to the right and the plot on the left may tend to bow up to the left.

What about the normality and equal variance assumptions? The normal probability plot of the residuals in the left panel of Figure 7.13 is certainly not straight, but neither is it systematically curved. The plot of the studentized residuals versus the fitted values in the right panel of Figure 7.13 shows no clear signs of unequal variance. Thus, all three of the assumptions appear to be reasonably met.

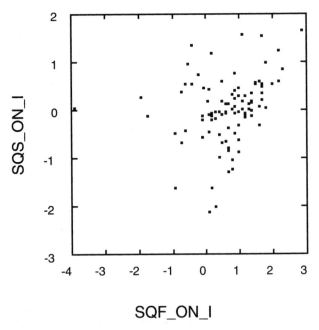

Figure 7.9 Added variable plot of the square root of family size in the regression of the square root of shoes on income and the square root of family size. SQS = square root of shoes; SQF = square root of family size.

There is one remaining difficulty, however. The regression output for this analysis is shown in Table 7.5, and it indicates that we have a collinearity problem (see section 6.3). One of the condition indices is over 30, and it involves at least the constant and the logged income variable and perhaps the square root of the family size variable.

When variables are collinear with the constant, this suggests that *centering* the variables, i.e., subtracting their means from each data point so that they have mean zero, will help reduce the collinearity. In my data set, log income is labelled LOG_INC and the square root of family size is labelled SQR_FAM. Their means are shown in Table 7.6, which was generated by the commands

```
>STATS
>STATISTICS LOG_INC SQR_FAM / MEAN N
```

With these means you can now center the variables with the following commands,

```
>LET LOG_I_CN = LOG_INC - 4.197
>LET SQR_F_CN = SQR_FAM - 1.53
```

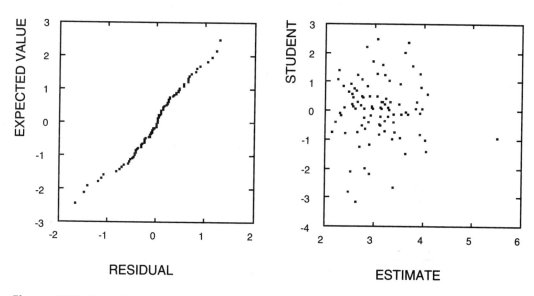

Figure 7.10 The left panel shows the normal probability plot of the residuals and the right panel shows the scatterplot of the studentized residuals versus the fitted values; both come from the regression of the square root of shoes on income and the square root of family size.

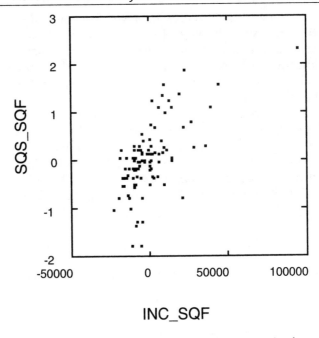

Figure 7.11 Added variable plot for income in the regression of the square root of shoes on income and the square root of family size. SQS = square root of shoes; SQF = square root of family size.

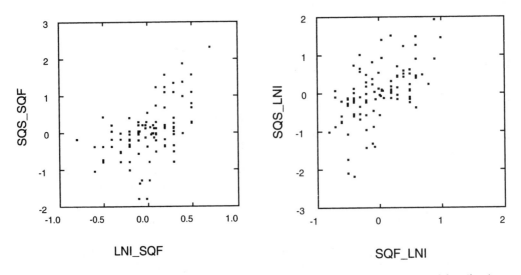

Figure 7.12 Added variable plots for log income (LNI) and square root of family size (SQF) in the regression of the square root of shoes (SQS) on those two predictors.

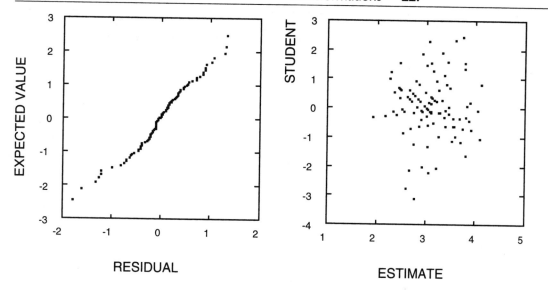

Figure 7.13 The left panel shows the normal probability plot of the residuals and the right panel shows the scatterplot of the studentized residuals versus the fitted values; both come from the regression of the square root of shoes on log income and the square root of family size.

where the suffix *cn* is just a reminder that the variable is centered. The new regression of the square root of shoe size on these centered variables is shown in Table 7.7. As you can see, the estimate of the constant has changed and its standard error it greatly reduced. However, the other coefficients have not changed at all. Most important, this analysis indicates that there was only one near collinearity among these variables, given that the collinearity has all but disappeared after centering, as evidenced by a large condition index of 1.640. If you are only interested in showing the significance of the slopes, therefore, the uncentered variables will suffice. If you are interested also in the estimate of the constant, and its confidence interval, then you might want to use the centered variables.

To summarize, we have found that taking the square root of the outcome variable, shoes, and taking the log of the income and the square-root of family size, results in a regression for which the normality, equal variance, and linearity assumptions are reasonably well met. In addition, the one near dependency among the predictors is almost entirely through the constant, and so can be eliminated by centering the

Table 7.5 Regression of the square root of shoes (SQR_SHOE) on log income (LOG_INC) and the square root of family size (SQR_FAM) in the SHOES data set.

```
EIGENVALUES OF UNIT SCALED X'X
                              1           2           3
                          2.931       0.067       0.002

CONDITION INDICES
                              1           2           3
                          1.000       6.633      37.292

VARIANCE PROPORTIONS
                              1           2           3
    CONSTANT          .430337E-03       0.006       0.994
    LOG_INC           .554050E-03       0.017       0.982
    SQR_FAM                 0.007       0.669       0.323
```

DEP VAR: SQR_SHOE   N: 100   MULTIPLE R: 0.624   SQUARED MULTIPLE R: 0.389
ADJUSTED SQUARED MULTIPLE R: 0.377   STANDARD ERROR OF ESTIMATE: 0.600

VARIABLE	COEFFICIENT	STD ERROR	STD COEF	TOLERANCE	T	P(2 TAIL)
CONSTANT	-4.024	0.984	0.000	.	-4.091	.89E-04
LOG_INC	1.331	0.205	0.580	0.790	6.492	.36E-08
SQR_FAM	1.001	0.145	0.617	0.790	6.914	.50E-09

CORRELATION MATRIX OF REGRESSION COEFFICIENTS

	CONSTANT	LOG_INC	SQR_FAM
CONSTANT	1.000		
LOG_INC	-0.978	1.000	
SQR_FAM	-0.626	0.458	1.000

ANALYSIS OF VARIANCE

SOURCE	SUM-OF-SQUARES	DF	MEAN-SQUARE	F-RATIO	P
REGRESSION	22.280	2	11.140	30.899	.414543E-10
RESIDUAL	34.972	97	0.361		

Table 7.6 Means of log income (LOG_INC) and the square root of family size (SQR_FAM) in the SHOES data set.

```
TOTAL OBSERVATIONS:    100

                      LOG_INC      SQR_FAM

  N OF CASES            100          100
  MEAN                 4.197        1.530
```

predictors. Is this, then, the regression that you should report? The answer is not easy. Although the assumptions may be *slightly* better met than in the regression using the untransformed variables (see section 6.5), this was obtained at the cost of changing the scales of the shoes and family size variables by taking square roots. These square root scales may have no intuitive meaning for us, and readers of your report may think you are engaging in silly statistical hocus-pocus to get your results. The economist Ronald Coase once said "If you torture the data long enough, it will confess." Simplicity should not be underemphasized, and in this situation many researchers (including myself) would probably not think the square root scales were appropriate (I would still pursue log income, however).

7.2.2 Polynomial Regression

When the outcome variable reaches its maximum (or minimum) within the range of the predictors, to satisfy linearity you may need to add polynomials to the model; that is, instead of *replacing* X_1 with X_1^2, you now have a model that contains $X_1 + X_1^2$. In other cases you may need to add interaction terms like $X_1 X_2$ to the model. These changes are less desirable than transformations because they add parameters to the model, which may have undesirable consequences for a scientific theory (not necessarily for prediction, however). Theorists try to minimize the number of free parameters in a model because of simplicity of interpretation and because the more parameters, the better the model will fit, almost no matter what those parameters are. So, if we add polynomials, we will increase the proportion of variance accounted for, whether the polynomials help with linearity or not.

Table 7.7 Regression of the square root of shoes (SQR_SHOE) on log income centered (LOG_I_CN) and the square root of family size centered (SQR_F_CN) in the SHOES data set.

```
EIGENVALUES OF UNIT SCALED X'X
                            1           2           3
                         1.458       1.000       0.542

CONDITION INDICES
                            1           2           3
                         1.000       1.207       1.640

VARIANCE PROPORTIONS
                            1           2           3
   CONSTANT          .138597E-05    1.000     .503902E-05
   LOG_I_CN              0.271    .373393E-05    0.729
   SQR_F_CN              0.271    .210954E-07    0.729

DEP VAR:SQR_SHOE    N: 100   MULTIPLE R: 0.624   SQUARED MULTIPLE R: 0.389
ADJUSTED SQUARED MULTIPLE R: 0.377      STANDARD ERROR OF ESTIMATE: 0.600

VARIABLE    COEFFICIENT   STD ERROR   STD COEF   TOLERANCE     T     P(2 TAIL)

CONSTANT        3.093        0.060      0.000        .       51.506   .10E-14
LOG_I_CN        1.331        0.205      0.580      0.790      6.492   .36E-08
SQR_F_CN        1.001        0.145      0.617      0.790      6.914   .50E-09


CORRELATION MATRIX OF REGRESSION COEFFICIENTS

                    CONSTANT      LOG_I_CN     SQR_F_CN
    CONSTANT         1.000
    LOG_I_CN      -.597147E-03    1.000
    SQR_F_CN        -0.001        0.458        1.000


                    ANALYSIS OF VARIANCE

SOURCE       SUM-OF-SQUARES   DF   MEAN-SQUARE    F-RATIO          P

REGRESSION       22.280        2      11.140      30.899    .414543E-10
RESIDUAL         34.972       97       0.361
```

Since we know that the added variable plot of family size in the regression of the square root of shoes on family size and income (Figure 7.6) is curved, let's try adding family size squared to the regression to see whether it helps. Before we do this, however, there is one little trick that can be very useful when adding polynomials. If we add family size squared to a model that already contains family size, these two variables will be highly correlated. We can greatly decrease this correlation by first centering the family size variable and then squaring the centered values. This will preserve that polynomial relationship while reducing collinearity.

The mean of the family size variable (FAMSIZE in my data set) can be found with the command

```
>STATS
>STATISTICS FAMSIZE / MEAN
```

as shown in Table 7.8. With this value we can now center the family size variable with the command

```
>LET FAM_CN = FAMSIZE - 2.56
```

again, using the *cn* suffix to indicate centering. Now we are in a position to compute the polynomial values with the command

```
>LET FAM_CN_2 = FAM_CN^2
```

With these two new columns, family size centered (FAM_CN) and family size centered squared (FAM_CN_2), we can compute the polynomial regression. The output is shown in Table 7.9 and was generated by the commands

```
>MGLH
>MODEL SQR_SHOE=CONSTANT+INCOME+FAM_CN+FAM_CN_2
>SAVE POLYNOM.RES /MODEL
>ESTIMATE
```

The two new variables are still correlated, but they pose no collinearity problems, as evidenced by the largest condition index of only 3.222 in Table 7.9.

Should we keep the squared term in the model? One criterion, of coarse, is to look at the added variable plots to see whether they have straightened out any. If they have, then you might try replacing *family size + family size squared* with some transformation of family size that does the same job. Then you will be back in the situation discussed in the previous section.

Table 7.8 Mean of the family size variable in the SHOES data set.

TOTAL OBSERVATIONS:	100
	FAMSIZE
N OF CASES	100
MEAN	2.560

Another criterion is to ask whether the new variable makes a substantial improvement in the proportion of variance accounted for in the model, whether due to improving linearity or not. Rather than comparing R^2s for models with and without the polynomial term, however, a better criterion is offered by a statistic called *Mallows' C_p*, which measures the difference in the magnitudes of the residuals between the models with and without the squared term. (Mallows' C_p can be used more generally to help eliminate any number of terms from a model, but this type of model building is beyond the scope of this book.) The reasons for preferring C_p over direct comparison of the R^2s are that (1) you don't have to compute both regressions, and (2) C_p penalizes each model for the number of parameters they contain, so that simply adding predictors does not necessarily increase the value of C_p, as it would for R^2.

The model including all predictors, i.e., with the polynomial term, is called the *full* model, and models with fewer predictors are called *restricted* models. The formula for the C_p statistic is

$$C_p = (k' - p)(t_p^2 - 1) + p \qquad (7.3)$$

where k' is the number of predictors in the full model (including the constant), p is the number of predictors in the restricted model (including the constant), and t_p^2 is the square of the *t*-test of the coefficient of the predictor of interest in the full model. When the coefficients of the predictors that you are considering leaving out of the model are equal to zero, then the expected value of $C_p = p$, and C_p is roughly distributed around this value as *F*. Thus, Mallows (1973) suggests that good models have C_p less than, or approximately equal to, p.

For example, consider the family size (centered) squared variable in Table 7.9. The full model including this variable has $k' = 4$, whereas the restricted model excluding this variable

Table 7.9 Regression of the square root of shoes on income, family size centered (FAM_CN), and family size centered squared (FAM_CN_2) in the SHOES data set.

```
EIGENVALUES OF UNIT SCALED X'X
                          1         2         3         4
                      2.053     1.442     0.307     0.198

CONDITION INDICES
                          1         2         3         4
                      1.000     1.193     2.588     3.222

VARIANCE PROPORTIONS
                          1         2         3         4
     CONSTANT         0.067     0.008     0.069     0.856
     INCOME           0.057     0.038     0.051     0.855
     FAM_CN_2         0.061     0.079     0.845     0.015
     FAM_CN           0.005     0.233     0.705     0.057

DEP VAR:SQR_SHOE    N: 100   MULTIPLE R: 0.707   SQUARED MULTIPLE R: 0.500
ADJUSTED SQUARED MULTIPLE R: 0.485      STANDARD ERROR OF ESTIMATE:  0.546

VARIABLE   COEFFICIENT    STD ERROR    STD COEF  TOLERANCE     T     P(2 TAIL)

CONSTANT       2.502        0.094        0.000       .       26.716   .10E-14
INCOME    .283848E-04  .341022E-05       0.643    0.872       8.323   .58E-12
FAM_CN_2       0.002        0.015        0.010    0.640       0.113    0.910
FAM_CN         0.284        0.046        0.585    0.573       6.145   .18E-07


CORRELATION MATRIX OF REGRESSION COEFFICIENTS

                   CONSTANT      INCOME    FAM_CN_2     FAM_CN
     CONSTANT        1.000
     INCOME         -0.718       1.000
     FAM_CN_2       -0.314      -0.092      1.000
     FAM_CN         -0.025       0.333     -0.590       1.000


                       ANALYSIS OF VARIANCE

SOURCE      SUM-OF-SQUARES    DF    MEAN-SQUARE      F-RATIO        P

REGRESSION       28.649        3       9.550         32.051   .192069E-13
RESIDUAL         28.603       96       0.298
```

would have $p = 3$. The value of t for this variable, from Table 7.9, is 0.113. Therefore, the value of C_p for the model excluding family size (centered) squared, is

$$C_p = (4 - 3)(.113^2 - 1) + 3 = 2.01$$

This value is less than $p = 3$, so this is a good model according to our criterion. Therefore, in terms of variance accounted for, we can feel comfortable leaving out the polynomial predictor.

For comparison, suppose that we considered dropping family size (centered) itself. Again, the full model including this variable has $k' = 4$, whereas the restricted model excluding family size would have $p = 3$. The value of t for family size (centered) in Table 7.9 is 6.145. Thus, the value of C_p is

$$C_p = (4 - 3)(6.145^2 - 1) + 3 = 39.76$$

This value is much greater than $p = 3$, so the model without family size would not be considered a good model.

All of these criteria, curvature in added variable plots, C_p, and the interpretation of the variables, may be important in deciding whether or not to move to a polynomial regression model. Fortunately, SYSTAT makes it easy to obtain all of the information you need for assessing the first two. Ensuring that you end up with meaningful predictors is up to you.

8

One-Way Analysis of Variance

A "one-way" analysis of variance decomposes the total variance in a set of data into a component corresponding to the variance of the means of the groups defined by a single categorical variable and a component corresponding to "error." The former is usually called the *between group* variance and the latter the *within group* variance. These variances can then be used to assess the plausibility that the groups all have the same mean, to put confidence intervals around the means, and so on. Data sets with only two groups can also be analyzed using SYSTAT's *t-test* procedure; data sets with three or more groups should always be analyzed using contrast analysis or post-hoc tests.

The analysis described in this chapter consists of looking at the raw data to check for unusual cases, computing the one-way ANOVA, computing contrasts and possibly performing post-hoc analyses, computing standard errors and confidence intervals, assessing the assumptions underlying the ANOVA, and graphing the results of the analysis.

8.1 The Example Data Set

The example that will be used in this chapter is a one-way design with three levels, which was generated in SYSTAT using the commands described in section 8.9. The example data set is shown in Table 8.1. This data set contains 15 cases in group 1, 10 cases in group 2, and 12 cases in group 3. Suppose that the scores represent test scores on a school achievement test and the groups represent three private tutoring methods to which students were randomly assigned. In order to use this data set in SYSTAT, you need to enter it into a data file with at least two columns: one column containing codes for the categorical tutoring variable, one column with the dependent variable, and any other information columns that you want, such as subject names. Note that if your categorical variable is a character variable, you must first recode it as a numeric variable to generate some of the plots used below. For example, suppose that the three tutorial groups were in a column labelled "TUTORIAL$," and were coded "interact," "lecture," and "computer" in the data set. You could create a new numeric variable called "GROUP" with three levels by recoding the tutorial groups with the following commands (see section 2.2.2):

```
> IF TUTORIAL$ = "INTERACT" THEN LET GROUP = 1
> IF TUTORIAL$ = "LECTURE" THEN LET GROUP = 2
> IF TUTORIAL$ = "COMPUTER" THEN LET GROUP = 3
```

Procedures that require numerical variables will be noted below. The complete, properly coded data file is shown in section 8.9.

Table 8.1 The TEST SCORES data set—test scores for students in three tutoring conditions.

Group 1	Group 2	Group 3
94.4	95.3	98.1
75.7	117.2	101.2
88.1	97.9	120.1
108.7	82.7	77.5
94.8	105.3	124.7
130.6	85.2	136.1
121.1	86.7	132.6
82.9	104.8	130.5
112.0	67.9	130.0
85.2	106.1	138.8
98.7		105.8
50.1		70.4
86.1		
99.8		
121.8		

8.2 Looking at the Data

8.2.1 Scatterplots

As in any analysis, the first step should be to examine the raw data to check for extreme observations and to get a feel for the shape of the data. For categorical data, it is usually best to plot the data by group because you are interested in detecting observations that are extreme within a group—even though they might not be extreme for the data as a whole. To create a scatterplot of the raw data by category, type

```
>GRAPH
>PLOT SCORE*GROUP/SMOOTH=LOWESS
```

(Note that this plot requires a numerical variable for the categories; see section 8.1. See section 3.2.2 for more on plots.) The option *smooth=lowess* simply draws a line connecting the means of the groups in this plot. The resulting scatterplot is shown in Figure 8.1. Extreme cases in this plot should be examined individually to ensure that they were recorded properly and do not differ from the rest of the data in some other exceptional way. For example, the student in group 1 with a score around 50 should be examined: Did this student differ from the rest of the class in some way?

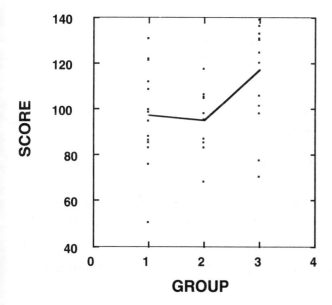

Figure 8.1 Scatterplot of the raw data in the TEST SCORES data set.

8.2.2 Box Plots

A second type of plot that is very useful in this context is a box plot of the scores by groups. This plot is shown in Figure 8.2. The plot is generated by the command

```
>BOX SCORE*GROUP/SORT
```

These three plots do not indicate any outliers within any of the three groups. If there were any outliers, you should examine those cases individually.

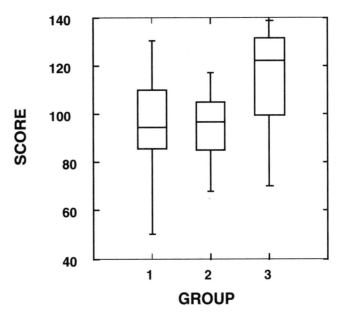

Figure 8.2 Box plots of the scores for each group in the TEST SCORES data set.

8.3 One-Way ANOVA

To compute summary statistics for each of the groups, the ANOVA, and Bartlett's test for homogeneity of variance, sort the data of the grouping variable, select the *by group* option, and then select the statistics you want to compute. This can be done using the following commands (see also sections 1.1.3, 2.1.1, and 3.1 for saving, sorting, and stats, respectively):

```
>SYSTAT
>SAVE TESTSCOR.SRT
```

```
>SORT SCORE
>EDIT "TESTSCOR.SRT"
>BY GROUP
>STATISTICS SCORE/ MAXIMUM MEAN MINIMUM N SD,
SEM VARIANCE MEDIAN
```

Recall that SCORE is the dependent variable and GROUP is the independent variable in the example data set. The output of this series of commands is shown in Table 8.2. (Note the earlier versions of SYSTAT did not have the median option.)

Bartlett's test in Table 8.2 is a significance test of the homogeneity of two or more variances that employs a χ^2 with df equal to the number of groups minus 1. In the example data this test is not significant, $\chi^2 = 1.98$, $p = .37$. In general, this test is too sensitive, and will indicate that the group variances are different too often, i.e., report p-values that are too small (Snedecor & Cochran, 1980, pp. 252–253). This test is discussed below in testing the homogeneity of variance assumption, section 8.6.2.4.

The one-way ANOVA provided by the statistics command contains most of the information that you need. However, another method of analyzing the data in SYSTAT allows you to save the residuals within groups, which can then be used to assess the assumptions underlying the ANOVA. This ANOVA procedure also gives some new information that may be interesting to you. The commands are:

Table 8.2 Summary statistics by groups for the TEST SCORES data set.

```
THE FOLLOWING RESULTS ARE FOR:
         GROUP      =        1.00000

TOTAL OBSERVATIONS:      15

                    SCORE
  N OF CASES              15
  MINIMUM            50.10000
  MAXIMUM           130.60000
  MEAN               96.66667
  VARIANCE          421.19524
  STANDARD DEV       20.52304
  STD. ERROR          5.29903
  MEDIAN             94.80000
```

Table 8.2 *(continued)*

```
THE FOLLOWING RESULTS ARE FOR:
        GROUP    =    2.00000

TOTAL OBSERVATIONS:    10

                SCORE
  N OF CASES         10
  MINIMUM        67.90000
  MAXIMUM       117.20000
  MEAN           94.91000
  VARIANCE      208.58100
  STANDARD DEV   14.44233
  STD. ERROR      4.56707
  MEDIAN         96.60000


THE FOLLOWING RESULTS ARE FOR:
        GROUP    =    3.00000

TOTAL OBSERVATIONS:    12

                SCORE
  N OF CASES         12
  MINIMUM        70.40000
  MAXIMUM       138.80000
  MEAN          113.81667
  VARIANCE      531.45970
  STANDARD DEV   23.05341
  STD. ERROR      6.65495
  MEDIAN        122.40000
```

SUMMARY STATISTICS FOR   SCORE

BARTLETT TEST FOR HOMOGENEITY OF GROUP VARIANCES

CHI-SQUARE = 1.98103   DF = 2   PROBABILITY = 0.37138

ANALYSIS OF VARIANCE

SOURCE	SUM OF SQUARES	DF	MEAN SQUARE	F	PROBABILITY
BETWEEN GROUPS	2602.71289	2	1301.35645	3.24861	0.05116
WITHIN GROUPS	13620.01900	34	400.58879		

```
>MGLH
>CATEGORY GROUP
>ANOVA SCORE
>PRINT LONG
>SAVE "TESTSCOR.RES"/MODEL
>ESTIMATE
```

The output is shown in Table 8.3.

The *estimates of effects* in Table 8.3 are the regression coefficients from an analysis on the effects coded dependent variable. The value called *constant* is the mean of the group means (the unweighted grand mean), and the values for the groups correspond to the estimated effects in the analysis of variance—the differences between the group means and the mean of the group means. The effect for the last group is never given, but can be found either by subtracting the mean of the last group from the constant (mean of the group means) or by adding up the effects for the other groups and changing the sign. This works because the effects for all groups must sum to zero.

The AVOVA table in Table 8.3 is identical to that in Table 8.2. Underneath this table are the least squares means and standard errors (LS MEAN and SE, respectively). The least squares means are simply the means of each group. The standard errors of those means use the pooled variance from all groups, MEAN-SQUARE$_{ERROR}$ from the ANOVA table, divided by the number of cases in each group. The *Durbin-Watson Statistic* and *first-order autocorrelation* are almost never relevant in ANOVA (see section 5.4.3.3 for a discussion). Note that the former is a sort of significance test of the latter, which is the correlation between each consecutive pair of numbers in the data set; this correlation is meaningless in the present example because the data were not ordered in any meaningful way within groups.

WARNING: CHECK THE NUMBER OF LEVELS

In versions prior to 5.0 for DOS and 5.2 for the Macintosh, SYSTAT needed to be told how many levels there were of each categorical variable, and the data set needed to be sorted by that variable. In the new versions, SYSTAT simply reads the data file to determine the number of levels of categorical variables, so you should always check to make sure that the number of levels is correct. The number of levels is given at the beginning of the ANOVA output, as shown at the top of Table 8.3.

MACINTOSH
① Choose the menu item **Stats/MGLH/Fully Factorial (M)ANOVA...**
② Select SCORE from the **Dependent variable(s)** list.
③ Select GROUP from the **Factor(s)** list.
④ Click on **More...**
⑤ Turn on the **Means & Std. Errors, Extended Output,** and **Save Residuals** check boxes.
⑥ Click **OK**, then type the save filename, e.g., "testscor.res" into the **Save Residuals as** box, and click **Save**.
⑦ Select **Residuals and Diagnostics** and click OK.

MS-DOS
① Choose the menu item **Statistics/MGLH/ANOVA/ Category/** and select GROUP from the variables list.
② Choose the menu item **Statistics/MGLH/ANOVA/ ANOVA/Variables** and select SCORE from the variables list.
③ Choose the menu item **Statistics/MGLH/ANOVA/ Estimate/Save/** and hit <esc>, then type the save filename, e.g., "testscor.res", hit <ret>, and select **Residuals** and hit <esc>.
④ Choose the menu item **Statistics/MGLH/ANOVA/ Estimate/Go!**

Table 8.3 One-way ANOVA on the TEST SCORES data set using the standard ANOVA procedures and showing extended results.

```
LEVELS ENCOUNTERED DURING PROCESSING ARE:
GROUP
       1.00000        2.00000        3.00000

DEP VAR: SCORE     N: 37   MULTIPLE R: 0.401  SQUARED MULTIPLE R: 0.160
                                  -1
ESTIMATES OF EFFECTS    B = (X'X)   X'Y

                              SCORE

    CONSTANT                101.79778
       GROUP     1.00000     -5.13111
       GROUP     2.00000     -6.88778


                      ANALYSIS OF VARIANCE

SOURCE         SUM-OF-SQUARES   DF   MEAN-SQUARE    F-RATIO         P
GROUP              2602.71289    2    1301.35645    3.24861   0.05116
ERROR             13620.01900   34     400.58879

LEAST SQUARES MEANS.
                              LS MEAN           SE           N
       GROUP  =   1.00000    96.66667      5.16778          15
       GROUP  =   2.00000    94.91000      6.32921          10
       GROUP  =   3.00000   113.81667      5.77775          12


DURBIN-WATSON D STATISTIC       0.411
FIRST ORDER AUTOCORRELATION     0.692

RESIDUALS HAVE BEEN SAVED
```

8.4 Contrasts

The F test for the GROUP effect in Table 8.3 is not quite significant at the traditional 0.05 level of significance. As I hope this example shows, you would probably not want to conclude that the experimental manipulations produced no differences between groups. F tests with more than one degree of freedom in the numerator, such as those in Table 8.3 are of little scientific interest because (1) when they are significant they only tell us that some differences exist, not which means are different, and (2) when they are not significant, this does not mean that none of the means are different or that there are no significant trends among the means. Experiments are almost always conducted with specific predictions in mind, and these predictions should always be tested directly. *F* tests with more than one degree of freedom in the numerator are virtually useless.

> **HINT: TESTS MUST *FOLLOW* AN ANOVA**
>
> All contrasts, comparisons, and post-hoc tests in SYSTAT must be performed immediately after the overall ANOVA has been computed. If you perform other operations, or try to test a hypothesis that results in an error, you may have to recompute the ANOVA before proceeding with specific tests. So if you get an error message like "You must save a file first" or "SYSTAT command expected about here ^," recompute the ANOVA and try again.

8.4.1 Planned Comparisons

SYSTAT offers a couple of methods for testing specific hypotheses, or *contrasts*. Suppose that your theory specifies that the mean of group 3 in the TEST SCORES data set should be higher than the means for groups 1 and 2, which should be approximately equal. This hypothesis can be expressed as

$$\bar{G}_3 - \frac{\bar{G}_1 + \bar{G}_2}{2} > 0 \qquad (8.1a)$$

which is equivalent to

$$(1) \quad \bar{G}_3 + (-.5)\bar{G}_1 + (-.5)\bar{G}_2 > 0 \qquad (8.1b)$$

Thus, the hypothesis is a simple linear combination of the three means using the contrast weights shown in parentheses in Equation 8.1b. You can multiply both sides of the inequality in Equation 8.1b by 2 to get rid of the decimal places, which gives

$$(2) \quad \bar{G}_3 + (-1)\bar{G}_1 + (-1)\bar{G}_2 > 0 \qquad (8.1c)$$

MACINTOSH

① Choose **Stats/MGLH/Test of Effects...**
② Select GROUP from the **Between Subjects** list.
③ Turn on **Coefficients** and type "–1 –1 2" into the coefficients box.
④ Click **OK**.

MS-DOS

① Choose **Statistics/MGLH/ MGLM/Hypothesis/ Effect/**, select GROUP from the variables list, and hit <esc>.
② Choose **Statistics/MGLH/ MGLM/Hypothesis/ Contrast/Matrix/**, type "–1 –1 2" into the numeric array box, and hit <esc>.
③ Choose **Statistics/MGLH/ MGLM/Hypothesis/Go (Test)!**

MACINTOSH

① Choose the menu item **Stats/MGLH/User Defined Contrasts...**
② Type "2*group[3] + group[1] + group[2] = 0" into the **Contrast** box.
③ Click **OK**.

MS-DOS

① Choose the menu item **Statistics/MGLH/MGLM/ Hypothesis/Specify/ Equation/**, type "2*group[3] + group[1] + group[2] = 0" into the box, and hit <esc>.
② Choose the menu item **Statistics/MGLH/MGLM/ Hypothesis/Go (Test)!**

All three expressions, Equations. 8.1a, 8.1b, and 8.1c are equivalent, but Equation 8.1c gives the contrast weights in the simplest form. The contrast weights for the hypothesis are –1, –1, and +2 for the means of groups 1, 2, and 3, respectively.

These contrast weights can be entered into SYSTAT directly to test the desired hypothesis. To test this hypothesis, first compute the one-way ANOVA as was done in Table 8.3. Then type the commands

▶ >HYPOTHESIS
>EFFECT = GROUP
>CONTRAST
>-1 -1 2
>TEST

The output from this procedure is shown in Table 8.4. An equivalent method for testing contrasts is offered by SYSTAT that allows you to enter directly expressions in the form of Equation 8.1c. (Leave out means that are multiplied by zero.) The commands are

>SPECIFY / POOLED
>2*GROUP[3] + GROUP[1] + GROUP[2] = 0
▶ >TEST

The numbers in the output are identical to those in Table 8.4. Which procedure you use is simply a matter of personal preference.

This hypothesis in Table 8.4 is statistically significant. Note that SYSTAT always reports "nondirectional," or "two-tailed," p-values. However, the present hypothesis was very specifically directional, so to find the directional p-value you can divide the given p value by-two. This gives $p = .008$ from Table 8.4, which is highly significant.

We rarely conduct research with the purpose of estimating probability values from F tests; rather, the goal of research is almost always to estimate the size of some specific effect, and the probability values simply give us information about how likely that effect was due to chance. There are a number of effect size estimates that one might use, but one of the simplest and most useful is the Pearson product-moment correlation r. To find the effect size r for the above contrast, plug the F and df values from Table 8.4 into the following formula:

Table 8.4 Test of the contrast predicting that the mean of group 3 would be higher than the means of groups 1 and 2 in the TEST SCORES data set.

TEST FOR EFFECT CALLED: GROUP

A MATRIX

	1	2	3
	0.00000	-3.00000	-3.00000

TEST OF HYPOTHESIS

SOURCE	SS	DF	MS	F	P
HYPOTHESIS	2600.16642	1	2600.16642	6.49086	0.01553
ERROR	13620.01900	34	400.58879		

$$r = \sqrt{\frac{F}{F + df_{\text{denom}}}} = \sqrt{\frac{6.491}{6.491 + 34}} = 0.40 \quad (8.2)$$

See Rosenthal and Rosnow (1991) for more on computing and interpreting the effect size r.

There remains one loose end that must be tied up. We have determined that our prediction was highly significant, but a reader of this result would not know whether there was also significant *deviation* from our prediction among the means. That is, our prediction may account for a significant amount of the variance among the means but this does not mean that there is no significant variance remaining. The size of the deviations from the predicted values are indexed by the sum of squares remaining between groups after the sum of squares for the prediction has been removed (this is only approximate with unequal n's). Therefore, from Table 8.3 and Table 8.4:

$$SS_{\text{residual between}} = SS_{\text{between}} - SS_{\text{contrast}} \quad (8.3)$$

$$= 2602.713 - 2600.166$$

$$= 2.547$$

This sum of squares is based on 1 df, so an F test of its significance is simply the sum of squares for the residual between, divided by the mean square within from Table 8.3. This $F_{(1,34)}$ is near zero, so we can conclude that the deviation from our prediction in the data is not significant. In fact, our hypothesis accounted for essentially all of the variance between groups:

$$\% \ SS_{between} \text{ accounted for} = \frac{SS_{contrast}}{SS_{between}}$$

$$= \frac{2600.166}{2602.713}$$

$$= .999 \qquad (8.4)$$

8.4.2 Polynomials

Sometimes the specific hypotheses that you want to test may be sets of polynomials. For example, you might have predicted that there would be a linearly increasing trend in the means across groups 1, 2, and 3 in the example data set. You could test this using the procedures described in the previous section using the contrast weights –1, 0, and +1 for the means of groups 1, 2, and 3, respectively (SYSTAT will automatically take into account the unequal sample sizes in a one-way ANOVA). Similarly, a test of the nonlinear (quadratic) trend across groups would use the contrast weights –1, 2, and –1 for the means of groups 1, 2, and 3, respectively. SYSTAT gives an easier method for computing such trends. Simply type

```
>HYPOTHESIS
>EFFECT = GROUP
>CONTRAST / POLYNOMIAL ORDER = 1
>TEST
```

for the linear trend, and type

```
>HYPOTHESIS
>EFFECT = GROUP
>CONTRAST / POLYNOMIAL ORDER = 2
>TEST
```

for the quadratic trend, and so on. The outputs from these tests are shown in Table 8.5. The linear trend is significant, $p = .034$, but the quadratic trend is not. As always, you should not

---

MACINTOSH

① Choose the menu item **Stats/MGLH/Test of Effects...**

② Select GROUP from the **Between Subjects** list.

③ Turn on **Polynomial** and type "1" into the **Order** box.

④ Click **OK**.

MS-DOS

① Choose the menu item **Statistics/MGLH/MGLM/ Hypothesis/Effect/** and select GROUP from the variables list.

② Choose the menu item **Statistics/MGLH/MGLM/ Hypothesis/Contrast/ Polynomial/ Order/**, type "1" into the box, and hit <esc>.

③ Choose the menu item **Statistics/MGLH/MGLM/ Hypothesis/Go (Test)!**

conclude from this that the linear trend exists but the quadratic trend does not : Even though the quadratic trend is not significant, its effect size is not significantly different from the effect size for the linear trend. The effect sizes for these trends are found in the same manner as those in Equation 8.2:

$$r = \sqrt{\frac{F}{F + df_{\text{denom}}}} = \sqrt{\frac{4.895}{4.895 + 34}} = 0.35 \tag{8.5}$$

$$r = \sqrt{\frac{F}{F + df_{\text{denom}}}} = \sqrt{\frac{1.938}{1.938 + 34}} = 0.23 \tag{8.6}$$

The linear trend accounts for around 75% of the variance between groups, whereas the nonsignificant quadratic trend accounts for only around 30% of the variance between groups (see Equation 8.4). The reason these trends together account for more than 100% of the variance is that they are not orthogonal due to the unequal sample sizes: With equal sample sizes these percentages would add up to 100%.

The linear trend above used the contrast weights –1, 0, and +1 for the means of groups 1, 2, and 3, respectively. Because the weight of zero on the mean of group 2 prevents the size of that mean from affecting the contrast, this is equivalent to a contrast comparing only groups 1 and 3, that is, using the contrast weights –1 and +1 for the means of groups 1 and 3, respectively. The test of this contrast will be identical to the output shown in Table 8.5. However, these two contrasts differ in an important way: The linear trend predicts that the mean of the second group will fall halfway between groups 1 and 3, whereas the contrast comparing only groups 1 and 3 makes no prediction about group 2. This difference should be taken into account in two ways: (1) whenever the linear trend in tested, the quadratic (or nonlinear residual, as in Equation 8.3) trend should also be tested to determine whether group 2 does indeed lie on the line between groups 1 and 3; and (2) the effect size for the contrast between groups 1 and 3 should not include the degrees of freedom for the second group. The effect size for the linear trend used pooled degrees of freedom from all three groups: 14 + 9 + 11 = 34. The effect size for the contrast between groups 1 and 3 should only pool the degrees of freedom from those two groups: 14 + 11 = 25. Thus, the effect size for the contrast between groups 1 and 3 is

$$r = \sqrt{\frac{F}{F + df_{\text{denom}}}} = \sqrt{\frac{4.895}{4.895 + 25}} = 0.40 \qquad (8.7)$$

Intuitively, the effect size for this contrast (.40) should be larger than the effect size for the linear trend (.35, Equation 8.5), because the size of the contrast between groups 1 and 3 alone represents only the endpoints of the linear trend.

That the same contrast, for example, –1, 0, +1, can at once be used to test both a linear trend and a difference between two groups, is often very confusing to students. Theoretically, the two contrasts test different effects: Calling an effect a "linear trend" is not only a description of differences between groups but also a claim about what would happen between groups (thus, the term "linear trend" is only appropriate when the independent variable is on a meaningful scale; see also section 8.8), whereas a "difference" makes no assumptions about what goes on "between" the groups being compared. What students need to realize is that a difference can be a *component* of a linear trend, and that with some numbers of levels of the independent variable that component is the only part of the linear trend that can be tested.

8.4.3 Error Terms

8.4.3.1 Pooled Error Terms

The default error term for contrasts in SYSTAT is the *pooled*, or "global," error term. This error term pools together the variance from all of the cells in the analysis and is equal to the mean square error from the overall one-way ANOVA (for example, in Table 8.3). The validity of pooling together sources of variance from multiple sources, however, is based on one of the assumptions underlying ANOVA: the assumption that these variances all represent estimates of the same variance σ^2. When this assumption appears to be substantially violated (see section 8.6.2), you should consider either transforming the data or using the *local* error estimate or *separate* error estimates described in the next two sections.

8.4.3.2 Local Error Terms

Often a contrast is tested that involves only a subset of the means in the design, that is, uses some contrast weights of zero.

Table 8.5 Test of the linear and quadratic trends across groups in the TEST SCORES data set.

Linear Trend

```
TEST FOR EFFECT CALLED:      GROUP

A MATRIX
                   1             2             3
               0.00000       -1.41421      -0.70711

TEST OF HYPOTHESIS

     SOURCE         SS         DF         MS          F           P

 HYPOTHESIS   1960.81667       1     1960.81667    4.89484     0.03376
      ERROR  13620.01900      34      400.58879
```

Quadratic Trend

```
TEST FOR EFFECT CALLED:      GROUP

A MATRIX
                   1             2             3
               0.00000    -.111022E-15     -1.22474

TEST OF HYPOTHESIS

     SOURCE         SS         DF         MS          F           P

 HYPOTHESIS    776.31517       1      776.31517    1.93794     0.17293
      ERROR  13620.01900      34      400.58879
```

In such cases it is sometimes wise to use as the estimate of error variance only the pooled variance from the cell actually used in the contrast. This error term is "local" to the particular contrast being tested and, hence, is sometimes called the *local error term*. Normally, you would want to use the variance pooled from all cells of the experiment because this is based on more data points and is, therefore, a better estimate of the true population variance, when the assumption of equal variances is met (see section 8.6.2). However, when the variances of different cells are not equal, it is often safer to use the local error term.

For example, in section 8.4.1 I discussed a contrast comparing the mean of group 1 to the mean of group 3 in the TEST SCORES data set. In Table 8.2 you can see that the variances in groups 1 and 3 are much closer in magnitude than either of them are to the variance in group 2. In such a case you might want to test the contrast using only the pooled error variance from first and third groups. The formula for pooling variances is

$$s^2_{pooled} = \frac{\Sigma(df_i s_i^2)}{\Sigma df_i} \tag{8.8}$$

where the s_i^2 are the variances from the groups used in the contrast and the df_i are the degrees of freedom for those groups. In the example data set, plugging in the variances from Table 8.2 and the degrees of freedom to Equation 8.8 gives

$$s^2_{pooled} = \frac{\Sigma(df_i s_i^2)}{\Sigma df_i}$$

$$= \frac{(14)(421.195) + (11)(531.460)}{(14 + 11)}$$

$$= 467.712 \tag{8.9}$$

This is somewhat larger than the mean square within used to test the contrast in Table 8.5. The sum of squares for the contrast given in Table 8.5 is 1960.817. Thus, the test of the contrast is $F_{(1,25)} = 1960.817/467.712 = 4.174$, $p = .052$. This contrast does not quite reach the traditional level of significance! Clearly, the choice of error terms can affect the outcome of the hypothesis testing. However, the effect size of this contrast is

$$r = \sqrt{\frac{F}{F + df_{denom}}} = \sqrt{\frac{4.174}{4.174 + 25}} = 0.38 \tag{8.10}$$

which is nearly identical to the effect size in Equation 8.5.

8.4.3.3 Adjusted Error Terms Based on Separate Variances

Often you want to compare groups or use contrasts on means from groups with unequal variances. For example, groups 2 and 3 in Table 8.2 have very different variances, so when testing a contrast comparing those groups you might worry about the validity of the equal variance assumption (see section 8.6.2). The polynomial contrasts in Table 8.5 also compared groups with unequal variances. If the deviation from equality is substantial (in the example data set the differences are arguably not large enough to worry about), you should consider either transforming the data or using an error term that adjusts for this inequality.

SYSTAT provides an error term adjustment using Satterthwaite's (1946) adjustment. Essentially, this adjustment penalizes you for having unequal variance by reducing the pooled error estimate and by reducing the degrees of freedom for that estimate. The commands for testing the linear contrast using the adjusted error estimate are

```
>HYPOTHESIS
>SPECIFY / SEPARATE
>GROUP[3] - GROUP[1] = 0
>TEST
```

The output for this test is shown in Table 8.6. The hypothesis is not quite significant at the traditional .05 level.

Error term adjustments such as this decrease the power of your tests and should generally only be used when the variances are quite different (see section 8.6.2) and when transformations of the dependent variable are either not desired or will not correct the problem.

8.5 Post-Hoc Tests

Frequently, investigators want to test hypotheses that are suggested by patterns in the data. Such tests are usually called *post-hoc*, or *unplanned*, comparisons. They are perfectly all right; in fact, one would be remiss in allowing unexpected effects to go unreported. However, you should be aware that the probability of finding significant results due to chance increases with the number of tests performed. SYSTAT provides a number of standard methods for assessing the significance of all pairwise comparisons among a set of means.

MACINTOSH
① Choose the menu item **Stats/MGLH/User Defined Contrasts...**
② Type "group[3] – group[1] = 0" into the **Contrast** box.
③ Turn on **Separate** under **Error Term Estimate**.
④ Click **OK**.

MS-DOS
① Choose the menu item **Statistics/MGLH/MGLM/ Hypothesis/Specify/ Equation/**, type "group[3] – group[1] = 0" into box, and hit <esc>.
② Choose the menu item **Statistics/MGLH/MGLM/ Hypothesis/Specify/ Options/ Separate** and hit <ret>.
③ Choose the menu item **Statistics/MGLH/MGLM/ Hypothesis/Go (Test)!**

Table 8.6 Linear contrast across groups 1, 2, and 3 in the TEST SCORES data set using the separate variances estimate for the error term.

```
USING SEPARATE VARIANCES ESTIMATE FOR ERROR TERM.

HYPOTHESIS.

A MATRIX
                            1              2              3
                      0.00000        2.00000        1.00000

NULL HYPOTHESIS VALUE FOR D
                      0.00000

TEST OF HYPOTHESIS

SOURCE          SS              DF             MS             F              P

HYPOTH     1960.81667            1       1960.81667        4.06426        0.05599
ERROR     10768.58922        22.32048     482.45327
```

8.5.1 Fisher's Least Significant Difference Test

One of SYSTAT's post-hoc procedures follows the spirit of Fisher's least significant difference (LSD) test. Traditionally, this test involved taking an error term, its degrees of freedom, and the critical value of t needed to obtain significance with that number of df, and then solving backwards to find the minimum difference between a pair of means that would be significant at the desired level. This difference was the *least significant difference*. You could then inspect a table of pairwise difference between means and pick out those differences that exceeded the LSD. Since computational work is now a lot cheaper than it was in Fisher's day, SYSTAT simply computes the p-values for all pairs of differences. The commands for computing these probabilities for the example data set are

```
>HYPOTHESIS
>POST GROUP / LSD
>TEST
```

The output is shown in Table 8.7. Notice that the probability for the difference between means of groups 1 and 3, .03376, is identical to the p-value for the contrast between those groups in the top panel of Table 8.7. Computing the probabilities of

MACINTOSH

① Choose the menu item **Stats/MGLH/Post Hoc Test...**
② Select GROUP from the **Effect** list.
③ Turn on **Fisher's LSD**.
④ Click **OK**.

MS-DOS

① Choose the menu item **Statistics/MGLH/MGLM/ Hypothesis/Post/ Options/LSD/** and hit <ret>.
② Choose the menu item **Statistics/MGLH/MGLM/ Hypothesis/Post/ Expression/** and select GROUP from the variables list.
③ Choose the menu item **Statistics/MGLH/MGLM/ Hypothesis/Go (Test)!**

all pairwise differences using the LSD command is identical to computing contrasts that compare all pairs of means and finding the probabilities from those contrasts.

The danger of this procedure is that it does not adjust the probabilities for the number of tests performed. Fisher warned that such comparisons should not be trusted unless the overall F from the one-way ANOVA was significant (in the example data it is not quite—see Table 8.3). When the overall F is significant, the largest pairwise difference between means is said to be "protected" at that significance level; that is, the largest difference is at least as significant as the overall F. A more direct approach, however, is simply to adjust the probabilities by multiplying them by the number of tests performed. This is called the *Bonferroni* adjustment, and is presented in the next section.

Table 8.7 Post-hoc comparisons following Fisher's least significant difference approach.

```
USING MODEL MSE OF       400.589 WITH       34. DF.
MATRIX OF PAIRWISE MEAN DIFFERENCES:

                           1              2              3
           1           0.00000
           2          -1.75667        0.00000
           3          17.15000       18.90667        0.00000


FISHER'S LEAST-SIGNIFICANT-DIFFERENCE TEST.
MATRIX OF PAIRWISE COMPARISON PROBABILITIES:

                           1              2              3
           1           1.00000
           2           0.83106        1.00000
           3           0.03376        0.03423        1.00000
```

8.5.2 Bonferroni Adjusted Pairwise Differences

If you compute contrasts on all of the pairwise differences between means in your experiment, you increase the probability of finding significant results due to chance. To adjust the probabilities so that you do not capitalize on chance, you can simply multiply the probabilities by the number of tests performed. This is called the Bonferroni adjustment, and SYSTAT will do it for you if you type the following commands

MACINTOSH

① Choose the menu item **Data/Select Cases...**

② Enter the criteria for the levels that you want to compare, e.g., "GROUP=1 or GROUP=2".

③ Click **OK**.

④ Choose the menu item **Stats/Stats/t-test...**

⑤ Click on the **Independent** button if it is not already selected.

⑥ Select the dependent variable from the left **Variables** list, e.g., SCORE.

⑦ Select the independent, or *grouping*, variable from the **Group** list, e.g., GROUP.

⑧ Click **OK**.

MS-DOS

① Choose the menu item **Data/Select/Select/** and hit <esc>.

② Type "group<3" and hit <ret>.

③ Choose the menu item **Statistics/Stats/Ttest/ Independent/Variables/ Dependent** and select SCORE from the variables list.

④ Choose the menu item **Statistics/Stats/Ttest/ Independent/Variables/ Grouping/** and select GROUP from the variables list.

⑤ Choose the menu item **Statistics/Stats/Ttest/ Independent/Go!**

(see section 8.5.1 for menu equivalents):

```
>HYPOTHESIS
>POST GROUP / BONFERRONI
>TEST
```

The output from this procedure is shown in Table 8.8. It is identical to the output in Table 8.7, except that all of the probabilities have been multiplied by three. This was done because there are three pairwise differences being tested by this post-hoc test.

Each of these pairwise comparisons corresponds to a *two-sample*, or *independent*, *t*-test. The is just a special case of a one-way ANOVA with only two levels. You could compute these tests singly using SYSTAT's independent *t*-test command. For example, to compare group 1 with group 2, simply type

```
>SELECT GROUP=1 OR GROUP=2
>STATS
>TTEST SCORE * GROUP
```

The " * " between the two variables tells SYSTAT that the second variable is a grouping variable and not another column of data. This command will give the *t* and unadjusted *p*-value for this particular comparison. If you only want to compute selected pairwise differences, this method is equivalent to a contrast between the two groups using the local error term and does not require that you run the ANOVA in advance.

The Bonferroni adjustment is quite conservative, i.e., the *p*-values will be too big, but it has two advantages over other post-hoc methods: (1) it is very easy to understand, which is a quality that should never be underrated, and (2) you can vary the size of the adjustment according to the number of tests in which you are interested. For example, if you are really interested only in two comparisons, between the means of groups 1 and 2 and between groups 2 and 3, you can use the Bonferroni adjustment but only adjust for two tests. SYSTAT allows you to specify the number of tests to adjust for after the "bonferroni" option on the post command (see section 8.5.1 for menu equivalents):

```
>HYPOTHESIS
>POST GROUP / BONFERRONI = 2
>TEST
```

This output is shown in Table 8.9. This table is identical to Table 8.7 except that the probabilities have been multiplied by two. Recall that the only two comparisons that we are interested in are between groups 1 and 2 and between groups 2 and

Table 8.8 Post-hoc comparisons of pairwise differences between means in the TEST SCORES data set, using the Bonferroni adjustment for three comparisons.

```
USING MODEL MSE OF        400.589 WITH       34. DF.
MATRIX OF PAIRWISE MEAN DIFFERENCES:

                          1              2              3

              1       0.00000
              2      -1.75667        0.00000
              3      17.15000       18.90667        0.00000


BONFERRONI ADJUSTMENT.
MATRIX OF PAIRWISE COMPARISON PROBABILITIES:

                          1              2              3

              1       1.00000
              2       1.00000        1.00000
              3       0.10127        0.10268        1.00000
```

Table 8.9 Post-hoc comparisons of pairwise differences between means in the TEST SCORES data set, using the Bonferroni adjustment for two comparisons.

```
USING MODEL MSE OF        400.589 WITH       34. DF.
MATRIX OF PAIRWISE MEAN DIFFERENCES:

                          1              2              3
              1       0.00000
              2      -1.75667        0.00000
              3      17.15000       18.90667        0.00000

BONFERRONI ADJUSTMENT.
PROBABILITIES SCALED FOR     2 COMPARISONS.
MATRIX OF PAIRWISE COMPARISON PROBABILITIES:

                          1              2              3
              1       1.00000
              2       1.00000        1.00000
              3       0.06751        0.06845        1.00000
```

3—the probability given for the difference between groups 1 and 3, .06751, should be ignored. If it is of interest, you should use the probabilities adjusted for three tests, as in Table 8.8.

8.5.3 Tukey-Kramer Honestly Significant Differences

The Tukey-Kramer *honestly significant difference* (HSD) method provides another way of adjusting the *p*-values for post-hoc comparisons. This method adjusts the confidence intervals around the mean differences among a set of means so that all of the confidence intervals simultaneously contain their respective population values of the mean differences with an overall probability equal to $1-\alpha$. SYSTAT will report the probabilities associated with each of these mean differences using this approach. The commands are (see section 8.5.1 for menu equivalents):

```
>HYPOTHESIS
>POST GROUP / TUKEY
>TEST
```

The output from this procedure for the example data is shown in Table 8.10. The Tukey method is slightly less conservative than the Bonferroni adjustment, but at the expense of intelligibility. When the sample sizes are unequal, the Scheffé method in the next section is usually preferred to the HSD. Generally, it would be better to pick out a priori only those comparisons that you are really interested in, and then use the Bonferroni procedure to adjust for that number of comparisons. This will be more powerful than using the HSD method and examining all possible pairwise differences.

Table 8.10 Post-hoc comparisons of pairwise differences between means in the TEST SCORES data set, using the Tukey-Kramer honestly significant difference method.

```
USING MODEL MSE OF        400.589 WITH       34. DF.
MATRIX OF PAIRWISE MEAN DIFFERENCES:

                        1              2              3
          1         0.00000
          2        -1.75667        0.00000
          3        17.15000       18.90667        0.00000


TUKEY HSD MULTIPLE COMPARISONS.
MATRIX OF PAIRWISE COMPARISON PROBABILITIES:

                        1              2              3
          1         1.00000
          2         0.97494        1.00000
          3         0.08350        0.08458        1.00000
```

8.5.4 Scheffé

The Scheffé method is probably the best known and most conservative of the post-hoc procedures, and it takes into account the number of tests that could possibly be performed. It is preferred over the Tukey HSD method when the sample sizes are unequal. This method provides a family of confidence intervals for evaluating all possible contrasts, including pairwise differences, that maintain an overall probability equal to $1 - \alpha$. SYSTAT will report the probabilities associated with each of the pairwise differences using the Scheffé approach. The commands are

```
>HYPOTHESIS
>POST GROUP / SCHEFFE
>TEST
```

(see section 8.5.1 for menu equivalents). The output from this procedure for the example data is shown in Table 8.11. The Scheffé method is slightly more conservative than the Tukey HSD method because it adjusts for more than just pairwise comparisons. Unless you are really just snooping in the data for any possible difference or trend, the Scheffé will be inappropriately lacking in power. It would be much better to pick out a priori only those comparisons that you are really interested in, and then use the Bonferroni procedure to adjust for that number of comparisons.

Table 8.11 Post-hoc comparisons of pairwise differences between means in the TEST SCORES data set, using the Scheffé method.

```
USING MODEL MSE OF       400.589 WITH      34. DF.
MATRIX OF PAIRWISE MEAN DIFFERENCES:

                     1              2              3
       1          0.00000
       2         -1.75667        0.00000
       3         17.15000       18.90667        0.00000


SCHEFFE TEST.
MATRIX OF PAIRWISE COMPARISON PROBABILITIES:

                     1              2              3
       1          1.00000
       2          0.97717        1.00000
       3          0.10162        0.10285        1.00000
```

8.5.5 Dunnett's Test

The Dunnett method is used when you are not interested in all possible pairwise differences, but rather are interested in comparing the mean of a control group with all other means. For example, in the TEST SCORES data set suppose that group 1 is a control group, and you are interested in the differences between this group and the means of groups 2 and 3. The commands for Dunnett's method are

```
>HYPOTHESIS
>POST GROUP / DUNNETT CONTROL = 1 TWO
>TEST
```

The "1" after the control statement gives the label of the control group, and the "two" tells SYSTAT to compute a two-tailed test (you can also specify "one" if you have an a priori prediction about the direction of each of the differences between the control group and other groups). The output from this procedure for the example data is shown in Table 8.12. This test is slightly more powerful than using a Bonferroni procedure to adjust for the number of comparisons between the control and other groups (Table 8.8), but, again, the Bonferroni procedure is easier to understand and describe.

8.6 Checking Assumptions

ANOVA makes four assumptions about the distribution of errors within each cell: They are normally distributed, independent, have constant variance, and are unbiased. Each of these assumptions is examined below, and SYSTAT procedures are illustrated for assessing their validity. These assumptions should always be assessed before reporting the results of an analysis of variance.

8.6.1 Normality

The normality assumption states that, within each cell of the design, the errors have a normal distribution. The estimated values of effects and means do not depend on this assumption, but the F and t tests, and corresponding confidence intervals, do. Fortunately, F's and t's are fairly robust to violations of the normality assumption, but serious violations may call for a transformation of some sort to make the errors look more normal. It turns out that in real data, transformations that make the errors more normal also often make them have more constant variance (see section 8.6.3).

MACINTOSH

① Choose the menu item **Stats/MGLH/Post Hoc Test...**
② Select GROUP from the **Effect** list.
③ Turn on **Dunnett**, type "1" into the box, and turn on **Two Tail**.
④ Click OK.

MS-DOS

① Choose the menu item **Statistics/MGLH/MGLM/Hypothesis/Post/Expression/** and select GROUP from the variables list.
② Choose **Statistics/MGLH/MGLM/Hypothesis/Post/Options/Dunnett/Control,** type "1" into the box, and hit <ret>.
③ Choose **Statistics/MGLH/MGLM/Hypothesis/Post/Options/Dunnett/Two,** hit <ret>, and hit <esc>.
④ Choose **Statistics/MGLH/MGLM/Hypothesis/Go (Test)!**

Table 8.12 Post-hoc comparisons of differences between the mean of a control group (group 1 in the TEST SCORES data set) and the means of other groups using Dunnett's method.

```
USING MODEL MSE OF        400.589 WITH     34. DF.
MATRIX OF MEAN DIFFERENCES FROM CONTROL:

           1          0.00000
           2         -1.75667
           3         17.15000

DUNNETT TWO SIDED TEST.
MATRIX OF PAIRWISE COMPARISON PROBABILITIES:

           1          1.00000
           2          0.96818
           3          0.06285
```

8.6.1.1 Stem and Leaf Displays and Box Plots

We do not get to see the real errors in our data, but must instead use the residuals within groups from the ANOVA. These residuals were saved into a file called "testscor.res" when the ANOVA in Table 8.3 was generated. You can get some idea of the distribution of errors within each group by looking at plots of the raw data, such as Figure 8.1 and Figure 8.2. Although the distributions of cases within groups in those plots are not perfectly symmetric, this is expected for small sample sizes. These plots certainly show no evidence for any substantial departure from normality.

When the equal variance assumption (see below) is approximately met and the sample sizes within groups are small, you may get a better picture of the distribution of the residuals by creating stem and leaf displays and box plots that ignore which groups the residuals come from. These plots are produced with the commands

```
>GRAPH
>STEM RESIDUAL
>BOX RESIDUAL
```

The resulting plots are shown in Figure 8.3. Both plots indicate a very slight negative skew, but no outliers. If anything, these plots reveal a slight tendency towards being short-tailed, but not meaningfully so.

```
STEM AND LEAF PLOT OF VARIABLE: RESIDUAL, N = 37

MINIMUM IS:          -47.
LOWER HINGE IS:       -11.
MEDIAN IS:              2.
UPPER HINGE IS:        15.
MAXIMUM IS:            34.

 -4     63
 -3     6
 -2     70
 -1   H 532210
 -0     988821
  0   M 022369
  1   H 00125668
  2     22445
  3     3
```

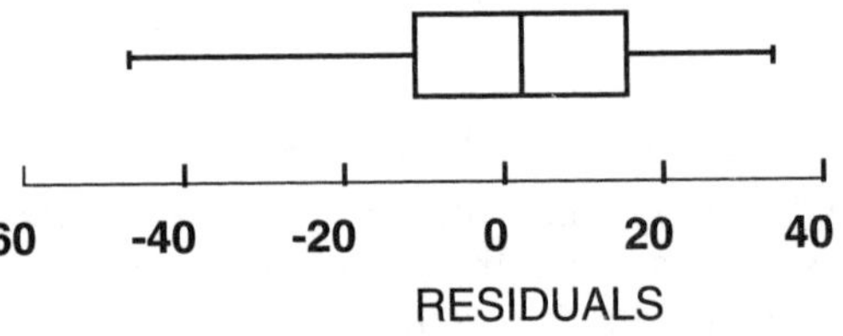

Figure 8.3 Stem and leaf display and box plot of the residuals from an ANOVA on the TEST SCORES data set in Table 8.3.

8.6.1.2 Skewness and Kurtosis

When plots of the data do not clearly reveal which, if any, deviations from normality exist, you can compute the skewness and kurtosis of the residuals directly. Simply type:

```
>STATS
>STATISTICS RESIDUALS/SKEWNESS KURTOSIS N
```

The output is shown in Table 8.13. Skewness is positive when the distribution is positively skewed and negative when the distribution is negatively skewed. With sample sizes greater than about 150, skewness is approximately normally distributed with mean 0 and standard deviation $\sqrt{6/N}$. By dividing the value of skewness by its standard deviation to obtain a Z, you can find an approximate probability of getting a skewness of that magnitude or larger due to chance. For sample sizes less than 100 this gives an inaccurate value of the probability. More accurate values for sample sizes up to 500 can be found in Snedecor and Cochran (1980, Appendix 20i). In the example data in Table 8.13, skewness for the residuals is –0.54, with a standard deviation of $\sqrt{6/37} = 0.40$. Dividing the former by the latter to obtain an approximate value of Z gives –1.35, which is obviously not reliably different from zero.

Kurtosis is positive when a distribution has "longer" tails than a normal distribution. Distributions with negative kurtosis are "flat-topped." For sample sizes greater than 1,000, kurtosis is approximately normally distributed with mean 0 and standard deviation $\sqrt{24/N}$, and can be compared to a Z distribution to find approximate probabilities. For smaller sample sizes probabilities obtained in this way will be somewhat inaccurate, more so for negative than positive values of kurtosis. More accurate values for small sample sizes can be found in Snedecor and Cochran (1980, Appendix 20ii). In the example data in Table 8.13, the value of kurtosis for the residuals is –0.15, with a standard deviation of $\sqrt{24/37} = 0.81$. Dividing kurtosis by its standard deviation gives $Z = -0.19$, again, clearly not different from zero.

Table 8.13 Skewness and kurtosis for the residuals from an ANOVA on the TEST SCORES data set in Table 8.3.

```
TOTAL OBSERVATIONS:         37

                     RESIDUAL

  N OF CASES                    37
  SKEWNESS(G1)             -0.54363
  KURTOSIS(G2)             -0.15149
```

8.6.1.3 Normal Probability Plots

Perhaps the best way to assess the normality assumption is by generating a normal probability plot of the residuals from the ANOVA. To generate the normal probability plot, simply open this file and type the commands

```
>GRAPH
>PPLOT RESIDUAL/NORMAL
```

In this plot, if the residuals are drawn from a normal distribution they should fall on a straight line. In Figure 8.4, no points are systematically off of the line formed by the rest of the data points. (See section 3.2.1.4 for more on normal probability plots.) Figure 8.4 is not straight, but it does not clearly indicate any interpretable deviations.

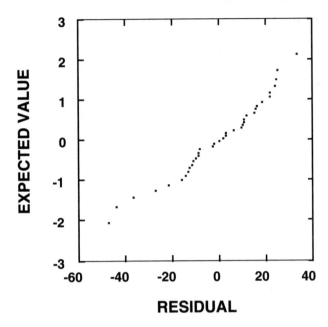

Figure 8.4 Normal probability plot of the residuals from an ANOVA on the TEST SCORES data set in Table 8.3.

To summarize, none of the diagnostics provide much evidence that the residuals are non-normally distributed. The stem and leaf display and box plot, and the values of skewness and kurtosis, suggest that the residuals might be slightly short-tailed and slightly negatively skewed, but none of these deviations are bigger than we would expect by chance. Thus, the normality assumption is adequately met in the TEST SCORES data set (which is comforting, considering that these data were generated from a normal distribution using SYSTAT's random number generator).

8.6.1.4 Remedies

If the residuals are not normally distributed, the p-values from the F tests and t tests in the regression output may be too small. Fortunately, F's and t's are fairly robust to violations of the normality assumption, so moderate departures from normality can often be ignored if the other assumptions are met. In such a case, the reader should be warned that the p-values are not accurate. If the deviation from normality is substantial (and you must rely on your experience or consult an applied textbook for what should count as "substantial"), the most common remedies are (1) to transform the dependent variable, or (2) perform the analysis on the medians rather than the means, which is beyond the scope of this book. Transforming dependent variables proceeds in the same fashion as transforming the outcome variable covered in Chapter 7, using the ladder of transformations.

8.6.2 Equal Variance

The second assumption underlying ANOVA is that the residuals have the same variance within each cell of the design. Equal variance is sometimes called *homoscedasticity*. In real data, the residuals often increase in magnitude as the mean of the cell increases. The reverse can also happen in theory, but is rarely seen in practice. Violations of this assumption can often be seen in a plot of the residuals for each group or cell, and formal tests are also available. *Bartlett's test* is provided automatically by SYSTAT and is discussed below. *Levene's test* is one of the best tests available in terms of power and resistance to Type I error, so it is described in full below.

The variances of the three groups in the example TEST SCORES data set are shown in Table 8.2. They are approximately 421, 209, and 531 for groups 1, 2, and 3, respectively. A rough rule of thumb is not to worry about unequal variance unless the ratio of the largest to the smallest is greater than 2.0 (some authors suggest 5.0 as the criterion). Certainly, if the ratio is less than 2.0, you should not worry about this assumption for any practical purpose.

8.6.2.1 Box Plots

Box plots are very useful for assessing the equal variance assumption. Such a plot was already shown in Figure 8.2. This plot shows that the variance of the second group is smaller than the other two, but not substantially so. Generally, this plot gives a quick way of locating possible deviations from equality by eye, which can then be explored by more formal means.

8.6.2.2 Plotting Residuals Against the Cell Means

Plots of the residuals for each group give much of the information contained in plots of the residuals against the cell means, but the latter is more useful for suggesting possible transformations when the assumption is violated. To plot the residuals against the cell means, open the file containing the residuals (saved during the computation of the ANOVA in Table 8.3) and type

```
>GRAPH
>PLOT RESIDUAL*ESTIMATE
```

The resulting plot for the example data is shown in Figure 8.5. The variance does increase with the magnitude of the cell mean in this plot. However, most of the increase is between the first two groups, and it occurs over a very small change in the magnitude of the cell means

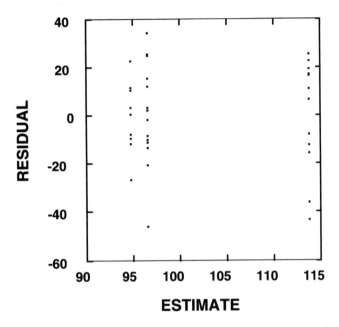

Figure 8.5 Plot of the residuals from the ANOVA on the TEST SCORES data set against the cell means.

8.6.2.3 Bartlett's Test for Homogeneity of Variance

Bartlett's test in Table 8.2 is a significance test of the homogeneity of two or more variances that employs a χ^2 with degrees of freedom equal to the number of groups minus 1. The formula for Bartlett's test is usually computed in steps:

$$M = \Sigma df_i \ln MS_{within} - \Sigma(df_i \ln S_i^2) \qquad (8.11)$$

$$C = 1 + \left[\frac{1}{3(a-1)}\right]\left[\Sigma\left(\frac{1}{df_i}\right) - \frac{1}{\Sigma df_i}\right] \qquad (8.12)$$

where the df_i are the degrees of freedom for each group, MS_{within} is the mean square within from the one-way ANOVA, the s_i^2 are the variances within each group, and a is the number of groups. The test statistic then is

$$\chi^2 = \frac{M}{C} \quad \text{with } a-1 \text{ degrees of freedom} \qquad (8.13)$$

In the example data this test is not significant, $\chi^2 = 1.98$, $df = 2$, $p = .37$. In general, this test is too sensitive, and will indicate that the group variances are different too often, i.e., report p-values that are too small (Snedecor and Cochran, 1980, pp. 252–253). So when this test is not significant, you usually do not have anything to worry about.

8.6.2.4 Levene's Test for Homogeneity of Variance

Levene's test is perhaps the best test for homogeneity of variance. As the variance within a group increases, so do the absolute magnitudes of the residuals within that group. Levene's test compares the average absolute size of the residuals within each group. Traditionally, Levene's test was computed using the deviations around the mean, but the mean is pulled in the direction of extreme values, which attenuates the size of the deviations of those extreme values from the mean. A better method is to use deviations around the medians of each group, which produces a more robust test (Conover et al., 1981).

Levene's test can be performed in four steps. First, find the medians for each group. This was done in Table 8.2 (in older versions of SYSTAT, medians can be found from the stem and leaf display output). The medians of the three groups in Table 8.2 are 94.8, 96.6, and 122.4, respectively. Second, open the original data set and create a new column of medians in which the median for a given group is paired with each score in that group. This can be easily accomplished, for example, with the commands

```
> IF GROUP = 1 THEN LET MEDIAN = 94.8
> IF GROUP = 2 THEN LET MEDIAN = 96.6
> IF GROUP = 3 THEN LET MEDIAN = 122.4
```

Third, compute the absolute deviations of the scores from their medians, using SYSTAT's ABS() function to set a new column called, say, "DEVIAT," equal to raw data minus their group medians. The appropriate command in the example data set is

```
>LET DEVIAT = ABS(SCORE - MEDIAN)
```

Fourth, and finally, compute a one-way ANOVA on the deviations using SYSTAT's usual ANOVA routine (see section 8.3 for menu descriptions):

```
>MGLH
>CATEGORY GROUP
>ANOVA DEVIAT
>PRINT SHORT
>ESTIMATE
```

The ANOVA results for the TEST SCORES data set are shown in Table 8.14. The F test in this ANOVA is Levene's test: If the average deviations from the medians of the groups do not differ significantly, one can conclude that there is not good evidence for heterogeneity of variance. The nonsignificant test in Table 8.14 confirms what Bartlett's test and other considerations above have already suggested: that the variances in the three groups do not differ significantly.

Table 8.14 Levene's test for homogeneity of variance for the TEST SCORES example data set.

```
LEVELS ENCOUNTERED DURING PROCESSING ARE:
GROUP
       1.00000      2.00000      3.00000

DEP VAR: DEVIAT      N: 37   MULTIPLE R: 0.210   SQUARED MULTIPLE R: 0.044

                      ANALYSIS OF VARIANCE

SOURCE         SUM-OF-SQUARES    DF    MEAN-SQUARE     F-RATIO          P

GROUP              263.84577      2      131.92289     0.78071    0.46612
ERROR             5745.22233     34      168.97713
```

8.6.2.5 Remedies

As a rule, if you can't see unequal variance in plots of the residuals Figures 8.1, 8.2, and 8.5), you don't have a serious problem. When unequal variance is detected in a plot, Levene's test should be employed to verify that it is not likely due to chance. For moderate departures from equal variance, F's and t's are fairly robust to violations of this assumption. When the departure from equality is substantial, the most common remedy is to perform a *variance stabilizing transformation* on the dependent variable (Snedecor & Cochran, 1980, pp. 287–292) Transforming variables is covered in Chapter 7.

8.6.3 Independence

The third assumption underlying ANOVA is that the errors are uncorrelated with one another. When the normality assumption is met, uncorrelated errors are also *independent*, so this assumption is often called the independence assumption. When errors are correlated, the estimates of means and effects are still unbiased, but the population variance is underestimated. This produces F's and t's that are too large and confidence intervals that are too small.

In ANOVA, correlated errors are usually impossible to detect in the data. Whether one should worry about correlated errors depends almost entirely on the experimental design. For example, if the subjects in each group are tested individually with no opportunity to influence each other, then their errors are probably uncorrelated. The most common violations of this assumption arise when either (1) subjects (or families, or twins, or plots of land) are measured more than once, that is, they contribute more than one case to the data, or (2) the experimental manipulation is administered to a group of subjects *as a group*, and these subjects within a group influence the effectiveness of the manipulation for each other. Both types of violation can often be dealt with using some type of repeated measures analysis (see Chapters 11 and 12).

8.6.4 Errors Are Unbiased

The fourth assumption underlying ANOVA is that the errors are unbiased, that is, that the expected value of the errors within each cell is zero. For the purposes of almost all ANOVAs, and certainly for the type of analyses covered in this book, this assumption can be considered true by definition. Because the cell means are unbiased estimates of the population values, and the errors are defined as the differences between the obtained scores and the cell means, in the long run the errors do have an expected value of zero.

8.7 Confidence Intervals

SYSTAT does not directly provide confidence intervals in ANOVA, but they easily can be obtained from estimates of means and variances that SYSTAT does provide. In the following sections I illustrate three types of confidence intervals using the TEST SCORES data set. The formula for confidence intervals in general is

$$X \pm t_{(\alpha/2,\, df)} SE_X \qquad (8.14)$$

where X is the thing that you want to place confidence intervals around, SE_X is the standard error of that thing, and df is the degrees of freedom of SE_X. The confidence intervals below differ in the way they estimate SE_X and the values of t that are employed.

8.7.1 Confidence Intervals Using Pooled Variance

The easiest confidence intervals to construct from SYSTAT's ANOVA output are confidence intervals around the group means using the mean squared error—the pooled variance from all the groups—as a common estimate of variance. The appropriate means and their standard errors are shown in Table 8.3 and are reproduced here:

```
LEAST SQUARES MEANS.
                          LS MEAN          SE         N
     GROUP    =  1.00000   96.66667      5.16778      15
     GROUP    =  2.00000   94.91000      6.32921      10
     GROUP    =  3.00000  113.81667      5.77775      12
```

The standard errors in this output are simply the square roots of the mean square error from the ANOVA table divided by the n in each group. For example, the mean square error in Table 8.3 is 400.589, so the estimate of the standard error for group 1 is $\sqrt{400.589/15} = 5.16778$, as shown in the above output. The degrees of freedom for the mean square error is 34, so the critical value of t corresponding to the .025 level (one-tailed) is 2.03. Plugging these values along with the means from the above output gives the following confidence intervals on the means:

group i: $\quad \bar{G}_i \pm t_{(.025, 34)} \, SE_{\bar{G}_i}$ (8.15)

group 1: $\quad 96.66667 \pm 2.03 \times 5.16778$ (8.15a)
$\quad\quad\quad\quad\quad \pm 10.5$

group 2: $\quad 94.91000 \pm 2.03 \times 6.32921$ (8.15b)
$\quad\quad\quad\quad\quad \pm 12.8$

group 3: $\quad 113.81667 \pm 2.03 \times 5.77775$ (8.15c)
$\quad\quad\quad\quad\quad \pm 11.7$

So, for example, the interval 86.2 to 107.2 has about a 95% chance of having captured the population mean for group 1. Note that, when the groups have equal n's, the confidence intervals computed in this manner will all be of the same size.

8.7.2 Confidence Intervals Using Separate Variances

When the assumption of equal variance is met, the mean square error from the ANOVA table gives a better estimate of the sample variance than do the separate variances for each of the individual groups. When you are concerned that the variances are different across groups, however, it is sometimes safer, and more informative to the reader, to compute confidence intervals using the separate variances in each group. The means of each group, along with their standard errors (Std. Error), are shown in Table 8.2, and are reproduced here:

```
Mean         Std. Error
96.6667         5.29903
94.9100         4.56707
113.8167        6.65495
```

These numbers can be entered directly into the formula for confidence intervals, along with the critical values of t associated with each group. Note that these t's should be based on the df within each group, 14, 9, and 11, for groups 1, 2, and 3 in the example design. The corresponding t's are 2.15, 2.26, and 2.20. The confidence intervals, then, are

group i: $\quad \bar{G}_i \pm t_{(.025, df_i)} \, SE_{\bar{G}_i}$ (8.16)

group 1: $\quad 96.66667 \pm 2.15 \times 5.29903$ (8.16a)
$\quad\quad\quad\quad\quad \pm 11.4$

group 2: $94.91000 \pm 2.26 \times 4.56707$ (8.16)
± 10.3

group 3: $113.81667 \pm 2.20 \times 6.65495$ (8.16c)
± 14.6

So, for example, there is about a 95% chance that the interval 85.3 to 108.1 is an interval that captures the population mean for group 1. The confidence intervals for groups 1 and 3 are slightly wider using separate variances than they were using the pooled variance in the previous section. Because of the smaller variance in group 2, however, the confidence interval for group 2 is actually smaller using its own separate variance.

8.7.3 Simultaneous Confidence Intervals

When you construct a 95% confidence interval around a mean, you hope that about 95% of such intervals will contain the true population value of that mean. However, when you construct confidence intervals around more than one mean, there is more than a 5% chance that one of the true population means will fall outside its respective confidence intervals. The situation is analogous to conducting multiple tests and can be remedied in much the same way using a Bonferroni-type adjustment. When finding the value of t, divide the α level by the number of confidence intervals that you wish to construct. For example, to construct three confidence intervals while maintaining an overall probability of .95 that all of those intervals will contain their respective means, divide .05 by 3 to get a new α' level of .017. Look up the upper-tail value of t corresponding to $\alpha'/2 = .0083$. For 34 degrees of freedom this value of t is 2.52. Plugging this value into Equation 8.16, one finds

group 1: $96.66667 \pm 2.52 \times 5.16778$ (8.17a)
± 13.0

group 2: $94.91000 \pm 2.52 \times 6.32921$ (8.17b)
± 15.9

group 3: $113.81667 \pm 2.52 \times 5.77775$ (8.17c)
± 14.6

About 95% of trios of confidence intervals generated in this way will consist of three confidence intervals that simultaneously contain all three of their respective population means.

8.8 Graphing the Results

Data presentation is very important, and SYSTAT offers a flexible graphics program for generating plots of the results of an ANOVA. Normally, you will want to present a plot of the group means that shows some type of error bars around those means. The most common type of error bars are ± 1 standard error and ± the 95% confidence interval (see sections 3.2.2.4 and 3.2.2.5 for more on graphing error bars). These graphs are illustrated below for both nominal independent variables and independent variables on interval or ratio scales.

8.8.1 Nominal Independent Variables

When the independent variable is categorical and does not have a meaningful scale, that is, the ordering of the categories is arbitrary, it is standard to plot the means of the categories using a bar graph. To plot the means along with errors bars showing ±1 standard error of the mean using each group's own variance for the variance estimate (section 8.6.2), type

```
>GRAPH
>BAR SCORE * GROUP/YMIN=80 YMAX=130 SERROR SORT
```

The *ymin* and *ymax* values were chosen so that the error bars fit completely on the graph. The option *SERROR* tells SYSTAT to include error bars based on the standard error of the mean for each group. The option *SORT* plots the data in order of the individual groups. The resulting graph is shown in Figure 8.6.

Often, it is more informative to the reader to show the confidence intervals around the group means, rather than just the standard errors. In general, you can generate any error bars you wish by creating a new column of data that contains the magnitude of the error bars for each group. For example, to put confidence intervals around the means for each group in the TEST SCORES data set, first create a new column of data that contains those confidence limits by typing

```
>IF GROUP=1 THEN LET CL=11.4
>IF GROUP=2 THEN LET CL=10.3
>IF GROUP=3 THEN LET CL=14.6
```

where "CL" is just my label for the column of confidence limits. In general, you can use this procedure to set the error bars to anything you want. These confidence limits are the 95% confidence limits found in section 8.7.1 based on the pooled variance estimate. To put these confidence limits around the means in the bar graph, type

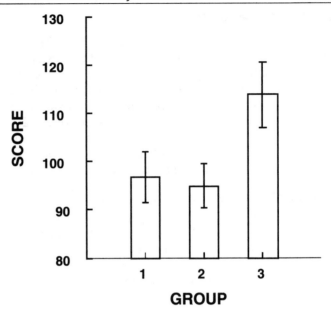

Figure 8.6 Bar graph showing the mean of the three groups in the TEST SCORES data set, along with error bars of ± 1 SE of the mean, where the standard errors are based on each group's separate variance.

```
>BAR SCORE*GROUP/YMIN=80 YMAX=130 ERROR=CL SORT
```

The option ERROR=CL tells SYSTAT to look in the data column CL to find the magnitudes of the errors bars in the graph. The resulting graph is shown in Figure 8.7.

8.8.2 Independent Variables on a Meaningful Scale

8.8.2.1 Equally Spaced Groups

When the independent variable is on a meaningful scale, it is standard to plot the means of the categories using a line graph. The options for choosing error bars are the same as those discussed in sections 3.2.2.5 and 8.8.1 and will not be repeated here. To create a line graph for the means in the TEST SCORES data set showing the 95% confidence intervals around the means, type

```
>GRAPH
>CPLOT SCORE * GROUP/YMIN=80 YMAX=130,
ERROR=CL LINE
```

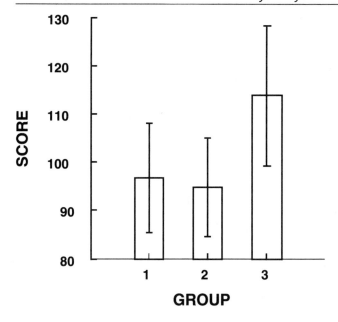

Figure 8.7 Bar graph showing the mean of the three groups in the TEST SCORES data set, along with error bars showing the 95% confidence intervals around each mean, with standard errors based on the pooled variance.

where "CL" is the column in the data set that contains the values of the confidence limits, as discussed in section 8.8.1. The *ymin* and *ymax* values were chosen so that the error bars fit completely on the graph. The option *LINE* tells SYSTAT to draw the line connecting the means of the groups. This graph is shown in Figure 8.8.

8.8.2.2 Unequally Spaced Groups

Sometimes the independent variable is on an interval or ratio scale, which means that the groups may not be equally spaced on the *X*-axis. To change this spacing in the graph, it is necessary to open a new data worksheet and create a new column containing the values of the independent variable and a second column with the means of the groups. If you desire, you can create a third column containing the values of the error bars for each of the means. These columns of data can then be used to generate a graph using SYSTAT's *plot* command.

For example, suppose that the three groups of subjects in the TEST SCORES data set were each administered different quan-

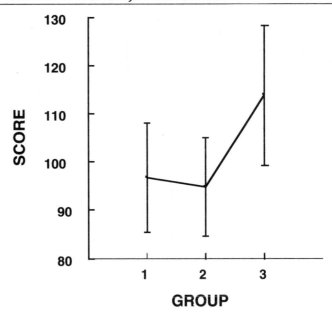

Figure 8.8 Line graph showing the mean of the three groups in the TEST SCORES data set, along with error bars showing the 95% confidence intervals around each mean, with standard errors based on the pooled variance.

tities of adrenalin to assess its effects on test-taking performance. Suppose that the dosages were 100 units, 200 units, and 600 units for groups 1, 2, and 3, respectively. To plot the results of the analysis, create a new file containing the three dosage levels in one column, the three group means in another column, and the magnitudes of the confidence intervals in a third column. The entries in this new file are as follows:

GROUP	SCORE	CL
100	96.7	11.4
200	94.8	10.3
600	113.8	14.6

The column labelled "CL" contains the confidence limits for each group based on the pooled error variance from the ANOVA (see section 8.7.1). To generate the plot of the means and error bars, type

```
>GRAPH
>PLOT  SCORE * GROUP/YMIN=70 YMAX=130,
ERROR=CL SMOOTH=LOWESS
```

The *ymin* and *ymax* options were set to ensure that the error bars would fit completely on the graph. The *ERROR=CL* option tells SYSTAT to use the values in the column CL as the magnitudes of the error bars. Finally, the option *SMOOTH=LOWESS* causes SYSTAT to draw a line connecting the means of the groups. The resulting graph is shown in Figure 8.9. This graph preserves the proper spacing between groups on the X-axis.

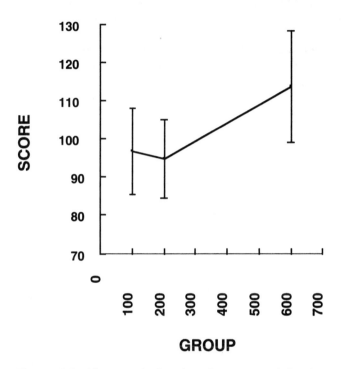

Figure 8.9 Line graph showing the means of the three groups in the TEST SCORES data set spaced according to three hypothetical drug dosage levels, along with error bars showing the 95% confidence intervals around each mean, with standard errors based on the pooled error variance.

8.9 Test Scores Data Set

The data in the TEST SCORES example data set in Table 8.1 were generated in SYSTAT by starting a new data column labelled "GROUP" and then using the following commands:

```
>REPEAT 37
>IF CASE <= 15 THEN LET GROUP=1
>IF CASE >15 AND CASE <= 25 THEN LET GROUP=2
>IF CASE >25 AND CASE <= 37 THEN LET GROUP=3
>IF GROUP=1 THEN LET SCORE=94+20*ZRN
>IF GROUP=2 THEN LET SCORE=100+20*ZRN
>IF GROUP=3 THEN LET SCORE=112+20*ZRN
```

To truncate the decimal places after the first, I set the number of decimal places to one, cut the entire data set, and then pasted it back in. Macintosh users should be able to reconstruct these numbers exactly by following this method. I'm afraid DOS users will have to type in the data directly if they want to use the same numbers. The complete data file is shown in Table 8.15.

Table 8.15 The TEST SCORES data set.

Case	Group	Score
1	1.0	94.4
2	1.0	75.7
3	1.0	88.1
4	1.0	108.7
5	1.0	94.8
6	1.0	130.6
7	1.0	121.1
8	1.0	82.9
9	1.0	112.0
10	1.0	85.2
11	1.0	98.7
12	1.0	50.1
13	1.0	86.1
14	1.0	99.8
15	1.0	121.8
16	2.0	95.3
17	2.0	117.2
18	2.0	97.9
19	2.0	82.7
20	2.0	105.3
21	2.0	85.2
22	2.0	86.7
23	2.0	104.8
24	2.0	67.9
25	2.0	106.1
26	3.0	98.1
27	3.0	101.2
28	3.0	120.1
29	3.0	77.5
30	3.0	124.7
31	3.0	136.1
32	3.0	132.6
33	3.0	130.5
34	3.0	130.0
35	3.0	138.8
36	3.0	105.8
37	3.0	70.4

9
Fixed Factors Factorial ANOVA

This chapter shows how to perform an analysis of variance in SYSTAT with two or more, fully-crossed, *fixed* factors. This is a very straightforward extension of the one-way ANOVA procedures described in the last chapter, so much of the explanatory discussion of these procedures will be found there. As usual, the analysis described in this chapter begins by looking at the raw data to check for unusual cases, computing the ANOVA, computing contrasts and possibly performing post-hoc analyses, computing standard errors and confidence intervals, assessing the assumptions underlying the ANOVA, and graphing the results of the analysis.

9.1 The Example Data Set

The example that will be used in this chapter is a two-way design with three levels of a factor called GROUP, the experimental treatments, and a factor called GENDER with two levels. Each group-by-gender combination has five observations, each taken from different subjects. (In fact, these are just the first ten cases from each group of the TEST SCORES data set shown in Table 8.1, which was generated in SYSTAT using the commands described in section 8.9.) The example data set is shown in Table 9.1. As in Chapter 8, suppose that the scores represent test scores on a school achievement test, and the groups represent three private tutoring methods to which students were randomly assigned.

Table 9.1 The TEST SCORES data set—test scores for students in three tutoring conditions.

	Group 1	Group 2	Group 3
Males	94.4	95.3	98.1
	75.7	117.2	101.2
	88.1	97.9	120.1
	108.7	82.7	77.5
	94.8	105.3	124.7
Females	130.6	85.2	136.1
	121.1	86.7	132.6
	82.9	104.8	130.5
	112.0	67.9	130.0
	85.2	106.1	138.8

To use this data set in SYSTAT, you need to enter it or read it into a data editor window with at least three columns (you may also have descriptive columns, such as subject names, and so on). One column will contain a code for the gender variable, one will contain a code for the group variable, and one will contain the dependent variable—the test scores. You have some options in coding the categorical variables (see section 8.1), but the most general method is to simply code each variable as numerical with the codes 1 through k, where k is the number of categories. So, for example, you might code the gender variable "1" for male and "2" for female, and code the group variable "1" for group 1, and so on. The fully coded data set corresponding to Table 9.1 is shown in Table 9.2.

Finally, it will be useful in some of the procedures below to have a fourth column containing a separate code for each cell, as though the data were from a one-way ANOVA. There are six cells total in the example data set, so the cells are shown coded 1 through 6 in the last column of Table 9.2. I entered these codes by hand, but if the data set were too large for that to be practical, you could enter the codes with the following six commands:

```
>IF GENDER=1 AND GROUP=1 THEN LET CELL=1
>IF GENDER=1 AND GROUP=2 THEN LET CELL=2
>IF GENDER=1 AND GROUP=3 THEN LET CELL=3
>IF GENDER=2 AND GROUP=1 THEN LET CELL=4
>IF GENDER=2 AND GROUP=2 THEN LET CELL=5
>IF GENDER=2 AND GROUP=3 THEN LET CELL=6
```

(See section 2.2.2 for more on conditional coding.) Depending on how you have coded your factors, you may also be able to come up with fewer commands that will do the same job. For example, the following command will create the same codes as the previous six commands:

```
>LET CELL = (GENDER-1)*3 + GROUP
```

Especially when the number of factors becomes large, it is often worth spending a couple of minutes working out a coding scheme such as this, rather than typing large numbers of *let* commands.

Table 9.2 The TEST SCORES data set from Table 9.1 coded into SYSTAT.

	GENDER	GROUP	SCORE	CELL
1	1.000	1.000	94.400	1
2	1.000	1.000	75.700	1
3	1.000	1.000	88.100	1
4	1.000	1.000	108.700	1
5	1.000	1.000	94.800	1
6	1.000	2.000	95.300	2
7	1.000	2.000	117.200	2
8	1.000	2.000	97.900	2
9	1.000	2.000	82.700	2
10	1.000	2.000	105.300	2
11	1.000	3.000	98.100	3
12	1.000	3.000	101.200	3
13	1.000	3.000	120.100	3
14	1.000	3.000	77.500	3
15	1.000	3.000	124.700	3
16	2.000	1.000	130.600	4
17	2.000	1.000	121.100	4
18	2.000	1.000	82.900	4
19	2.000	1.000	112.000	4
20	2.000	1.000	85.200	4
21	2.000	2.000	85.200	5
22	2.000	2.000	86.700	5
23	2.000	2.000	104.800	5
24	2.000	2.000	67.900	5
25	2.000	2.000	106.100	5
26	2.000	3.000	136.100	6
27	2.000	3.000	132.600	6
28	2.000	3.000	130.500	6
29	2.000	3.000	130.000	6
30	2.000	3.000	138.800	6

9.2 Looking at the Data

9.2.1 Scatterplots

In factorial designs, you should examine the raw data to check for observations that are extreme relative to their own cells, not necessarily to the data set as a whole. Thus, for categorical data, it is usually best to plot the data by each cell, ignoring the factorial structure of the data. To create a scatterplot of the raw data by cell, you can use the column of cell labels shown in the last column of Table 9.2:

```
>GRAPH
>PLOT SCORE*CELL/SMOOTH=LOWESS
```

(Note that this plot requires a numerical variable for the categories; see section 8.1.) The option *smooth=lowess* draws a line connecting the means of the groups in this plot. The resulting scatterplot is shown in Figure 9.1. Extreme cases in this plot should be examined individually to ensure that they were recorded properly and do not differ from the rest of the data in some other exceptional way. None of the data points in this plot appear to stand apart from other cases in their cells. However, the last cell appears to have much smaller variance than the rest; we will return to this in section 9.6.2.

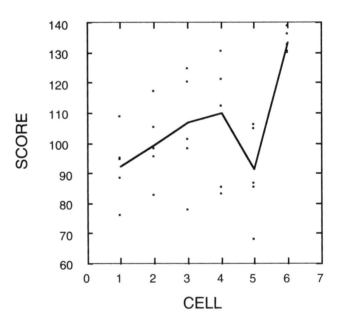

Figure 9.1 Scatterplot of the raw data in the TEST SCORES data set.

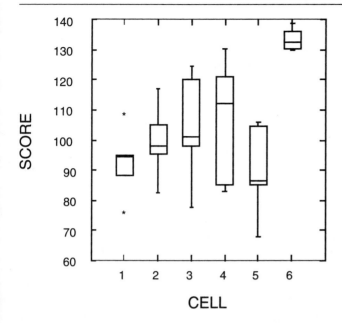

Figure 9.2 Box plots of the scores for each group in the TEST SCORES data set.

9.2.2 Box Plots

A second type of plot that is very useful in this context is a box plot of the scores by cells. This plot is shown in Figure 9.2. The plot is generated by the command

```
>BOX SCORE*CELL/SORT
```

The box plot for the first cell suggests that it has two outliers, but you should not take this seriously due to the small number of data points (five) in each plot. When the number of data points in the box plot is small, the main use of the plot is to give you a rough idea of the spread of the data. If there were any extreme outliers, you should examine those cases individually.

9.3 Factorial ANOVA

You can perform ANOVAs using SYSTAT's usual *model* command, but a simpler method that also gives output more appropriate to analysis of variance is, unsurprisingly, the *anova* command. Thus, for a fully crossed factorial design, the commands are

MACINTOSH

① Choose the menu item **Stats/MGLH/Fully Factorial (M)ANOVA**.

② Select SCORE from the **Dependent Variable(s)** list.

③ Select GENDER and GROUP from the **Factor(s)** list.

④ Click on **More...**

⑤ Turn on **Means & Std. Errors, Extended Output,** and **Save Residuals**.

⑥ Click **OK**.

⑦ Type "twoway.res" into the **Save residuals as** box and click **OK**.

MS-DOS

① Choose the menu item **Statistics/MGLH/ANOVA/ ANOVA/Variables/** and select SCORE from the variables list.

② Choose the menu item **Statistics/MGLH/ANOVA/ Category/** and select GENDER and GROUP from the variables list.

③ Choose the menu item **Statistics/MGLH/ANOVA/ Estimate/Save/**, hit <esc>, and type the save filename, e.g., "twoway.res" into the box, hit <ret>, and select **Residuals**.

④ Choose the menu item **Statistics/MGLH/ANOVA/ Estimate/Go!**

```
>MGLH
>CATEGORY GENDER GROUP
>ANOVA SCORE
>PRINT LONG
>SAVE    TWOWAY.RES / MODEL
>ESTIMATE
```

The results, using extended output (*print=long*) are shown in Table 9.3.

For balanced designs, such as the present example, the *estimates of effects* in Table 9.3 are simply unweighted estimates of (1) the grand mean (constant), (2) the differences between the marginal means and the grand mean (for the main effects), and (3) the residuals corresponding to the interaction within the cells. More generally, these are the regression coefficients from a regression of the dependent variable on the effects coded independent variables. These effects can be used to reconstruct an entire table of cell and marginal effects, sometimes called a *mean polish table* (Rosenthal & Rosnow, 1991, pp. 370–372). Although SYSTAT does not give all of the values for such a table, because the rows, columns, and interactions in mean polish tables always add to zero, you can easily compute the values that SYSTAT does not report. For example, the effect for males (gender=1) in Table 9.3 is –5.627; therefore, the effect for females (gender=2) must be +5.627. Similarly, the effects for the first two groups are –5.057 and –9.497, so the effect for the third group must be 0–(–5.097–9.457) = +14.554.

For the interaction effects, the rows and columns also add to zero. So the effect for females in group 1 must be zero minus the effect for males in group 1: 0–(–1.383) = +1.383. Similarly, the effect for females in group 2 must be zero minus the effect for males in group 2: 0–(10.397) = –10.397. The effect for males in group 3 can be found by subtracting the effects for males in groups 1 and 2 from zero: 0–(–1.383+10.397) = –9.014. Finally, the effect for females in group 3 can be found by subtracting this value from zero: 0–(–9.014) = +9.014.

The completed mean polish table is shown in Table 9.4. The values in italics in this table are the values given in the SYSTAT output in Table 9.3. The remainder of the values were computed in the previous two paragraphs. The values in this table are particularly important for graphing the interaction effects, which is discussed in section 9.8.3.

At the bottom of the output in Table 9.3 are the least squares means and standard errors (LS MEAN and SE, respectively) for each row, column, and cell of the design. With balanced designs, the least squares means are simply the means of each

Table 9.3 Two-way ANOVA on the TEST SCORES data set using the standard ANOVA procedures and showing extended results.

```
LEVELS ENCOUNTERED DURING PROCESSING ARE:
GENDER        1.000        2.000
GROUP         1.000        2.000        3.000

DEP VAR: SCORE     N: 30   MULTIPLE R: 0.725   SQUARED MULTIPLE R: 0.525
                             -1
ESTIMATES OF EFFECTS  B = (X'X)   X'Y
                                  SCORE

   CONSTANT                       104.407

      GENDER         1.000         -5.627

       GROUP         1.000         -5.057
       GROUP         2.000         -9.497

      GENDER         1.000
       GROUP         1.000         -1.383

      GENDER         1.000
       GROUP         2.000         10.397

                    ANALYSIS OF VARIANCE
SOURCE          SUM-OF-SQUARES   DF   MEAN-SQUARE    F-RATIO      P

GENDER              949.781       1      949.781      4.110     0.054
GROUP              3275.561       2     1637.780      7.087     0.004
GENDER*GROUP       1912.445       2      956.222      4.138     0.029

ERROR              5546.232      24      231.093

LEAST SQUARES MEANS.
                              LS MEAN         SE         N
     GENDER    =    1.000      98.780       3.925       15

     GENDER    =    2.000     110.033       3.925       15

      GROUP    =    1.000      99.350       4.807       10

      GROUP    =    2.000      94.910       4.807       10

      GROUP    =    3.000     118.960       4.807       10

     GENDER    =    1.000
      GROUP    =    1.000      92.340       6.798        5

     GENDER    =    1.000
      GROUP    =    2.000      99.680       6.798        5

     GENDER    =    1.000
      GROUP    =    3.000     104.320       6.798        5

     GENDER    =    2.000
      GROUP    =    1.000     106.360       6.798        5

     GENDER    =    2.000
      GROUP    =    2.000      90.140       6.798        5

     GENDER    =    2.000
      GROUP    =    3.000     133.600       6.798        5

DURBIN-WATSON D STATISTIC       2.548
FIRST ORDER AUTOCORRELATION     0.277

RESIDUALS HAVE BEEN SAVED
```

Table 9.4 Mean polish table derived from the ANOVA output in Table 9.3. The values in italics were given by SYSTAT, the rest were derived by hand (see text).

	Group 1	Group 2	Group 3	Row Effects
Gender 1	*−1.383*	*10.397*	*−9.014*	*−5.627*
Gender 2	+1.383	−10.397	+9.014	+5.627
Col. Effects	*−5.057*	*−9.497*	*+14.554*	GM= *104.407*

group. The standard errors of those means use the pooled variance from all groups: MEAN-SQUARE$_{ERROR}$ from the ANOVA table divided by the number of cases in each group.

The *Durbin-Watson Statistic* and *first-order autocorrelation* are almost never relevant in ANOVA (see section 5.4.3.3 for discussion).

9.4 Contrasts

The *F* test for the gender effect in Table 9.3 is not quite significant at the traditional 0.05 level of significance. Because this *F* test has only onedegree of freedom in the numerator, it is a contrast and no further analysis is necessary unless you wish to test simple effects of gender within each group (see section 9.4.1.2).

On the other hand, the *F* test of group is significant at the .05 level, which means there exist some differences among the groups, but we don't know where. *F* tests with more than onedegree of freedom in the numerator, like the group and gender*group tests in Table 9.3, are of little scientific interest because (1) if they are significant they only tell us that some differences exist, not which means are different, and (2) if they are not significant, this does not mean that none of the means are different or that there are no significant trends among the means. Experiments are almost always conducted with specific predictions in mind, and these predictions should always be tested directly.

9.4.1 Planned Comparisons

9.4.1.1 Main Effects

SYSTAT offers a couple of methods for testing specific hypotheses, or *contrasts*. Suppose that your theory specifies that group 3 in the TEST SCORES data set should be higher than the means for group 1 and group 2, which should be approximately equal. This hypothesis can be expressed using the contrast weights

WARNING: CHECK THE DIRECTION OF YOUR CONTRASTS!

Because SYSTAT always computes nondirectional F tests for contrasts, it is impossible to tell whether a given contrast is in the direction you predicted or in exactly the opposite direction from the F test alone. Both would be equally significant! However, if you specified your contrast weights properly, you can make this determination by looking at the "null hypothesis contrast" line in the extended output: If this value is negative, then the contrast came out in the direction opposite the prediction made by your contrast weights. It is also important to examine plots of the effects that your contrast is testing (see section 9.8) as a double-check to ensure that you found what you think you found with your contrast.

$-1, -1$, and $+2$, respectively, for the means of the three groups (see section 8.4.1 for the derivation of these weights). These contrast weights can be entered into SYSTAT directly to test the desired hypothesis. To test this hypothesis, first compute the two-way ANOVA as was done in Table 9.3, and then type the commands

```
>HYP
>EFFECT = GROUP
>CONTRAST
>-1 -1 2
>TEST
```

(See section 8.4.1 for menu descriptions.) The output from this procedure is shown in Table 9.5. The value of the contrast C is given in the output as "null hypothesis contrast AB" ("AB" refers to two of the matrices involved in computing such a contrast, but you don't really need to know this). This value, 43.66, is equal to $\Sigma \lambda_j \bar{X}_j$, where the λ_j are the contrast weights and the $\bar{X}_j$ are the means of the three groups. For equal cell sizes, the sum of squares (and hence the mean squares) for the contrast is just

$$SS_C = \frac{nC^2}{\Sigma \lambda_j^2} = \frac{10 \times 43.66^2}{(-1)^2 + (-1)^2 + 2^2} = 3176.993 \qquad (9.1)$$

where n is the number of cases entering into each of the means being tested by the contrast. This value is shown in the output in Table 9.5.

An equivalent method for testing contrasts is offered by SYSTAT that allows you to compare combinations of means directly. (Leave out any means that are multiplied by zero.) For example, to test the same contrast as in Table 9.5, the commands are

```
>HYP
>SPECIFY / POOLED
>GROUP[1] + GROUP[2] - 2*GROUP[3] = 0
>TEST
```

(See section 8.4.1 for menu descriptions.) The numbers in the output are identical to those in Table 9.5. Which procedure you use is simply a matter of personal preference. However, the latter procedure has a bit more generality and is used in the next two sections.

The hypothesis in Table 9.5 is statistically significant. Note that SYSTAT always reports "nondirectional," or "two-tailed," p-values. However, the present hypothesis was very specifically directional, so to find the directional p-value you can divide the given p-value by two. This gives $p = .0005$ from Table 9.5, which is highly significant.

Table 9.5 Test of the contrast predicting that the mean of group 3 would be higher than the means of groups 1 and 2 in the TEST SCORES data set.

```
TEST FOR EFFECT CALLED:      GROUP

A MATRIX
   1            2            3            4            5            6
   0.000        0.000       -3.000       -3.000        0.000        0.000

NULL HYPOTHESIS CONTRAST AB
                43.660

                            -1
INVERSE CONTRAST A(X'X)  A'
                 0.600

TEST OF HYPOTHESIS
       SOURCE          SS           DF           MS              F            P

   HYPOTHESIS       3176.993         1        3176.993        13.748        0.001
        ERROR       5546.232        24         231.093
```

To find the effect size r for the contrast, plug the F and df values from Table 9.5 into the following formula:

$$r = \sqrt{\frac{F}{F + df_{\text{denom}}}} = \sqrt{\frac{13.748}{13.748 + 24}} = 0.60 \qquad (9.2)$$

There remains one loose end that must be tied up. We have determined that our prediction was highly significant, but a reader of this result would not know whether there was significant deviation from our prediction among the means; that is, our prediction may account for a significant amount of the variance among the means, but this does not mean that there is no significant variance remaining. The size of the deviations from the predicted values is indexed by the sum of squares remaining between groups after the sum of squares for the prediction has been removed. Therefore, from Table 9.3 and Table 9.5:

$$SS_{\text{residual between}} = SS_{\text{between}} - SS_{\text{contrast}} \qquad (9.3)$$

$$= 3275.561 - 3176.993$$

$$= 98.568$$

This sum of squares is based on 1 df, so an F test of its significance is simply the sum of squares for the residual between, divided by the mean square within from Table 9.3. This $F_{(1,24)}$ is near zero, so we can conclude that the deviation from our prediction in the data is not significant. In fact, our hypothesis accounted for most of the variance between groups:

$$\% \ SS_{\text{between}} \text{ accounted for} = \frac{SS_{\text{contrast}}}{SS_{\text{between}}}$$

$$= \frac{3176.993}{3275.561} \qquad (9.4)$$

$$= .97$$

MACINTOSH

After you have computed the overall ANOVA:

① Choose the menu item **Stats/MGLH/User Defined Contrasts...**

② Type "gender[1]group[1] + gender[1]group[2] − 2*gender[1]group[3] = 0" into the **Contrast** box.

③ Click **OK**.

MS-DOS

After you have computed the overall ANOVA:

① Choose the menu item **Statistics/MGLH/MGLM/ Hypothesis/Specify/ Equation/**, type "gender[1]group[1] + gender[1]group[2] − 2*gender[1]group[3] = 0" into the box and hit <esc>.

② Choose the menu item **Statistics/MGLH/MGLM/ Hypothesis/Go (Test)!**

9.4.1.2 Simple Effects

Sometimes you may want to ask a specific question about an effect within a single level of one factor. These are called simple effects. For example, you might ask whether the contrast computed above, that group 3 was equal to the average of groups 1 and 2, is significant for males or females separately. You can test this effect, and any contrast among cell means in the example data set, by designating the specific cell means in user-defined contrasts using the notation GENDER[i]GROUP[j]. So, to test the contrast from above on males only, type

```
>HYP
>SPECIFY / POOLED
>GENDER[1]GROUP[1] + GENDER[1]GROUP[2],
- 2*GENDER[1]GROUP[3] = 0
>TEST
```

The output is shown in the top panel of Table 9.6. You can test the same effect for females only by substituting gender[2] in place of all of the occurrences of gender[1] in the command line. The output for females is shown in the bottom panel of Table 9.6. As you can see, only the simple effect for females is significant, and the effect size for females is larger than that for males:

$$\text{males:} \quad r = \sqrt{\frac{F}{F + df_{\text{denom}}}} = \sqrt{\frac{0.996}{0.996 + 24}} = 0.20 \quad (9.5)$$

$$\text{females:} \quad r = \sqrt{\frac{F}{F + df_{\text{denom}}}} = \sqrt{\frac{18.025}{18.025 + 24}} = 0.65 \quad (9.6)$$

However, we cannot tell from this whether the effect for males is significantly smaller than the effect for females. The question of whether two simple effects differ is a question about the interaction and must be tested directly. This is shown in the next section.

Both of these contrasts used the pooled error term from the ANOVA in Table 9.3. If you want to use the local error term for each simple effect you can do so; see section 9.4.3.2.

9.4.1.3 Interaction Effects

As with main effects, tests of interaction effects with more than 1 df in the numerator do not test any specific hypothesis. They

Table 9.6 Test of the simple contrast effects for males and females separately, predicting that the mean of group 3 would be higher than the means of groups 1 and 2 in the TEST SCORES data set.

Simple Effect at Gender 1 (males)

```
A MATRIX
     1           2           3           4           5           6
   0.000       0.000      -3.000      -3.000      -3.000      -3.000

NULL HYPOTHESIS VALUE FOR D
                 0.000

NULL HYPOTHESIS CONTRAST AB-D
                16.620

                     -1
INVERSE CONTRAST A(X'X)   A'
                 1.200

TEST OF HYPOTHESIS
     SOURCE        SS          DF         MS          F           P

   HYPOTHESIS    230.187        1       230.187     0.996       0.328
     ERROR      5546.232       24       231.093
```

Simple Effect at Gender 2 (females)

```
A MATRIX
     1           2           3           4           5           6
   0.000       0.000      -3.000      -3.000       3.000       3.000

NULL HYPOTHESIS VALUE FOR D
                 0.000

NULL HYPOTHESIS CONTRAST AB-D
                70.700

                     -1
INVERSE CONTRAST A(X'X)   A'
                 1.200

TEST OF HYPOTHESIS
     SOURCE        SS          DF         MS          F           P

   HYPOTHESIS   4165.408        1      4165.408    18.025     .282826E-03
     ERROR      5546.232       24       231.093
```

tell you that there are some differences among the interaction effects, but do not tell you where. Fortunately, as with main effects, you can test contrasts on interactions. Suppose, for example, you were interested in determining whether the contrast tested in section 9.4.1.1 was larger, or "more true," for women than for men; that is, is the contrast on the simple effect significantly greater for females than for males? SYSTAT will not allow you to compute a contrast on the interaction directly. However, when the cell sizes are all equal and the contrast weights sum to zero in both the rows and the columns, as they do for this contrast, a contrast on the means is identical to a contrast on the interaction (see section 9.9 for treatment of unequal cell sizes). Thus, you can compute the contrast on the interaction by typing the commands

```
>HYP
>SPECIFY / POOLED
>GENDER[1]GROUP[1] + GENDER[1]GROUP[2],
-2*GENDER[1]GROUP[3] = GENDER[2]GROUP[1],
+ GENDER[2]GROUP[2] -2*GENDER[2]GROUP[3]
>TEST
```

(See section 9.4.1.2 for menu descriptions.) The contrast output is shown in Table 9.7. The contrast is significant, which means that the contrasts on the simple effects for males and females computed in the last section are significantly different. The effect size r for this interaction contrast can be computed in the usual manner:

$$r = \sqrt{\frac{F}{F + df_{\text{denom}}}} = \sqrt{\frac{5.273}{5.273 + 24}} = 0.42 \qquad (9.7)$$

> **HINT: VARIABLE LABELS AND CONTRASTS**
>
> If you are planning to perform a number of user-defined contrasts, you will save yourself a lot of typing by giving your variables very short labels. For example, if I had used the letter "G" for group and the letter "S" (sex) for gender, I could have saved typing 54 letters in the command that generated the contrast in Table 9.7.

This contrast is purely on the interaction: You could have gotten the same result by using the interaction effects in Table 9.4 rather than the cell means in Table 9.3 (which is what

Table 9.7 Test of the contrast predicting that the mean of group 3 would be higher than the means of group 1 and group 2 for females, and the opposite would be true for males.

```
A MATRIX
         1          2          3          4          5          6
         0.000      0.000      0.000      0.000     -6.000     -6.000

NULL HYPOTHESIS VALUE FOR D
                    0.000

NULL HYPOTHESIS CONTRAST AB-D
                  -54.080

                      -1
INVERSE CONTRAST A(X'X)   A'
                    2.400

TEST OF HYPOTHESIS
     SOURCE        SS         DF         MS           F          P

   HYPOTHESIS    1218.603      1       1218.603     5.273      0.031
   ERROR         5546.232     24        231.093
```

SYSTAT does). Some people don't believe this, so here goes. To find the contrast weights, you take the weights used above, $-1, -1,$ and $+2$ and multiply them by -1 for the males and $+1$ for the females. This gives the six contrast weights shown in Table 9.8. Then you multiply these weights by the interaction effects using the formulas in Equation 9.1:

$$SS_C = \frac{5[(1)(-1.383)+(1)(10.397)+(-2)(-9.014)+(-1)(1.383)+(-1)(-10.397)+(2)(9.014)]^2}{1^2+1^2+(-2)^2+(-1)^2+(-1)^2+2^2}$$

$$= 1218.783$$

which is within rounding error of the value shown in Table 9.7. If you have unequal cell sizes and you really want your contrast to be on estimates of the interaction alone, you must compute the contrast on the interaction effects by hand as done here, but using the formula for sums of squares given in Equation 9.11 in section 9.9.

Table 9.8 Contrast weights predicting that the mean of group 3 would be higher than the means of groups 1 and 2 for females, and the opposite would be true for males.

		Group 1	Group 2	Group 3
	λ	−1	−1	+2
Males		−1	+1	+1	−2
Females		+1	−1	−1	+2

9.4.2 Polynomials

Sometimes the specific hypotheses that you want to test may be sets of orthogonal polynomials. For example, suppose that for theoretical reasons you predicted that there would be a linearly increasing trend in the means across groups 1, 2, and 3 in the example data set. You could test this using the procedures described in the previous section using the contrast weights −1, 0, and +1 for the means of groups 1, 2, and 3, respectively. Similarly, a test of the nonlinear (quadratic) trend across groups would use the contrast weights −1, 2, and −1 for the means of groups 1, 2, and 3, respectively. SYSTAT gives an easier method for computing such trends. Simply type

```
>HYPOTHESIS
>EFFECT = GROUP
>CONTRAST /  POLYNOMIAL ORDER = 1
>TEST
```

for the linear trend, and type

```
>HYPOTHESIS
>EFFECT = GROUP
>CONTRAST /  POLYNOMIAL ORDER = 2
>TEST
```

for the quadratic trend, and so on. (See section 8.4.2 for menu descriptions.) The outputs from these tests are shown in Table 9.9.

The linear trend is significant, $p = .008$, and the quadratic is significant, $p = .023$, indicating that the differences among the means of the three groups contain both linear and quadratic components (assuming the groups are equally spaced on a meaningful scale; see section 8.8.2). The effect sizes for these trends are found in the same manner as those in Equation 9.7:

Table 9.9 Tests of the contrasts predicting linear and quadratic trends across the three groups in the TEST SCORES example.

Linear Trend

A MATRIX

1	2	3	4	5	6
0.000	0.000	-1.414	-0.707	0.000	0.000

NULL HYPOTHESIS CONTRAST AB
   13.866

INVERSE CONTRAST $A(X'X)^{-1}A'$
   0.100

TEST OF HYPOTHESIS

SOURCE	SS	DF	MS	F	P
HYPOTHESIS	1922.761	1	1922.761	8.320	0.008
ERROR	5546.232	24	231.093		

Quadratic Trend

A MATRIX

1	2	3	4	5	6
0.000	0.000	-.111022E-15	-1.225	0.000	0.000

NULL HYPOTHESIS CONTRAST AB
   11.631

INVERSE CONTRAST $A(X'X)^{-1}A'$
   0.100

TEST OF HYPOTHESIS

SOURCE	SS	DF	MS	F	P
HYPOTHESIS	1352.800	1	1352.800	5.854	0.023
ERROR	5546.232	24	231.093		

$$r = \sqrt{\frac{F}{F + df_{\text{denom}}}} = \sqrt{\frac{8.320}{8.320 + 24}} = 0.51 \qquad (9.8)$$

$$r = \sqrt{\frac{F}{F + df_{\text{denom}}}} = \sqrt{\frac{5.854}{5.854 + 24}} = 0.44 \qquad (9.9)$$

See section 8.4.2 for a discussion of the calculation and interpretation of effect sizes for contrasts that use weights of zero, as the linear contrast does.

9.4.3 Error Terms

9.4.3.1 Pooled Error Terms

The default error term for contrasts in SYSTAT is the *pooled*, or "global," error term. This error term pools together the variance from all of the cells in the analysis and is equal to the mean square error from the overall ANOVA (for example, in Table 9.3). The validity of pooling together sources of variance from multiple sources, however, is based on one of the assumptions underlying ANOVA—the assumption that these variances all represent estimates of the same variance σ^2. When this assumption appears to be substantially violated (see section 9.6.2), you should consider either transforming the data or using the *local* error estimate or *separate* error estimates described in the next two sections.

9.4.3.2 Local Error Terms

Often a contrast is tested that involves only a subset of the means in the design, that is, uses some contrast weights of zero. In such cases it is sometimes wise to use as the estimate of error variance only the pooled variance from the cell actually used in the contrast. This error term is "local" to the particular contrast being tested, and hence is sometimes called the *local error term*. Normally, you would want to use the variance pooled from all cells of the experiment because this is based on more data points and is therefore a better estimate of the true population variance, when the assumption of equal variances is met (see section 9.6.2). However, when the variances of different cells are not equal, it is often safer to use the local error term.

For example, in section 9.4.1.2 we computed the contrast on the simple effect for males separately, but we used the error term from both males and females. In Figure 9.1 you can see that the variance for the females in group 3 is smaller in magnitude than the rest of the groups. In such a case you might want to test the contrast for males using only the pooled error variance from the male groups. To find the variance for males only, you can either pool the variances yourself, or you can compute the ANOVA for males only using statistics by groups (see sections 3.1.2 and 8.3). This latter method is probably easier in the present context:

```
>SELECT GENDER=1
>BY GROUPS
>STATS
>STATISTICS SCORE
```

The ANOVA from this output is shown in Table 9.10. The mean squares within groups in this table is the pooled error estimate for males only, the local error term for the contrast on the simple effect of males. You can now proceed to test this contrast by hand, which is easy: $F_{(1,12)} = 230.187/221.176 = 1.041$, or you can ask SYSTAT to compute the contrast with an error term that you specify. To do this, you simply add the *serror* command, as shown here:

```
>HYPOTHESIS
>SPECIFY / POOLED
>GENDER[1]GROUP[1] + GENDER[1]GROUP[2],
- 2*GENDER[1]GROUP[3] =0
>SERROR = 221.176 (12)
[or just "ERROR=" in DOS]
>TEST
```

This contrast is shown in Table 9.11. The effect size of this contrast is somewhat larger than it was when we used the global error term,

$$r = \sqrt{\frac{F}{F + df_{denom}}} = \sqrt{\frac{1.041}{1.041 + 12}} = 0.28 \qquad (9.10)$$

but it is not reliably different from zero.

MACINTOSH

After you have computed the overall ANOVA:

① Choose the menu item **Stats/MGLH/User Defined Contrasts...**

② Type "gender[1]group[1] + gender[1]group[2] - 2*gender[1]group[3] = 0" into the **Contrast** box.

③ Click on the **MSE** button, type "221.176" into the MSE box, and "12" into the **df** box.

④ Click **OK**.

MS-DOS

After you have computed the overall ANOVA:

① Choose the menu item **Statistics/MGLH/MGLM/ Hypothesis/Specify/ Equation/**, type "gender[1]group[1] + gender[1]group[2] - 2*gender[1]group[3] = 0" into the box and hit <esc>.

② Choose the menu item **Statistics/MGLH/MGLM/ Hypothesis/Error**, hit <esc>, and type "221.176 (12)" into the box.

③ Choose the menu item **Statistics/MGLH/MGLM/ Hypothesis/Go (Test)**!

Table 9.10 One-way ANOVA on males only in the TEST SCORES data set.

```
                          ANALYSIS OF VARIANCE

SOURCE            SUM OF SQUARES   DF   MEAN SQUARE      F        PROBABILITY

BETWEEN GROUPS         364.876      2      182.438      0.825        0.462
WITHIN GROUPS         2654.108     12      221.176
```

Table 9.11 Contrast on males only using the local pooled error term for males only.

```
A MATRIX
      1            2            3            4            5            6
  0.000        0.000       -3.000       -3.000       -3.000       -3.000

NULL HYPOTHESIS VALUE FOR D
         0.000

TEST OF HYPOTHESIS
    SOURCE         SS          DF         MS           F            P

 HYPOTHESIS     230.187         1       230.187      1.041        0.328
      ERROR    2654.112        12       221.176
```

9.4.3.3 Adjusted Error Terms Based on Separate Variances

Often you want to compare groups or use contrasts on means from groups with unequal variances. For example, females in group 3 have very different variance from any of the other groups (see Figure 9.1), so when testing a contrast using females in group 3 you might worry about the validity of the equal variance assumption (see section 9.6.2). If the deviation from equality is substantial (in the example data set the differences are arguably not large enough to worry about), you should consider either transforming the data or using an error term that adjusts for this inequality using Satterthwaite's (1946) adjustment (see section 8.4.3.3). For example, suppose you were interested in testing whether the mean of females in group 3 is larger than the mean of females in group 1, using an adjusted error term. You could compute this contrast by typing

```
>HYPOTHESIS
>SPECIFY / SEPARATE
>GENDER[2]GROUP[3] - GENDER[2]GROUP[1] = 0
>TEST
```

(See section 8.4.3.3 for menu equivalents.) The output for this test is shown in Table 9.12. The hypothesis is significant at the traditional .05 level even though the error term has a very small number of degrees of freedom due to the variance adjustment. Error term adjustments such as this decrease the power of your tests and should generally only be used when the variances are quite different (see section 9.6.2) and when transformations of the dependent variable are either not desired or will not correct the problem.

Table 9.12 Contrast comparing females in groups 1 and 3 in the TEST SCORES data set using the separate variances estimate for the error term.

```
USING SEPARATE VARIANCES ESTIMATE FOR ERROR TERM.

HYPOTHESIS.

A MATRIX
         1          2          3          4          5          6
     0.000      0.000      2.000      1.000     -2.000     -1.000

NULL HYPOTHESIS VALUE FOR D
                            0.000

TEST OF HYPOTHESIS
SOURCE          SS         DF         MS           F          P

HYPOTH      1855.044        1       1855.044      7.845       0.046
ERROR       1004.382        4.248    236.454
```

9.5 Post-Hoc Tests

Frequently, investigators want to test hypotheses that are suggested by patterns in the data. Such tests are usually called *post-hoc*, or *unplanned*, comparisons. They are perfectly all right; in fact, one would be remiss in allowing unexpected effects to go unreported. However, you should be aware that the probability of finding significant results due to chance increases with the number of tests performed. SYSTAT provides a number of standard methods for assessing the significance of all pairwise comparisons among a set of means. These methods were discussed in section 8.5, and that discussion will not be repeated here.

9.5.1 Marginal Means

You can compute post-hoc tests on marginal means, the main effects, exactly as for a one-way ANOVA in section 8.5. For example, if you wanted to test all pairwise differences among the three groups using Tukey's HSD test (see section 8.5.3), you would type

```
>HYP
>POST GROUP / TUKEY
>TEST
```

The output is shown in Table 9.13. The top part of this output shows the labels of the three groups. This is followed by a table of pairwise differences between the means and a table of probabilities corresponding to those differences. Only groups 1 and 3 ($p = .022$) and groups 2 and 3 ($p = .005$) are significantly different using this test.

Table 9.13 Post-hoc comparisons of the three group means using Tukey's HSD test.

```
COL/
ROW     GROUP
  1      1.000
  2      2.000
  3      3.000


USING LEAST SQUARES MEANS.

POST HOC TEST OF      SCORE

USING MODEL MSE OF       231.093 WITH      24. DF.
MATRIX OF PAIRWISE MEAN DIFFERENCES:

                     1            2            3
       1           0.000
       2          -4.440        0.000
       3          19.610       24.050        0.000

TUKEY HSD MULTIPLE COMPARISONS.
MATRIX OF PAIRWISE COMPARISON PROBABILITIES:

                     1            2            3
       1           1.000
       2           0.793        1.000
       3           0.022        0.005        1.000
```

9.5.2 Cell Means

You can also test pairwise differences among cell means. For example, if you wanted to test all pairwise differences among the six cell means in Table 9.3 using a Bonferroni correction for the number of tests, type

```
>HYPOTHESIS
>POST GENDER*GROUP / BONFERRONI
>TEST
```

(See section 8.5.2 for menu descriptions.) The output from this procedure is shown in Table 9.14. The top part of this output shows the labels of the six cells. This is followed by a table of pairwise differences between the means and a table of probabilities corresponding to those differences. Because there are 15 pairwise differences being tested by this post-hoc test, all of the probabilities were multiplied by 15, with a maximum value of 1.0. As you can see, after adjusting for the number of tests the only significant differences among the cell means are between females in group 3 (cell 6) and three other cells: females in group 2 (cell 5), males in group 1 (cell 1), and males in group 2 (cell 2). Because the latter two differences cut across the gender and group variables, they may be difficult to interpret. See section 8.5.2 for more information on Bonferroni corrections.

9.6 Checking Assumptions

ANOVA makes four assumptions about the distribution of errors within each cell: They are normally distributed, have constant variance, are independent, and are unbiased. Each of these assumptions is discussed below, and SYSTAT procedures are illustrated for assessing the validity of the first and second. These assumptions should always be assessed before reporting the results of an analysis of variance. See section 8.6 for more information.

9.6.1 Normality

The normality assumption states that, within each cell of the design, the errors have a normal distribution. The validity of this assumption can be assessed both graphically and using summary statistics, as described in the following sections. When your cell sizes are large enough, you can assess normality within each cell, but when the cell sizes are small, you may want to collapse across cells. Unfortunately, collapsing across

Table 9.14 Post-hoc comparisons of pairwise differences between six cell means in the TEST SCORES data set, using the Bonferroni adjustment for 15 comparisons.

```
COL/
ROW    GENDER         GROUP
 1      1.000         1.000
 2      1.000         2.000
 3      1.000         3.000
 4      2.000         1.000
 5      2.000         2.000
 6      2.000         3.000
```

USING LEAST SQUARES MEANS.

POST HOC TEST OF   SCORE

USING MODEL MSE OF   231.093 WITH   24. DF.
MATRIX OF PAIRWISE MEAN DIFFERENCES:

	1	2	3	4	5	6
1	0.000					
2	7.340	0.000				
3	11.980	4.640	0.000			
4	14.020	6.680	2.040	0.000		
5	-2.200	-9.540	-14.180	-16.220	0.000	
6	41.260	33.920	29.280	27.240	43.460	0.000

BONFERRONI ADJUSTMENT.
MATRIX OF PAIRWISE COMPARISON PROBABILITIES:

	1	2	3	4	5	6
1	1.000					
2	1.000	1.000				
3	1.000	1.000	1.000			
4	1.000	1.000	1.000	1.000		
5	1.000	1.000	1.000	1.000	1.000	
6	0.004	0.026	0.084	0.138	0.002	1.000

cells can give non-normal looking results when the equal variance assumption is violated. You will often want to assess the normality assumption and the equal variance assumption together. For small cell sizes the normality assumption will rarely appear to be badly violated.

9.6.1.1 Stem and Leaf Displays and Box Plots

You can get some idea of the distribution of errors within each group by looking at plots of the raw data, such as Figures 9.1 and 9.2. Although the distributions of cases within groups in those plots are not perfectly symmetric, this is expected for very small cell sizes. These plots certainly show no evidence for any substantial departure from normality.

When the cell sizes are small, as they are in the present example, and the equal variance assumption is met, an even better way of assessing the normality assumption graphically is to look at plots of the studentized residuals from all of the cells in the ANOVA. These residuals can be saved into a file when the ANOVA is calculated by SYSTAT. This was done above in section 9.3 when we generated Table 9.3. Stem and leaf displays and box plots (see sections 3.2.1.2 and 3.2.1.3) are both good for getting a feel for the distribution of the residuals. Open the file containing the studentized residuals (e.g., "twoway.res"), and then you can produce these plots with the commands

```
>GRAPH
>STEM STUDENT
>BOX STUDENT
```

The resulting plots are shown in Figure 9.3. Both plots indicate a very slight negative skew, but no outliers. The deviation from normality is clearly not large enough to worry most reviewers.

9.6.1.2 Skewness and Kurtosis

When plots of the data do not clearly reveal whether any substantial deviations from normality exist, you can compute the skewness and kurtosis of the residuals directly (see section 3.1.1 for more information). Simply type:

```
>STATS
>STATISTICS RESIDUALS/SKEWNESS KURTOSIS N
```

The output is shown in Table 9.15. The value of skewness is $-.206$, which suggests, as does the box plot in Figure 9.3, that the distribution is negatively skewed. However, the standard deviation of skewness for the example is $\sqrt{6/30} = 0.45$, so the approximate Z is 0.46, which is not anywhere near significance.

The value of kurtosis in Table 9.15 is $-.722$, suggesting that the distribution is slightly flat-topped. However, the p-values for kurtosis with small sample sizes (Snedecor & Cochran,

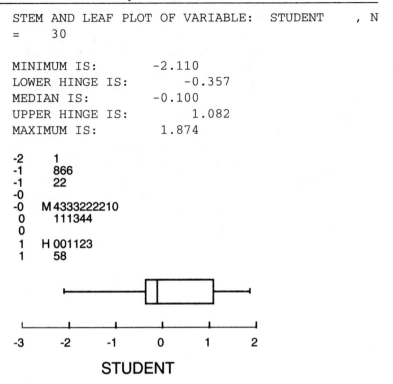

```
STEM AND LEAF PLOT OF VARIABLE:    STUDENT    , N
 =     30

MINIMUM IS:          -2.110
LOWER HINGE IS:       -0.357
MEDIAN IS:           -0.100
UPPER HINGE IS:        1.082
MAXIMUM IS:           1.874

 -2    1
 -1    866
 -1    22
 -0
 -0    M 4333222210
  0    111344
  0
  1    H 001123
  1    58
```

Figure 9.3 Stem and leaf display and box plot of the studentized residuals from the two-way ANOVA on the TEST SCORES data set in Table 9.3.

1982, Appendix 20ii) indicate that this value is not nearly significant

Thus, the values of skewness and kurtosis are consistent with what we can see in Figure 9.3: The distribution of residuals does not deviate substantially from normality.

Table 9.15 Skewness and kurtosis for the residuals from the two-way ANOVA on the TEST SCORES data set in Table 9.3.

TOTAL OBSERVATIONS:	30
	RESIDUAL
N OF CASES	30
SKEWNESS(G1)	-0.206
KURTOSIS(G2)	-0.722

9.6.1.3 Remedies

If the residuals are not normally distributed, the p-values from the F tests and t tests in the regression output may be too small. Fortunately, F's and t's are fairly robust to violations of the normality assumption, so moderate departures from normality can often be ignored if the other assumptions are met. In such a case, the reader should be warned that the p-values are not accurate. If the deviation from normality is substantial, the most common remedies are: (1) to transform the dependent variable, or (2) to perform the analysis on the medians rather than the means, which is beyond the scope of this book. Transforming dependent variables proceeds in the same fashion as transforming the outcome variable covered in Chapter 7.

9.6.2 Equal Variance

The second assumption underlying ANOVA is that the residuals have the same variance within each cell of the design (see section 8.6.2). You can assess this assumption both graphically and with formal tests. *Levene's test* is one of the best tests available in terms of power and resistance to Type I error, and it is described below.

9.6.2.1 Comparing Variances Among Cells

You can find the variances for each cell in the design by using the statistics command. For example, to find the variances for the six cells in the two-way TEST SCORES data set, type

```
>BY GENDER GROUP
>STATS
>STATISTICS SCORES/ VARIANCE
```

The variances of the six cells, along with the means and medians, are shown in Table 9.16. One rough rule of thumb is not to worry about unequal variance unless the ratio of the largest to the smallest is greater than 2.0 (some authors suggest ratios as high as 5.0 as the criterion). The largest variance in Table 9.16 is about 459 and the smallest is about 14, giving a ratio of about 33. Clearly, this assumption is violated, but the small numbers of data points in each group should caution us against making too much of this ratio alone.

9.6.2.2 Scatterplots and Box Plots

Scatterplots such as the one shown in Figure 9.1 and box plots such as the one in Figure 9.2 are very useful for assessing the equal variance assumption. These plots show that the variance

Table 9.16 N, mean, variance, and median for each cell in the two-way TEST SCORES data set.

```
THE FOLLOWING RESULTS ARE FOR:         THE FOLLOWING RESULTS ARE FOR:
        GENDER   =     1.000                   GENDER   =     2.000
        GROUP    =     1.000                   GROUP    =     1.000

TOTAL OBSERVATIONS:    5               TOTAL OBSERVATIONS:    5
                 SCORE                                   SCORE
   N OF CASES        5                    N OF CASES        5
   MEAN         92.340                    MEAN        106.360
   VARIANCE    143.203                    VARIANCE    458.693
   MEDIAN       94.400                    MEDIAN      112.000


THE FOLLOWING RESULTS ARE FOR:         THE FOLLOWING RESULTS ARE FOR:
        GENDER   =     1.000                   GENDER   =     2.000
        GROUP    =     2.000                   GROUP    =     2.000

TOTAL OBSERVATIONS:    5               TOTAL OBSERVATIONS:    5
                 SCORE                                   SCORE
   N OF CASES        5                    N OF CASES        5
   MEAN         99.680                    MEAN         90.140
   VARIANCE    162.302                    VARIANCE    250.123
   MEDIAN       97.900                    MEDIAN       86.700


THE FOLLOWING RESULTS ARE FOR:         THE FOLLOWING RESULTS ARE FOR:
        GENDER   =     1.000                   GENDER   =     2.000
        GROUP    =     3.000                   GROUP    =     3.000

TOTAL OBSERVATIONS:    5               TOTAL OBSERVATIONS:    5
                 SCORE                                   SCORE
   N OF CASES        5                    N OF CASES        5
   MEAN        104.320                    MEAN        133.600
   VARIANCE    358.022                    VARIANCE     14.215
   MEDIAN      101.200                    MEDIAN      132.600
```

of the sixth cell, the females in group 3, is smaller than the rest. These plots also make clear that the sixth cell has the highest mean.

9.6.2.3 Plotting Residuals Against the Cell Means

Plots of the residuals for each group give much of the information contained in plots of the residuals against the cell means, but the latter are more useful for suggesting possible transformations when the assumption is violated. To plot the residuals against the cell means, open the file containing the residuals (saved while computing the ANOVA in Table 9.3) and type

```
>GRAPH
>PLOT RESIDUAL*ESTIMATE
```

The resulting plot for the example data is shown in Figure 9.4. The variance appears to be fairly stable as the means of the cells increase, until the large decrease in the sixth cell, the females in group 3. Thus, the largest mean is associated with the smallest variance, an unusual occurrence.

9.6.2.4 Levene's Test for Homogeneity of Variance

Levene's test can be performed in four steps (see section 8.6.2.4). First, find the medians for each group. This was done

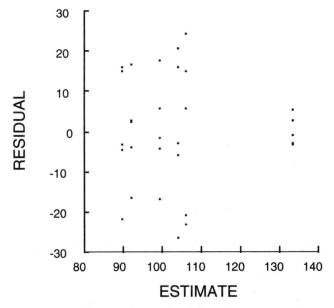

Figure 9.4 Plot of the residuals from the two-way ANOVA on the TEST SCORES data set against their cell means.

in Table 9.16. Second, open the original data set and create a new column of medians in which the median for a given group is paired with each score in that group. When the data set contains a column with codes for each cell, as was done in section 9.1, this can be easily accomplished with commands like the following (see also section 2.2.2):

```
> IF CELL = 1 THEN LET MEDIAN = 94.4
> IF CELL = 2 THEN LET MEDIAN = 97.9
> IF CELL = 3 THEN LET MEDIAN = 101.2
> IF CELL = 4 THEN LET MEDIAN = 112.0
> IF CELL = 5 THEN LET MEDIAN = 86.7
> IF CELL = 6 THEN LET MEDIAN = 132.6
```

Third, compute the absolute deviations of the scores from their medians, using SYSTAT's ABS() function to set a new column called, say, "DEVIAT," equal to raw data minus their group medians. The appropriate command in the example data set is

```
>LET DEVIAT = ABS(SCORE - MEDIAN)
```

Fourth, and finally, compute a one-way ANOVA on the deviations using SYSTAT's usual ANOVA routine:

```
>MGLH
>CATEGORY CELL
>ANOVA DEVIAT
>PRINT SHORT
>ESTIMATE
```

(See section 8.3 for menu descriptions.) The ANOVA results for the TEST SCORES data set are shown in the top panel of Table 9.17. The F test in this ANOVA is Levene's test: If the average deviations from the medians of the groups do not differ significantly, one can conclude that there is not good evidence for heterogeneity of variance. The nonsignificant test in Table 9.17 shows why we should not always trust our eyes: Even though the graphs above show that cell six has lower variance than the rest of the cells, such differences among cells with such small numbers of data points are quite likely due to chance.

There is one more thing to consider here, however. Levene's test with more than 1 df in the numerator of the F test suffers from the same problem as any multi-df test: Significant differences among the means can get lost. When the data suggests a trend among the variances, or if you have a theory about how the variances might differ, you can test these hypotheses with contrasts. For example, suppose that prior to looking at the data we had predicted that the females in group 3 would have lower variance than the other cells. The F test in the top panel of Table 9.17 is not a sufficient test of this hypothesis. Instead, we should test the contrast using the weights -1, -1, -1, -1, -1, $+5$, for the means of cells 1 through 6, respectively, which can be done with the following commands (see section 8.4.1 for menu descriptions):

```
>HYPOTHESIS
>SPECIFY / POOLED
>5*CELL[6] - CELL[1] - CELL[2] - CELL[3],
- CELL[4] - CELL[5] = 0
>TEST
```

Note that in this command I put the positive contrast weight at the beginning; SYSTAT seems to like this better (see Hint box below).

HINT: ENTERING USER-DEFINED CONTRASTS

SYSTAT is a bit finicky about the way contrasts must be entered in the commands and dialog boxes. Generally, it doesn't like minus signs at the beginning, and it will sometimes balk at the weights that you specify. If you get an error message, try rearranging the order of the means and try again.

The output from this contrast test is shown in the bottom panel of Table 9.17. The p-value in this output is nondirectional, but our a priori hypothesis was very clearly directional; so to find the directional p-value we must divide the value in the bottom panel of Table 9.17 by two. This gives $p = .031$, which is significant at the .05 level. Thus, even though the overall F in the top panel of Table 9.17 is not significant, our prediction is.

When the pattern of unequal variance is not predicted in advance, such contrasts capitalize on chance and should be corrected for the number of implicit tests performed. In reality, the smaller variance of the females in group 3 was not predicted in advance, so the contrast in the lower panel should not only employ a nondirectional p-value, but should also be corrected for the number of implicit tests. Clearly, this nondirectional test is not significant even before adjustment.

Table 9.17 Levene's test for homogeneity of variance for the two-way TEST SCORES data set.

One-way ANOVA on deviations from the median

```
LEVELS ENCOUNTERED DURING PROCESSING ARE:
CELL
        1.000      2.000      3.000      4.000      5.000      6.000
```

```
DEP VAR: DEVIAT    N: 30   MULTIPLE R: 0.470   SQUARED MULTIPLE R: 0.221
```

ANALYSIS OF VARIANCE

SOURCE	SUM-OF-SQUARES	DF	MEAN-SQUARE	F-RATIO	P
CELL	589.607	5	117.921	1.361	0.274
ERROR	2078.752	24	86.615		

Contrast comparing the variance in cell six to the other cells

```
HYPOTHESIS.

A MATRIX
     1         2         3         4         5         6
 0.000    -6.000    -6.000    -6.000    -6.000    -6.000

NULL HYPOTHESIS VALUE FOR D
              0.000
```

TEST OF HYPOTHESIS

SOURCE	SS	DF	MS	F	P
HYPOTHESIS	330.932	1	330.932	3.821	0.062
ERROR	2078.752	24	86.615		

9.6.2.5 Remedies

If you can't see unequal variance in plots of the residuals in Figures 9.3 or 9.4, you don't have a serious problem. If you can see unequal variance (or if you don't trust your eyes) Levene's test can be employed to determine whether the inequality is likely due to chance. With only five data points per cell, it would only take one or two residuals to be a little different for a different visual pattern of variances to emerge. Furthermore, for moderate departures from equal variance, F's and t's are fairly robust to violations of this assumption. When the departure from equality is substantial, the most common remedy is to perform a *variance stabilizing transformation* on the dependent variable (Snedecor & Cochran, 1982, pp. 287–292). Such transformations are discussed in section 7.1.1.2.

In the two-way TEST SCORES data set, because the variance is smallest in the group with the largest mean, if you did want to transform this data to equalize the variances, you would want to use a transformation "up the ladder" (see section 7.1.1), with an exponent greater than 1. For example, if you square the scores, the ratio of the largest to the smallest variance drops from about 33 to about 20; if you use scores cubed, the ratio drops to about 13. This is moving the variances in the right direction, but unless these transformations have theoretical justification (which is unlikely) you would be better off living with the unequal variance.

9.6.3 Independence

The third assumption underlying ANOVA is that the errors are uncorrelated with one another. When the normality assumption is met, uncorrelated errors are also *independent*, so this assumption is often called the independence assumption. When errors are correlated the estimates of means and effects are still unbiased, but the population variance is underestimated. This produces F's and t's that are too large, and confidence intervals that are too small.

As stated in the previous chapter, in ANOVA correlated errors are usually impossible to detect in the data. Whether one should worry about correlated errors depends almost entirely on the experimental design. For example, if the subjects in each group are tested individually with no opportunity to influence each other, then their errors are probably uncorrelated. The most common violations of this assumption arise when either (1) subjects (or families, or twins, or plots of land) are measured more than once, that is, they contribute more than one case to the data, or (2) the experimental manipulation is administered to a group of subjects *as a group*, and these subjects within a group influence the effectiveness of the manipulation for each other. Both types of violation sometimes can be dealt with using some type of repeated measures analysis (see Chapters 11 and 12).

9.6.4 Errors Are Unbiased

The fourth assumption underlying ANOVA is that the errors are unbiased, that is, that the expected value of the errors within each cell is zero. For the purposes of almost all ANOVAs, and certainly for the type of analyses covered in this book, this assumption can be considered true by definition. Because the cell means are unbiased estimates of the population values, and the errors are defined as the differences between the obtained scores and the cell means, in the long run the errors do have an expected value of zero.

9.7 Confidence Intervals

SYSTAT does not directly provide confidence intervals in ANOVA, but they can easily be obtained from estimates of means and variances that SYSTAT does provide. So long as the ANOVA has no random factors, the confidence intervals can be computed in the manner described in section 8.7, so that discussion will not be repeated here.

9.8 Graphing the Results

Normally, you will want to present a plot of each effect of interest in your data set, with, perhaps, some type of error bars around those means. This section shows graphs for the two-way TEST SCORES data set. More information and types of graphs can be found in section 8.8

9.8.1 Bar Chart with Error Bars

When the independent variable is categorical and does not have a meaningful scale, that is, the ordering of the categories is arbitrary, it is standard to plot the means of the categories using a bar graph with error bars. SYSTAT provides a few options for error bars. Unfortunately, the easiest one, using the standard error of the mean for each group, is not appropriate when you have more than one independent variable because SYSTAT will ignore all of the variables except the one that you are plotting. What you really want is an error estimate based on the variance within the smallest cells of the design. The easiest way to do this is to use the standard error based on the pooled error variance from the ANOVA output, as shown in Table 9.3. For example, the means for the two genders have standard errors equal to 3.925. So, to plot the means for males and females, along with errors bars showing ±1 standard error of the mean using the appropriate pooled variance for the variance estimate (see also section 8.8), you must first create a new column containing the standard errors. To do this type

```
>LET SE_GEND = 3.925
```

Now you can create the desired plot using the commands

```
>GRAPH
>BAR SCORE * GENDER/ERROR=SE_GEND SORT
```

The option *ERROR=SE_GEND* tells SYSTAT to include error bars based on the values found in the column SE_GEND. The option *SORT* plots the data in order of the individual groups. The resulting graph is shown in Figure 9.5. If you wanted to use the axis labels *male* and *female* instead of "1" and "2," you could create a character column containing those labels and use it in the plot. You can also use other types of error bars, using separate variance estimates, confidence limits, and so on. See section 3.2.2.5.

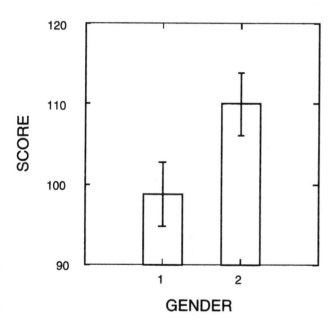

Figure 9.5 Bar graph showing the mean of males and females in the TEST SCORES data set, with error bars of ± 1 SE of the mean, where the standard errors are based on the pooled variance from the ANOVA.

9.8.2 Line Chart with Error Bars

When an independent variable is on a meaningful scale, it is common to graph the means using a line graph. As for bar graphs in the previous section, if you want to include error bars you will need to create a new column in your data set that contains the values for those error bars. For example, suppose that the *group* variable in the TEST SCORES data set is on a meaningful scale and you want to plot the means along with error bars based on the pooled error estimate. First, you must create a new column containing the values 4.807, which is the value of the standard errors shown in Table 9.3:

```
>LET SE_GROUP = 4.807
```

You can then plot the means for the groups, along with error bars, using the commands

```
>GRAPH
>CPLOT SCORE * GROUP/YMIN=80 YMAX=140,
ERROR=SE_GROUP LINE=1
```

The resulting graph is shown in Figure 9.6. The *ymin* and *ymax* option were adjusted by trial and error to ensure that the

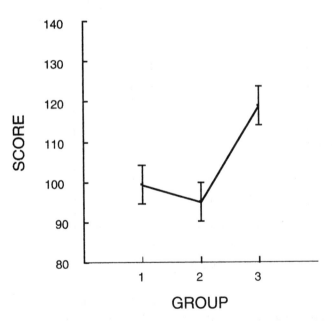

Figure 9.6 Line graph of the means of the three groups in the TEST SCORES data set, including standard error bars based on the pooled error variance from the ANOVA.

error bars fit completely on the graph. The *line=1* option tells SYSTAT to connect the means with a solid line, as shown in the graph.

9.8.3 Two-Way Plots of Means

SYSTAT allows you to plot more than one dependent variable in the same graph. For example, suppose that you want to create separate bars for men and women within each group in a bar plot. First, you must create two new columns in the data set containing only the test scores of men and women, respectively. This can be done easily using the commands

```
>IF GENDER = 1 THEN LET M = SCORE
>IF GENDER = 2 THEN LET F = SCORE
```

Each of these columns will have missing data, but this won't affect the graph. To include the appropriate error bars you need to create yet another column, this time containing the values of the standard error, for example. The standard errors for the means for males and females within each of the groups based on the pooled error variance is 6.798, as shown in Table 9.3. Thus, the new column can be created by typing

```
>LET SE_CELLS = 6.798
```

With these new columns you can now create the bar graph by typing, for example

```
>GRAPH
>BAR  M F * GROUP/ERROR1=SE_CELLS,SE_CELLS,
YMIN=80 YMAX=140 FILL=7,4 LLABEL="MALES",
"FEMALES" YLAB="SCORE"
```

The resulting bar graph is shown in Figure 9.7. The *error1* option does two things: (1) it tells SYSTAT to show only the upper limit on the error bar, which is useful when the bars in the graph are filled with some kind of color or pattern, and (2) it specifies the columns containing the error bars values for males and females, respectively, separated by a comma. In this case the error bar values are contained in the same column, but both still must be specified. The *ymin* and *ymax* options were set by trial and error to make the graph look good. The *fill* option specifies the numbers corresponding to the desired patterns used to fill the bars. The *llabel* option allows you to specify the names of the subgroups to be used in the legend. Make sure that the order of the labels corresponds to the order of the subgroups at the beginning of the bar command. Finally, the *ylab* option allows you to specify the label of the Y-

MACINTOSH

① Choose the menu item **Graph/Bar/Bar**.

② Select M and F from the left variable list and GROUP from the right variable list.

③ Click on **Error**, type "se_cells, se_cells" into the box, and turn on **One Way**.

④ Click on the **Fill** icon, type "7,4" into the box, and click **OK**.

⑤ Click on the **Axes** icon, type "80" into the **Y min** box and "140" into the **Y max** box, type "SCORE" into the **Y label** box, and click **OK**.

⑥ Click on the **Legend** icon, type "males" into the top box and "females" into the second box, and click **OK**.

⑦ Click **OK**.

axis. Because SYSTAT thinks you have two dependent variables in this plot, it does not automatically know what to call the Y-axis.

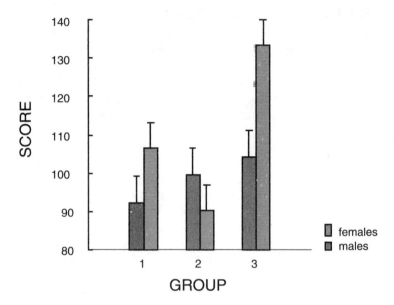

Figure 9.7 Bar graph showing the means of male and female subgroups in the two-way TEST SCORES data set.

If you prefer a line graph over a bar graph, you can generate one using the same columns for the males and females scores and error bars that you just created. Simply type

```
>CPLOT M F * GROUP/ERROR=SE_CELLS, SE_CELLS,
YMIN=80 YMAX=140 YLAB="SCORE" LLABEL= "MALES",
"FEMALES" LINE=1,11
```

The options are identical to those for the bar chart above, except that the *line* option replaces the *fill* option. The resulting graph is shown in Figure 9.8.

9.8.4 Plotting Interactions

The two-way plots in the previous section showed the means of the cells corresponding to each gender-by-group combination. Such plots are often mistakenly called "plots of the interaction," but, in fact, they contain both main effects as well as the interaction (Rosenthal & Rosnow, 1991, pp. 365–367). If you really want a plot of the interaction itself, you must use the residuals shown in Table 9.4. For example, to plot the interaction in the two-way TEST SCORES data set, open a new

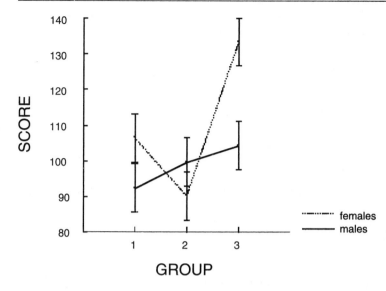

Figure 9.8 Line graph showing the means of male and female subgroups in the two-way TEST SCORES data set.

data worksheet and make one column with the group labels, one column with the interaction residuals corresponding to males, and one column with the interaction residuals corresponding to females. If you want to use error bars, you can enter another column or columns containing the error bar values. This data worksheet is shown in Table 9.18. With these columns you can now plot the interaction with the commands

```
>GRAPH
>CPLOT MALES FEMALES * GROUP/YLAB="RESIDUAL",
ERROR=SE,SE LINE=1,10
```

This plot is shown in Figure 9.9. Plotting the interaction residuals can help prevent common misinterpretations of the interaction, which often arise from looking at plots of the means, such as Figures 9.7 or 9.8, instead of plots of the interaction.

Table 9.18 A data file showing the columns to plot the interaction for the two-way TEST SCORES data set.

	GROUP	MALES	FEMALES	SE
1	1.000	−1.383	1.383	6.798
2	2.000	10.397	−10.397	6.798
3	3.000	−9.014	9.014	6.798

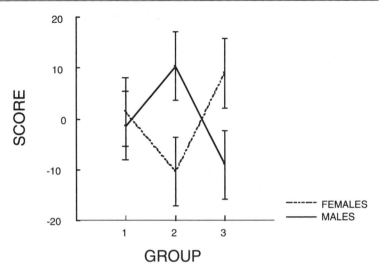

Figure 9.9 Plot of the interaction residuals for the two-way TEST SCORES analysis of variance.

9.9 Unequal Cell Sizes

There has been considerable controversy over the proper way to analyze factorial designs with unequal numbers of data points in the cells. There are two issues at stake here: (1) what is the thing you want to estimate, and (2) how do you compute the correct standard errors and significance tests. The first question is by far the most important, but it has received very little attention relative to the second question. As a result, statisticians have developed ways of getting the correct standard errors and significance tests to answer questions that few researchers would be interested in.

It would go beyond the scope of this book to discuss all of the possible approaches to analyzing data with unequal cell sizes. Fortunately, from a scientific standpoint, one method will serve to answer most questions of interest: the method of unweighted means. This method does not get the standard errors and significance tests exactly right, but it does estimate the things that researchers are usually interested in: the population cell and marginal means.

To illustrate, consider the data set in Table 9.19, which is the same as Table 9.1 except that there are a few additional subjects in each group. (This data set was also used in section 3.2.3.) Suppose you are interested in testing the significance of the group effect in this experiment, say the contrast −1, −1, and +2 across the three groups. Are you interested in estimating the relative proportions of males and females in each group in the population? Do you want to answer questions about means in a population that has these same unequal proportions of males and females? In some applied contexts the answer might be yes, but in almost any research setting you would have had equal cell sizes if you could have, and you are not interested in generalizing your results to a population that has these same inequalities. Therefore, most of the time you would like to act as though you really had equal cell sizes: This is the method of unweighted means.

Table 9.19 The TEST SCORES data set—test scores for students in three tutoring conditions.

	Group 1	Group 2	Group 3
Males	94.4	95.3	98.1
	75.7	117.2	101.2
	88.1	97.9	120.1
	108.7	82.7	77.5
	94.8	105.3	124.7
	130.6		136.1
	121.1		
Females	130.6	85.2	136.1
	121.1	86.7	132.6
	82.9	104.8	130.5
	112.0	67.9	130.0
	85.2	106.1	138.8
	86.1		70.4
	99.8		
	121.8		

MACINTOSH

① Choose the menu item **Stats/MGLH/Means Model...**
② Select SCORE from the **Dependent Variable(s)** list.
③ Select GENDER and GROUP from the **Factor(s)** list.
④ Click on **More...**
⑤ Turn on **Means & Std. Errors, Extended Output,** and **Save Residuals.**
⑥ Select *Unweighted means* from the **Means Model** list and click OK.
⑦ Type "unweight.res" into the **Save residuals as** box and click OK.

MS-DOS

① Choose the menu item **Statistics/MGLH/MGLM/ Model/Options/Means/** and hit <esc> (unweighted is the default; *do not* select **Weight**).
② Choose the menu item **Statistics/MGLH/MGLM/ Model/Variables/** and enter the model statement "SCORE = GENDER * GROUP".
③ Choose the menu item **Statistics/MGLH/MGLM/ Category/Variables/** and select GENDER and GROUP from the variables list.
④ Choose the menu item **Statistics/MGLH/MGLM/ Estimate/Save/,** hit <esc>, and type the save file- name, e.g., "unweight.res" into the box, hit <ret>, and select **Residuals.**
⑤ Choose the menu item **Statistics/MGLH/MGLM/ Estimate/Go!**

The unweighted means ANOVA can be computed in SYSTAT using the following commands:

```
>MGLH
>CATEGORY GROUP GENDER
>MEANS SCORE / UNWEIGHTED
  [leave off the '/ UNWEIGHTED' in DOS]
>PRINT LONG
>SAVE  UNWEIGHT.RES / RESID
>ESTIMATE
```

The output from this analysis is shown in Table 9.20. The means in this output are the best estimates of the population cell means, and if you simply average across the gender and

Table 9.20 Unweighted means ANOVA on the data in Table 9.19.

```
LEVELS ENCOUNTERED DURING PROCESSING ARE:
GROUP          1.000          2.000          3.000
GENDER         1.000          2.000
NUMBER OF CASES PROCESSED =      37
MEANS MODEL.

DEP VAR: SCORE       N: 37   MULTIPLE R: 0.458   SQUARED MULTIPLE R: 0.210

                    UNWEIGHTED MEANS MODEL
                     ANALYSIS OF VARIANCE
```

SOURCE	SUM-OF-SQUARES	DF	MEAN-SQUARE	F-RATIO	P
GROUP	2552.017	2	1276.009	3.086	0.060
GENDER	120.401	1	120.401	0.291	0.593
GROUP*GENDER	665.045	2	332.523	0.804	0.457
ERROR	12819.380	31	413.528		

LEAST SQUARES MEANS.

			LS MEAN	SE	N
GROUP	=	1.000			
GENDER	=	1.000	101.914	7.686	7
GROUP	=	1.000			
GENDER	=	2.000	92.075	7.190	8
GROUP	=	2.000			
GENDER	=	1.000	99.680	9.094	5
GROUP	=	2.000			
GENDER	=	2.000	90.140	9.094	5
GROUP	=	3.000			
GENDER	=	1.000	109.617	8.302	6
GROUP	=	3.000			
GENDER	=	2.000	118.017	8.302	6

```
WARNING: CASE   37 IS AN OUTLIER (STUDENTIZED RESIDUAL =       -2.843)

DURBIN-WATSON D STATISTIC      1.814
FIRST ORDER AUTOCORRELATION    0.002

RESIDUALS HAVE BEEN SAVED
```

group means, you can find the best estimates of the group and gender means, respectively. However, the significance tests in the ANOVA table are not based on exactly the correct standard errors of the effects being tested—instead, they estimate those standard errors by pretending that all of the cell sizes are equal, using the harmonic mean of the cell sizes as the n for each cell.

Fortunately, the inaccuracy of the F tests in Table 9.20 need not concern us, because overall F tests are usually not of much interest. Instead of overall F tests, compute contrasts using the formula

$$SS_C = \frac{C^2}{\Sigma \left(\frac{\lambda_{ij}^2}{n_{ij}} \right)} \tag{9.11}$$

where the contrast weights λ_{ij} are multiplied by each of the six cell means (not the marginal means) and summed to get C, and the n_{ij} are the individual cell sizes. This formula both (1) estimates the right estimand, and (2) gets the sum-of-squares right for that estimand. Fortunately, this is the formula SYSTAT uses to compute its contrasts when you use user-defined contrasts.

For example, to compute the linear trend in groups, first compute the unweighted means ANOVA and then type the following commands (see section 8.4.1 for menus):

```
>HYPOTHESIS
>SPECIFY / POOLED
>GROUP[1] - GROUP[3] = 0
>PRINT=SHORT
>TEST
```

(Recall that you leave out group 2 because it is multiplied by zero.) The output from these commands is shown in the top panel of Table 9.21. Similarly, the quadratic trend in groups can be tested with the commands

```
>HYP
>SPECIFY / POOLED
>GROUP[1] - 2*GROUP[2] + GROUP[3] =0
>TEST
```

This output is shown in the bottom panel of Table 9.21.

One of the problems that arises with unequal cell sizes can be demonstrated with the outputs in Tables 9.20 and 9.21. The sums of squares for the linear and quadratic contrasts on

Table 9.21 Linear and quadratic contrasts for the group factor using the unweighted means from Table 9.20.

Linear trend

```
CONTRASTING USING UNWEIGHTED MEANS.

HYPOTHESIS.

A MATRIX
       1           2           3           4           5           6
   -0.500      -0.500       1.000       1.000      -0.500      -0.500

NULL HYPOTHESIS VALUE FOR D
                0.000

TEST OF HYPOTHESIS

       SOURCE        SS         DF         MS           F           P

   HYPOTHESIS    1882.801        1      1882.801      4.553       0.041
        ERROR   12819.380       31       413.528
```

Quadratic trend

```
CONTRASTING USING UNWEIGHTED MEANS.

HYPOTHESIS.

A MATRIX
       1           2           3           4           5           6
    0.500       0.500      -1.000      -1.000       0.500       0.500

NULL HYPOTHESIS VALUE FOR D
                0.000

TEST OF HYPOTHESIS
       SOURCE        SS         DF         MS           F           P

   HYPOTHESIS     800.721        1       800.721      1.936       0.174
        ERROR   12819.380       31       413.528
```

groups in Table 9.21 sum to 2683.522, which is greater than the sum of squares for groups in Table 9.20. This is because the linear and quadratic contrasts are no longer orthogonal due to the unequal cell sizes.

Another, more worrisome problem is that the group and gender effects are not orthogonal, and neither is orthogonal to the interaction. This makes causal inference problematic. In regression we encounter correlated variables all the time, but usually are not bothered by it too much because with the lack of randomization in most regression analyses we already know better than to draw causal inference from the results alone. In ANOVA, however, random assignment to treatments often does make causal inference possible, and we hate to give that up because of unequal cell sizes. Unfortunately, there's not much choice. For example, because the group and gender effects are correlated in Table 9.20, you can't know for certain whether the treatments that the groups received *caused* the groups to differ, or whether the group sum-of-squares was produced spuriously through its correlation with the gender effect and the interaction. For more on interpreting effects in this situation, see Estes (1991, Chapter 10).

10

Nested Fixed Factors ANOVA

This chapter shows how to perform an analysis of variance in SYSTAT with two or more *nested* fixed factors. Any design that does not have fully crossed factors, where each level of one factor occurs in combination with each level of the other factors, has some type of nesting. For example, each school in your study may have a classroom, but each classroom occurs in only a single school. Classrooms are said to be nested in schools. Similarly, schools are nested in school districts, school districts nested within counties, and so on. This chapter analyzes a data set of this type in SYSTAT by looking at the raw data to check for unusual cases, computing an ANOVA, computing contrasts and performing post-hoc analyses, computing standard errors and confidence intervals, assessing the assumptions underlying the ANOVA, and graphing the results of the analysis.

10.1 The Example Data Set

The example that will be used in this chapter is a hypothetical study of the effect of an experimental teaching method on spatial reasoning test performance. Six schools were chosen to participate in this study, and within each school two classrooms were chosen. Finally, within each classroom four students were chosen at random, two of whom were randomly assigned to the experimental treatment and two of whom were assigned to a control condition. Thus, classrooms are nested in schools and pupils are nested in classrooms and treatments. Treatments are crossed with schools and classrooms. The example data set is shown in Table 10.1.

Table 10.1 The TEACHING METHOD data set: performance scores for 24 pupils in a new teaching method condition and 24 pupils in a control condition. Pupils are nested in teaching conditions and classrooms, and classrooms are nested in schools.

School		1		2		3		4		5		6	
Class		1	2	1	2	1	2	1	2	1	2	1	2
Treat	P1	67	69	76	72	81	75	74	65	82	69	85	83
	P2	62	78	70	75	76	76	69	77	81	70	79	85
Control	P3	71	81	78	86	78	85	74	85	89	83	89	92
	P4	76	71	66	78	83	76	79	84	80	84	79	89

To use this data set in SYSTAT, you need to enter it, or read it into a data editor window with at least four columns (you may also have descriptive columns, such as pupil names or numbers, but these are not necessary). One column will contain a code for the school, one will code for classroom, one will code for treatment, and one will contain the dependent variable—the performance scores. You have some options in coding the categorical variables (see section 8.1), but the most general method is to simply code each variable as numerical with the codes 1 through k, where k is the number of categories. For a nested factor, such as classroom, it makes no difference whether you code it from 1 to 12, with a different code for each classroom, or code it classrooms "1" and "2" within each school. The latter method is shown in the fully coded data set in Table 10.2. Although class "1" appears under all six schools, you should be aware that these are all different classrooms: Class 1 nested in school 1 is not the same as class 1 nested in school 2, and so on. Table 10.2 also shows a column with numerical labels for each student.

Table 10.2 The TEACHING METHOD data set from Table 10.1 coded into SYSTAT.

SCHOOL	CLASS	TREAT	PUPIL	SCORE
1	1	1	1	67
1	1	1	2	62
1	1	2	3	71
1	1	2	4	76
1	2	1	5	69
1	2	1	6	78
1	2	2	7	81
1	2	2	8	71

Table 10.2 *(continued)*

SCHOOL	CLASS	TREAT	PUPIL	SCORE
2	1	1	9	76
2	1	1	10	70
2	1	2	11	78
2	1	2	12	66
2	2	1	13	72
2	2	1	14	75
2	2	2	15	86
2	2	2	16	78
3	1	1	17	81
3	1	1	18	76
3	1	2	19	78
3	1	2	20	83
3	2	1	21	75
3	2	1	22	76
3	2	2	23	85
3	2	2	24	76
4	1	1	25	74
4	1	1	26	69
4	1	2	27	74
4	1	2	28	79
4	2	1	29	65
4	2	1	30	77
4	2	2	31	85
4	2	2	32	84
5	1	1	33	82
5	1	1	34	81
5	1	2	35	89
5	1	2	36	80
5	2	1	37	69
5	2	1	38	70
5	2	2	39	83
5	2	2	40	84
6	1	1	41	85
6	1	1	42	79
6	1	2	43	89
6	1	2	44	79
6	2	1	45	83
6	2	1	46	85
6	2	2	47	92
6	2	2	48	89

10.2 Looking at the Data
10.2.1 Scatterplots

In nested designs, you should examine the raw data to check for observations that are extreme relative to their own cells, not necessarily to the data set as a whole. You could do this, for example, by plotting the scores for the two students within each cell against the cell number. However, with only two students per cell, this would hardly be worth the trouble. A method that is almost as good is to sort the file by cell (as shown in Table 10.2), set the pupil's identification number equal to the case number (also as shown in Table 10.2), and then plot the scores against the pupils' identification numbers. In this plot, every even-odd pair of points on the X-axis corresponds to one cell of the design. Because the number of data points is so small, you would not be able to evaluate outliers within a cell, but you can still find points that are extreme relative to the entire data set. This plot is shown in Figure 10.1, and was created by typing

```
>GRAPH
>PLOT SCORE*PUPIL
```

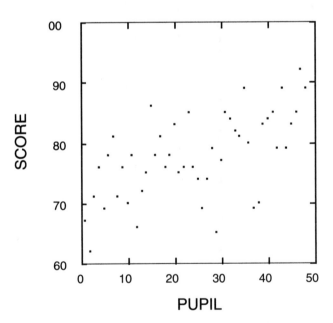

Figure 10.1 Scatterplot of the raw data in the TEACHING METHOD data set.

Extreme cases (vertically) in this plot should be examined individually to ensure that they were recorded properly and do not differ from the rest of the data in some other exceptional way. None of the data points in this plot appear to stand apart from other cases in the data set. The smallest score is pupil 2, but this score is not far from that of pupil 1 who is in the same cell. Similarly, the largest score is pupil 47, but this score is not far from that of pupil 48 who is in the same cell. Neither of these points stand apart from the rest of the data set.

10.2.2 Box Plots

Box plots can also be useful when the number of data points in the smallest cell is not too small. In the present example, however, box plots of pairs of points would be virtually useless. You might be interested in forming box plots of all of the scores within schools, for example, as shown in Figure 10.2, but you should keep in mind that each box in such a plot contains within it two classrooms and both experimental and control groups. The plot in Figure 10.2 was generated by the command

>BOX SCORE*SCHOOL/SORT

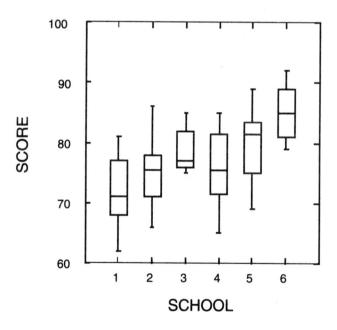

Figure 10.2 Box plots of the scores for each school in the TEACHING METHOD data set.

MACINTOSH

① Choose the menu item **Stats/MGLH/General Linear Model...**

② Select SCORE from the **Dependent Variable(s)** list.

③ Select SCHOOL, TREAT, and CLASS from the **Categorical Variable(s)** list.

④ Select SCHOOL and TREAT from the **Independent Variable(s)** list.

⑤ Type "CLASS{SCHOOL}", "SCHOOL*TREAT", and "TREAT*CLASS{SCHOOL}" into the **Interactions** box.

⑥ Click on **More...**

⑦ Turn on **Means & Std. Errors**, **Extended Output**, and **Save Residuals**.

⑧ Click **OK**.

⑨ Type "nested.res" into the **Save residuals as** box, and click **OK**.

MS-DOS

① Choose the menu item **Utilities/Output/Results/Long**.

② Choose the menu item **Statistics/MGLH/MGLM/Model/Variables/**, hit <esc>, and type "SCORE = CONSTANT + SCHOOL + TREAT + CLASS{SCHOOL} + TREAT*SCHOOL + TREAT*CLASS{SCHOOL}" into the box.

③ Choose the menu item **Statistics/MGLH/MGLM/Category/Variables** and select SCHOOL, TREAT, and CLASS from the variables list.

④ Choose the menu item **Statistics/MGLH/MGLM/Estimate/Save/**, hit <esc>, and type the save filename, e.g., "nested.res", into the box.

⑤ Choose the menu item **Statistics/MGLH/MGLM/Estimate/Go!**

When the number of data points in the box plot is small, the main use of the plot is to show the medians of the schools graphically and to give you a rough idea of the spread of the data. If there were any outliers, you might want to examine those cases individually, but because the boxes contain both experimental and control groups the data within them may be bimodal, or otherwise non-normal, and so some "outliers" might be expected.

10.3 Nested ANOVA

You can perform ANOVAs using SYSTAT's menus, but when you have nested variables it is usually easier to use commands. Categorical variables are entered into the model in the usual manner, but nested factors must indicate which factors they are nested within by listing those factors in braces ("curly brackets") after the nested factor's label. Thus, because classrooms are nested in schools in the example design, whenever CLASS appears in the model it must be followed by {SCHOOL} to indicate the nesting. You must do this for interactions as well as main effects, with the constraint that any non-nested factors in the interaction be listed first. So the treatment-by-classroom interaction would be entered into the model as TREAT*CLASS{SCHOOL} and not as CLASS{SCHOOL}*TREAT. Thus, the ANOVA commands for the example design are:

```
>MGLH
>CATEGORY SCHOOL CLASS TREAT
>MODEL SCORE=CONSTANT+SCHOOL+TREAT,
+CLASS{SCHOOL}+TREAT*SCHOOL,
+TREAT*CLASS{SCHOOL}
>PRINT LONG
>SAVE  NESTED.RES / MODEL
>ESTIMATE
```

The results, using extended output (*print=long*) are shown in Table 10.3. Because this design has 24 cells, it takes six pages to print all of the effects and means. The ANOVA is shown in the part of Table 10.3 beginning at the bottom of page 333.

HINT: ENTERING NESTED MODELS

You must use braces "{...}" to indicate nesting, not parentheses. If you get an error message about illegal subscripts, you probably have a parenthesis in your model somewhere. Simply edit the model and replace any parentheses with braces and try again.

HINT: NESTED DESIGNS (MS-DOS ONLY)

When you enter the model statement for nested designs using DOS menus, you can save yourself a good bit of typing (and mistyping) by selecting variables from the variables list and then editing the model statement. It is easiest to do the editing for each nested term as you come to it. For example, to enter the nested design above, first choose the menu item **Statistics/MGLH/MGLM/Model/Variables/** and then select SCORE, SCHOOL, TREAT, CLASS, and SCHOOL again from the variables list. This should give you the following model so far:

SCORE=CONSTANT+SCHOOL+TREAT+CLASS+SCHOOL

Now you can hit <esc>, which will put the cursor at the end of the model statement. With the left arrow key move back through the model to the "S" in the last SCHOOL, hit <del> until you have deleted all the characters back to the last "S" in CLASS, and then type "{". Then you can use the right arrow key to move to the end of the model and append a "}" after the final SCHOOL. This should produce the model statement

SCORE=CONSTANT+SCHOOL+TREAT+CLASS{SCHOOL}

which shows the proper nesting of CLASS. Now you can hit the <F8> key to bring back the variables list to enter "+ TREAT + SCHOOL", which you can then edit to read TREAT*SCHOOL and "+ TREAT + CLASS + SCHOOL", which you can edit to read TREAT*CLASS{SCHOOL}.

The mean of the control group (treatment 1) was 74.833 and the mean of the experimental group (treatment 2) was 80.667. The difference between them is highly significant, as can be seen in Table 10.3, indicating that the treatment improved

Table 10.3 ANOVA on the TEACHING METHOD data set showing standard output. Classrooms are nested in schools.

```
LEVELS ENCOUNTERED DURING PROCESSING ARE:
SCHOOL
      1.000     2.000     3.000     4.000     5.000     6.000
CLASS
      1.000     2.000
TREAT
      1.000     2.000
```

DEP VAR: SCORE   N: 48   MULTIPLE R: 0.878   SQUARED MULTIPLE R: 0.771

ESTIMATES OF EFFECTS   B = $(X'X)^{-1} X'Y$

		SCORE
CONSTANT		77.750
SCHOOL	1.000	−5.875
SCHOOL	2.000	−2.625
SCHOOL	3.000	1.000
SCHOOL	4.000	−1.875
SCHOOL	5.000	2.000
TREAT	1.000	−2.917
CLASS SCHOOL	1.000 1.000	−2.875
CLASS SCHOOL	1.000 2.000	−2.625
CLASS SCHOOL	1.000 3.000	0.750
CLASS SCHOOL	1.000 4.000	−1.875
CLASS SCHOOL	1.000 5.000	3.250
CLASS SCHOOL	1.000 6.000	−2.125

Table 10.3 (continued)

TREAT	1.000		
SCHOOL	1.000	0.042	
TREAT	1.000		
SCHOOL	2.000	1.042	
TREAT	1.000		
SCHOOL	3.000	1.167	
TREAT	1.000		
SCHOOL	4.000	-1.708	
TREAT	1.000		
SCHOOL	5.000	-1.333	
TREAT	1.000		
CLASS	1.000		
SCHOOL	1.000	-1.625	
TREAT	1.000		
CLASS	1.000		
SCHOOL	2.000	2.375	
TREAT	1.000		
CLASS	1.000		
SCHOOL	3.000	0.750	
TREAT	1.000		
CLASS	1.000		
SCHOOL	4.000	2.125	
TREAT	1.000		
CLASS	1.000		
SCHOOL	5.000	2.750	
TREAT	1.000		
CLASS	1.000		
SCHOOL	6.000	1.125	

ANALYSIS OF VARIANCE

SOURCE	SUM-OF-SQUARES	DF	MEAN-SQUARE	F-RATIO	P
SCHOOL	834.500	5	166.900	7.674	.198115E-03
TREAT	408.333	1	408.333	18.774	.226549E-03
CLASS {SCHOOL}	274.500	6	45.750	2.103	0.090
TREAT*SCHOOL	62.167	5	12.433	0.572	0.721
TREAT*CLASS {SCHOOL}	177.500	6	29.583	1.360	0.270
ERROR	522.000	24	21.750		

Table 10.3 *(continued)*

LEAST SQUARES MEANS.

			LS MEAN	SE	N
SCHOOL	=	1.000	71.875	1.649	8
SCHOOL	=	2.000	75.125	1.649	8
SCHOOL	=	3.000	78.750	1.649	8
SCHOOL	=	4.000	75.875	1.649	8
SCHOOL	=	5.000	79.750	1.649	8
SCHOOL	=	6.000	85.125	1.649	8
TREAT	=	1.000	74.833	0.952	24
TREAT	=	2.000	80.667	0.952	24
CLASS SCHOOL	= =	1.000 1.000	69.000	2.332	4
CLASS SCHOOL	= =	2.000 1.000	74.750	2.332	4
CLASS SCHOOL	= =	1.000 2.000	72.500	2.332	4
CLASS SCHOOL	= =	2.000 2.000	77.750	2.332	4
CLASS SCHOOL	= =	1.000 3.000	79.500	2.332	4
CLASS SCHOOL	= =	2.000 3.000	78.000	2.332	4
CLASS SCHOOL	= =	1.000 4.000	74.000	2.332	4
CLASS SCHOOL	= =	2.000 4.000	77.750	2.332	4
CLASS SCHOOL	= =	1.000 5.000	83.000	2.332	4
CLASS SCHOOL	= =	2.000 5.000	76.500	2.332	4
CLASS SCHOOL	= =	1.000 6.000	83.000	2.332	4
CLASS SCHOOL	= =	2.000 6.000	87.250	2.332	4

Table 10.3 *(continued)*

TREAT	=	1.000			
SCHOOL	=	1.000	69.000	2.332	4
TREAT	=	1.000			
SCHOOL	=	2.000	73.250	2.332	4
TREAT	=	1.000			
SCHOOL	=	3.000	77.000	2.332	4
TREAT	=	1.000			
SCHOOL	=	4.000	71.250	2.332	4
TREAT	=	1.000			
SCHOOL	=	5.000	75.500	2.332	4
TREAT	=	1.000			
SCHOOL	=	6.000	83.000	2.332	4
TREAT	=	2.000			
SCHOOL	=	1.000	74.750	2.332	4
TREAT	=	2.000			
SCHOOL	=	2.000	77.000	2.332	4
TREAT	=	2.000			
SCHOOL	=	3.000	80.500	2.332	4
TREAT	=	2.000			
SCHOOL	=	4.000	80.500	2.332	4
TREAT	=	2.000			
SCHOOL	=	5.000	84.000	2.332	4
TREAT	=	2.000			
SCHOOL	=	6.000	87.250	2.332	4
TREAT	=	1.000			
CLASS	=	1.000			
SCHOOL	=	1.000	64.500	3.298	2
TREAT	=	2.000			
CLASS	=	1.000			
SCHOOL	=	1.000	73.500	3.298	2
TREAT	=	1.000			
CLASS	=	2.000			
SCHOOL	=	1.000	73.500	3.298	2
TREAT	=	2.000			
CLASS	=	2.000			
SCHOOL	=	1.000	76.000	3.298	2

Table 10.3 (continued)

TREAT	=	1.000			
CLASS	=	1.000			
SCHOOL	=	2.000	73.000	3.298	2
TREAT	=	2.000			
CLASS	=	1.000			
SCHOOL	=	2.000	72.000	3.298	2
TREAT	=	1.000			
CLASS	=	2.000			
SCHOOL	=	2.000	73.500	3.298	2
TREAT	=	2.000			
CLASS	=	2.000			
SCHOOL	=	2.000	82.000	3.298	2
TREAT	=	1.000			
CLASS	=	1.000			
SCHOOL	=	3.000	78.500	3.298	2
TREAT	=	2.000			
CLASS	=	1.000			
SCHOOL	=	3.000	80.500	3.298	2
TREAT	=	1.000			
CLASS	=	2.000			
SCHOOL	=	3.000	75.500	3.298	2
TREAT	=	2.000			
CLASS	=	2.000			
SCHOOL	=	3.000	80.500	3.298	2
TREAT	=	1.000			
CLASS	=	1.000			
SCHOOL	=	4.000	71.500	3.298	2
TREAT	=	2.000			
CLASS	=	1.000			
SCHOOL	=	4.000	76.500	3.298	2
TREAT	=	1.000			
CLASS	=	2.000			
SCHOOL	=	4.000	71.000	3.298	2
TREAT	=	2.000			
CLASS	=	2.000			
SCHOOL	=	4.000	84.500	3.298	2
TREAT	=	1.000			
CLASS	=	1.000			
SCHOOL	=	5.000	81.500	3.298	2
TREAT	=	2.000			
CLASS	=	1.000			
SCHOOL	=	5.000	84.500	3.298	2

Table 10.3 *(continued)*

```
        TREAT   =   1.000
        CLASS   =   2.000
        SCHOOL  =   5.000        69.500       3.298        2

        TREAT   =   2.000
        CLASS   =   2.000
        SCHOOL  =   5.000        83.500       3.298        2

        TREAT   =   1.000
        CLASS   =   1.000
        SCHOOL  =   6.000        82.000       3.298        2

        TREAT   =   2.000
        CLASS   =   1.000
        SCHOOL  =   6.000        84.000       3.298        2

        TREAT   =   1.000
        CLASS   =   2.000
        SCHOOL  =   6.000        84.000       3.298        2

        TREAT   =   2.000
        CLASS   =   2.000
        SCHOOL  =   6.000        90.500       3.298        2

DURBIN-WATSON D STATISTIC        3.090
FIRST ORDER AUTOCORRELATION     -0.553

RESIDUALS HAVE BEEN SAVED
```

performance scores. The effect size *r* can be computed for the treatment effect in the usual manner:

$$r = \sqrt{\frac{F}{F + df_{\text{denom}}}} = \sqrt{\frac{18.774}{18.774 + 24}} = 0.66 \tag{10.1}$$

The remainder of the effects in this table are multi-*df*, so they should be examined more closely with contrasts. The *Durbin-Watson Statistic* and *first-order autocorrelation* are almost never relevant in ANOVA (see section 5.4.3.3 for a discussion).

10.4 Contrasts

All of the *F* tests in Table 10.3, except for the treatment effect, have more than one degree of freedom in the numerator, so they are of little use to us. First, if they are significant, they only tell us that some differences exist, not which means are different, and second, if they are not significant, this does not mean that none of the means are different or that there are no significant trends among the means. Experiments are almost always conducted with specific predictions in mind, and these predictions should always be tested directly with contrasts.

10.4.1 Planned Comparisons

10.4.1.1 Main Effects

When a main effect is not nested in any other factors, you can perform contrasts on it just as you would in a fully crossed design (see also section 9.4.1.1). Suppose, for example, that you had theoretical reasons for testing the difference between the first three and the last three schools in Table 10.1. The easiest method is to use the *effect* commands (see section 8.4.1 for menus):

```
>HYP
>EFFECT = SCHOOL
>CONTRAST
>-1 -1 -1 1 1 1
>TEST
```

The corresponding output is shown in Table 10.4. The contrast is in the predicted direction and is highly significant. The effect size r for the contrast is simply

$$r = \sqrt{\frac{F}{F + df_{\text{denom}}}} = \sqrt{\frac{13.793}{13.793 + 24}} = 0.60 \qquad (10.2)$$

See section 9.4.1.1 for a second method of computing contrasts on main effects and section 9.4.1.2 for how to compute contrasts on simple effects.

10.4.1.2 Interaction Effects

Tests of interaction effects between non-nested factors present a bigger problem for SYSTAT than do interactions in fully crossed designs (see section 10.4.1.3). SYSTAT will not allow you to test contrasts on effects for anything but main effects, and I have had bad luck trying to use contrasts on cell means. You could learn to compute contrasts using SYSTAT's *C matrix* procedures, but this requires as much manual labor as computing the contrasts by hand. So, if you get more than one or two "unable to understand what you mean" messages when trying

Table 10.4 Test of the contrast predicting that the mean of group 3 would be higher than the means of groups 1 and 2 in the TEST SCORES data set.

```
TEST FOR EFFECT CALLED:      SCHOOL

A MATRIX

   1        2        3        4        5        6        7        8        9
 0.000   -2.000   -2.000   -2.000    0.000    0.000    0.000    0.000    0.000

  10       11       12       13       14       15       16       17       18
 0.000    0.000    0.000    0.000    0.000    0.000    0.000    0.000    0.000

  19       20       21       22       23       24
 0.000    0.000    0.000    0.000    0.000    0.000

NULL HYPOTHESIS CONTRAST AB
                    15.000

                      -1
INVERSE CONTRAST A(X'X)    A'
                     0.750

TEST OF HYPOTHESIS
      SOURCE           SS          DF         MS              F            P

   HYPOTHESIS       300.000        1       300.000         13.793        0.001
        ERROR       522.000       24        21.750
```

to compute interaction contrasts on cell means in SYSTAT, you should probably bite the bullet and do it by hand. Here's how.

Suppose that the six schools in the TEACHING METHOD data set differ in the socioeconomic status (SES) of the communities in which they operate. School 1 is in the poorest community, school 6 is in the wealthiest community, and the other schools fall in order in between. Furthermore, suppose that you want to test the hypothesis that new experimental treatment is more effective as the SES of the school increases. The derivation of the appropriate contrast weights is shown in Table 10.5. Because the effectiveness of the treatment is measured relative to the control group, begin by writing the weights –1 and +1 at the beginning of the rows of the table for the control and experimental groups, respectively, as shown in Table 10.5. If the schools are roughly equally spaced in SES, the hypothesis that the treatment-control difference increases with SES can be tested using the equally spaced linear contrasts weights –5, –3, –1, 1, 3, and 5. Write these across the top margin for the columns, as shown in Table 10.5. Finally, to obtain the contrast weights for the 12 cells defined by the treatment * schools interaction, simply multiply the weights in the rows by the weights in the column, cell by cell. This results in weights of 5, 3, 1, –1, –3, and –5 for the control group, and –5, –3, –1, 1, 3, and 5 for the treatment group, as shown in the cells of the table in Table 10.5

Table 10.5 Contrast weights predicting that the benefit of the experimental teaching method increases as the SES of the school increases.

Treatment		Schools					
		1	2	3	4	5	6
	λ	–5	–3	–1	1	3	5
Controls	–1	5	3	1	–1	–3	–5
Experimentals	+1	–5	–3	–1	1	3	5

When the cells sizes are all equal and the contrast weights sum to zero in both the rows and the columns, as they are for this contrast, a contrast on the means is identical to a contrast on the interaction (a contrast on interaction residuals is illustrated in section 9.4.1.2). These means can be found in the extended ANOVA output in Table 10.3, and are reproduced here in Table 10.6.

Table 10.6 Means for the experimental and control conditions in each of the six schools in the TEACHING METHOD data set.

Treatment	Schools					
	1	2	3	4	5	6
Controls	69.00	73.25	77.00	71.25	75.50	83.00
Experimentals	74.75	77.00	80.50	80.50	84.00	87.25

The contrast can now be computed easily using standard contrast formulas. First, multiply each mean in Table 10.6 by its corresponding contrast weight in Table 10.5, and sum them up:

$$C = \Sigma \lambda_{ij} \overline{X}_{ij}$$

$$= (5)(69) + (3)(73.25) + (1)(77) + (-1)(71.25) +$$
$$(-3)(75.5) + (-5)(83) + (-5)(74.75) + (-3)(77) +$$
$$(-1)(80.5) + (1)(80.5) + (3)(84) + (5)(87.25)$$

$$= 12.5 \qquad (10.3)$$

The sum of squares of this contrast is

$$SS_C = \frac{nC^2}{\Sigma \lambda_{ij}^2} = \frac{4 \times 12.5^2}{(4)(1^2 + 3^2 + 5^2)} = 4.464 \qquad (10.4)$$

where n is the number of pupils in each treatment condition in each school. The test of the contrast using the pooled error term from the ANOVA (Table 10.3) is $F_{(1, 24)} = 4.464/21.750 = 0.21$, which is very far from significant. The effect size r for this interaction contrast can be computed in the usual manner:

$$r = \sqrt{\frac{F}{F + df_{\text{denom}}}} = \sqrt{\frac{0.21}{0.21 + 24}} = 0.09 \qquad (10.5)$$

10.4.1.3 An Aside: Nested "Main" and "Interaction" Effects

If you want to test a contrast on a factor that is nested in another factor, such as the classroom effect in the sample data set, you cannot use SYSTAT's normal contrast methods. If you try to use the effects method, you will get an error message like

YOU MAY COMPUTE CONTRASTS ONLY ON MAIN EFFECTS.

If you try to use the contrasts-on-cell-means method, you will get an error message like

MARGINAL OR CELL IDENTIFIED HAS ZERO OBSERVATIONS.

This is because the nesting implies that the classrooms are not ordered in a meaningful way within schools; that is, you could have just as easily called classroom 1 in school 1 "classroom 2," and the analysis would come out exactly the same. If the classrooms are ordered in a meaningful way within schools, then they should probably not be nested. For example, suppose that the first classroom in each school in Tables 10.1 and 10.2 is a classroom of first graders, and the second classroom is a class of second graders. Although classrooms are still nested within schools, *grade* is crossed with schools. The nested classroom effects and the grade effects are confounded in such a design, but if you are interested in the grade effect you should run the analysis as a fully crossed design.

For example, if you wanted to determine whether the second graders had higher performance scores overall than first graders, you could test this contrast by hand. The means for the first graders can be found by averaging the means for the first class in the six schools in Table 10.3: (69.0 + 72.5 + 79.5 + 74.0 + 83.0 + 83.0)/6 = 76.833. Similarly, the mean for the second graders is (74.75 + 77.75 + 78.0 + 77.75 + 76.5 + 87.25)/6 = 78.667. To compute the contrast you simply multiply the contrast weights –1 and +1 by their respective means:

$$C = \Sigma \lambda_j \overline{X}_{\cdot j}$$

$$= (-1)(76.833) + (1)(78.667) = 1.834 \qquad (10.6)$$

Then the sum-of-squares of the contrast is

$$SS_C = \frac{nC^2}{\Sigma \lambda_j^2} = \frac{24 \times 1.835^2}{(-1)^2 + 1^2} = 40.363 \qquad (10.7)$$

where n is the number of pupils in each grade. The test of the contrast using the pooled error term from the ANOVA (Table 10.3) is $F_{(1, 24)} = 40.363/21.750 = 1.86$, which is not significant, $p = .093$, one-tailed (see section 2.2.3.3 for how to find p-values in SYSTAT). Similarly, if you wanted to know whether the treatment was more effective for second graders than for first graders you could compute the interaction contrast by hand. Once again, you could find the means in question by averaging over the means in the output in Table 10.3 Then you could compute the contrast of those means using Equations 10.3 and 10.4. This sum of squares is 75.000, and the test is $F_{(1, 24)} = 75.000/21.750 = 3.448$, which is significant, $p = .038$, one-tailed.

However, both of these contrasts would be generated automatically if you analyzed the data as a fully crossed design. You could do this with the following commands (see section 9.3 for menus and information on factorial designs):

```
>MGLH
>CATEGORY SCHOOL CLASS TREAT
>ANOVA SCORE
>PRINT SHORT
>ESTIMATE
```

The output is shown in Table 10.7. Note that the mean squares for school, treatment, and error are exactly the same as they were in Table 10.3. But now the 6 *df* class{school} effect in Table 10.3 has been broken into a 1 *df* class effect and a 5 *df* school*class effect in Table 10.7. The sum-of-squares for the class effect is identical to the sum-of-squares for the contrast that we computed by hand in Equation 10.4. Similarly, the 6 *df* treat*class{school} effect in Table 10.3 has been broken into a 1 *df* class*treatment effect and a 5 *df* school*class*treatment effect in Table 10.7. The sum-of-squares for class*treatment is identical to the sum-of-squares for the interaction contrast that we computed in the previous paragraph.

To summarize, if the levels of the nested factor can be grouped meaningfully across "nests," then you should consider analyzing the factor as crossed rather than nested. This can save a lot of work computing contrasts by hand. If the levels cannot be meaningfully grouped across nests, then the nested design is more appropriate, and contrasts on the nested effects themselves will rarely be of interest.

10.4.2 Polynomials

Sometimes the specific hypotheses that you want to test may be sets of orthogonal polynomials. For example, suppose that the schools in Tables 10.1 and 10.2 are ordered by the SES of their communities, as we suggested above. To generate the contrast for the quadratic trend, for example, across the means of these six schools, type

```
>HYPOTHESIS
>EFFECT = SCHOOL
>CONTRAST /  POLYNOMIAL ORDER = 2
>TEST
```

Table 10.7 ANOVA on the TEACHING METHOD data set assuming a fully crossed design.

```
LEVELS ENCOUNTERED DURING PROCESSING ARE:
SCHOOL
      1.000      2.000      3.000      4.000      5.000      6.000
CLASS
      1.000      2.000
TREAT
      1.000      2.000
```

DEP VAR: SCORE   N: 48   MULTIPLE R: 0.878   SQUARED MULTIPLE R: 0.771

ANALYSIS OF VARIANCE

SOURCE	SUM-OF-SQUARES	DF	MEAN-SQUARE	F-RATIO	P
SCHOOL	834.500	5	166.900	7.674	.198115E-03
CLASS	40.333	1	40.333	1.854	0.186
TREAT	408.333	1	408.333	18.774	.226549E-03
SCHOOL*CLASS	234.167	5	46.833	2.153	0.093
SCHOOL*TREAT	62.167	5	12.433	0.572	0.721
CLASS*TREAT	75.000	1	75.000	3.448	0.076
SCHOOL*CLASS *TREAT	102.500	5	20.500	0.943	0.472
ERROR	522.000	24	21.750		

(See section 8.4.2.) The output is shown in Table 10.8. The quadratic trend is not significant, $p = .449$. The effect sizes for this trend are

$$r = \sqrt{\frac{F}{F + df_{\text{denom}}}} = \sqrt{\frac{0.592}{0.592 + 24}} = 0.16 \qquad (10.8)$$

This contrast assumed that the schools were equally spaced on the SES scale. Suppose instead that the spacing between the third and fourth schools was twice as large as between schools 1 and 2, schools 2 and 3, and schools 4 and 5, and that the space between schools 5 and 6 was three times as large. There are an infinite number of metrics that capture this spacing, for example, 1, 2, 3, 5, 6, 9. You can tell SYSTAT to use this spacing by specifying any metric that captures the desired spacing after the *metric* option, as shown in the following commands:

```
>HYP
>EFFECT = SCHOOL
>CONTRAST /POLYNOMIAL ORDER=2 METRIC,
= 1 2 3 5 6 9
>TEST
```

The output from this analysis is shown in Table 10.9. Notice that SYSTAT automatically adjusted the values in its A matrix to account for the unequal spacing among schools. Unfortunately, the effect size for this new quadratic trend has gotten even smaller

$$r = \sqrt{\frac{F}{F + df_{\text{denom}}}} = \sqrt{\frac{0.175}{0.175 + 24}} = 0.085 \qquad (10.9)$$

and is not significant, $p = .679$.

10.4.3 Error Terms

The default error term for contrasts in SYSTAT is the *pooled*, or "global," error term, which was used in all of the examples in this chapter. This error term pools together the variance from all of the cells in the analysis and is equal to the mean square error from the overall ANOVA (for example, in Table 10.3b). The validity of pooling together sources of variance from multiple sources, however, is based on one of the assumptions underlying ANOVA: the assumption that these variances all represent estimates of the same variance σ^2. When this assumption appears to be substantially violated (see section 10.6.2), you should consider either transforming the data or using the *local* error estimate or *separate* error estimates described in the next two sections. In the present example, with only two observations per cell, the within cell variances will differ greatly by chance, so the equal variance assumption will be difficult to evaluate. See section 8.4.3 for more on using error terms other than the pooled error term.

10.5 Post-Hoc Tests

Frequently, investigators want to test hypotheses that are suggested by patterns in the data. Such tests are usually called

MACINTOSH
① Choose the menu item **Stats/MGLH/Test of Effects...**
② Select SCHOOL from the **Between Subjects** list.
③ Turn on **Polynomial**, type "2" into the **Order** box, and type "1 2 3 5 6 9" into the **Metric** box.
④ Click **OK**.

MS-DOS
① Choose the menu item **Statistics/MGLH/MGLM/ Hypothesis/Effect/** and select SCHOOL from the variables list.
② Choose the menu item **Statistics/MGLH/MGLM/ Hypothesis/Contrast/ Polynomial/Order/**, type "2" into the box, and hit <esc>.
③ Choose the menu item **Statistics/MGLH/MGLM/ Hypothesis/Contrast/ Polynomial/Metric/**, type "1 2 3 5 6 9" into the numeric vector box, and hit <esc>.
④ Choose the menu item **Statistics/MGLH/MGLM/ Hypothesis/Go (Test)!**

Table 10.8 Test of the quadratic trend across schools in the TEACHING METHOD data set.

```
TEST FOR EFFECT CALLED:      SCHOOL

A MATRIX
    1         2         3         4         5         6         7         8         9
  0.000     0.000    -0.655    -0.982    -0.982    -0.655     0.000     0.000     0.000

   10        11        12        13        14        15        16        17        18
  0.000     0.000     0.000     0.000     0.000     0.000     0.000     0.000     0.000

   19        20        21        22        23        24
  0.000     0.000     0.000     0.000     0.000     0.000

TEST OF HYPOTHESIS
     SOURCE          SS          DF         MS              F            P

  HYPOTHESIS      12.871         1       12.871          0.592        0.449
       ERROR     522.000        24       21.750
```

Table 10.9 Test of the quadratic trend across schools in the TEACHING METHOD data set, with schools spaced according to the metric 1, 2, 3, 5, 6, 9.

```
TEST FOR EFFECT CALLED:      SCHOOL

A MATRIX
    1         2         3         4         5         6         7         8         9
  0.000    -0.038    -0.470    -0.777    -1.017    -0.950     0.000     0.000     0.000

   10        11        12        13        14        15        16        17        18
  0.000     0.000     0.000     0.000     0.000     0.000     0.000     0.000     0.000

   19        20        21        22        23        24
  0.000     0.000     0.000     0.000     0.000     0.000

TEST OF HYPOTHESIS
     SOURCE          SS          DF         MS              F            P

  HYPOTHESIS       3.805         1        3.805          0.175        0.679
       ERROR     522.000        24       21.750
```

post-hoc, or *unplanned*, comparisons. SYSTAT provides a number of standard methods for assessing the significance of all pairwise comparisons among a set of means. These methods were discussed in section 8.5, and that discussion will not be repeated here.

10.5.1 Marginal Means

You can compute post-hoc tests on marginal means, the main effects, exactly as for a one-way ANOVA in section 8.5. For example, if you wanted to test all pairwise differences among the six schools using the Bonferroni adjustment (see section 8.5.2), you would type

```
>HYP
>POST SCHOOL / BONFERRONI
>TEST
```

The output is shown in Table 10.10. Even after adjusting for the 15 comparisons performed, schools 1, 2, and 4 are significantly different from school 6, and school 1 is significantly different from school 5.

10.5.2 Cell Means

You can also test pairwise differences among cell means. For example, if you wanted to test all pairwise differences among the 12 cell means in Table 10.3 defined by the treatment*schools conditions using a Bonferroni correction for the number of tests, type

```
>HYPOTHESIS
>POST TREAT*SCHOOL / BONFERRONI
>TEST
```

(See section 8.5.2.) The output from this procedure is shown in Table 10.11. Because there are 66 pairwise differences being tested by this post-hoc test, all of the probabilities were multiplied by 66, with a maximum value of 1.0. As you can see, after adjusting for the number of tests, there are only five significant differences remaining among the cell means. See section 8.5.2 for more information on Bonferroni corrections. It should be pointed out that many of these pairwise differences will be hard to interpret because they confound different independent variables. Because the Bonferroni correction is very conservative when the number of tests is large, it is worthwhile to think about which pairwise differences you are really interested in and only test those.

Table 10.10 Post-hoc comparisons of the 12 treatment*school means, using a Bonferroni correction for the number of tests.

```
COL/
ROW   SCHOOL
  1          1.000
  2          2.000
  3          3.000
  4          4.000
  5          5.000
  6          6.000
```

USING LEAST SQUARES MEANS.

POST HOC TEST OF   SCORE

USING MODEL MSE OF   21.750 WITH   24. DF.
MATRIX OF PAIRWISE MEAN DIFFERENCES:

	1	2	3	4	5	6
1	0.000					
2	3.250	0.000				
3	6.875	3.625	0.000			
4	4.000	0.750	-2.875	0.000		
5	7.875	4.625	1.000	3.875	0.000	
6	13.250	10.000	6.375	9.250	5.375	0.000

BONFERRONI ADJUSTMENT.
MATRIX OF PAIRWISE COMPARISON PROBABILITIES:

	1	2	3	4	5	6
1	1.000					
2	1.000	1.000				
3	0.105	1.000	1.000			
4	1.000	1.000	1.000	1.000		
5	0.037	0.883	1.000	1.000	1.000	
6	0.000	0.004	0.174	0.009	0.452	1.000

10.6 Checking Assumptions

ANOVA makes four assumptions about the distribution of errors within each cell: They are normally distributed, have constant variance, are independent, and are unbiased. Each of these assumptions is discussed below, and SYSTAT procedures are illustrated for assessing the validity of the first and second.

Table 10.11 Post-hoc comparisons of pairwise differences between 12 cell means in the TEST SCORES data set, using the Bonferroni adjustment for 15 comparisons.

```
COL/
ROW         TREAT       SCHOOL
  1         1.000       1.000
  2         1.000       2.000
  3         1.000       3.000
  4         1.000       4.000
  5         1.000       5.000
  6         1.000       6.000
  7         2.000       1.000
  8         2.000       2.000
  9         2.000       3.000
 10         2.000       4.000
 11         2.000       5.000
 12         2.000       6.000
```

USING LEAST SQUARES MEANS.

POST HOC TEST OF   SCORE

USING MODEL MSE OF   21.750 WITH   24. DF.
MATRIX OF PAIRWISE MEAN DIFFERENCES:

```
        1       2       3       4       5       6       7       8       9      10      11      12
 1      0
 2   4.250     0
 3   8.000  3.750      0
 4   2.250 -2.000 -5.750      0
 5   6.500  2.250 -1.500  4.250      0
 6  14.000  9.750  6.000 11.750  7.500      0
 7   5.750  1.500 -2.250  3.500 -0.750 -8.250      0
 8   8.000  3.750  0.000  5.750  1.500 -6.000  2.250      0
 9  11.500  7.250  3.500  9.250  5.000 -2.500  5.750  3.500      0
10  11.500  7.250  3.500  9.250  5.000 -2.500  5.750  3.500  0.000      0
11  15.000 10.750  7.000 12.750  8.500  1.000  9.250  7.000  3.500  3.500      0
12  18.250 14.000 10.250 16.000 11.750  4.250 12.500 10.250  6.750  6.750  3.250      0
```

BONFERRONI ADJUSTMENT.
MATRIX OF PAIRWISE COMPARISON PROBABILITIES:

```
        1       2       3       4       5       6       7       8       9      10      11      12
 1  1.000
 2  1.000   1.000
 3  1.000   1.000   1.000
 4  1.000   1.000   1.000   1.000
 5  1.000   1.000   1.000   1.000   1.000
 6  0.019   0.454   1.000   0.104   1.000   1.000
 7  1.000   1.000   1.000   1.000   1.000   1.000   1.000
 8  1.000   1.000   1.000   1.000   1.000   1.000   1.000   1.000
 9  0.125   1.000   1.000   0.648   1.000   1.000   1.000   1.000   1.000
10  0.125   1.000   1.000   0.648   1.000   1.000   1.000   1.000   1.000   1.000
11  0.009   0.219   1.000   0.049   1.000   1.000   0.648   1.000   1.000   1.000   1.000
12  0.001   0.019   0.316   0.004   0.104   1.000   0.059   0.316   1.000   1.000   1.000   1.000
```

These assumptions should always be assessed before reporting the results of an analysis of variance. See section 8.6 for more information.

10.6.1 Normality

For very small cell sizes the normality assumption cannot be assessed within cells, but assuming equal variance, the assumption can be assessed across cells. The following sections do this using stem and leaf displays, box plots, and summary statistics.

10.6.1.1 Stem and Leaf Displays and Box Plots

When the cell sizes are small, as they are in the present example, and the equal variance assumption is approximately met, a good way of assessing the normality assumption graphically is to look at plots of the studentized residuals from all of the cells in the ANOVA. These residuals can be saved into a file when the ANOVA is calculated by SYSTAT. This was done above in section 10.3 when we generated Table 10.3. Stem and leaf displays and box plots (see sections 3.2.1.2 and 3.2.1.3) are both good for getting a feel for the distribution of the residuals. Open the file containing the studentized residuals (e.g., "nested.res") and then produce these plots with the commands

```
>GRAPH
>STEM STUDENT
>BOX STUDENT
```

The resulting plots are shown in Figure 10.3. Both plots indicate that the distribution is very symmetric and has no outliers. On the basis of these plots one would conclude that the normality assumption is adequately met. The distribution of residuals in the stem and leaf display appears to be somewhat flat-topped, and this will be further assessed in the next section.

10.6.1.2 Skewness and Kurtosis

You can compute the skewness and kurtosis of the residuals directly (see section 3.1.1 for more information). Simply type:

```
>STATS
>STATISTICS RESIDUALS/SKEWNESS KURTOSIS N
```

The output is shown in Table 10.12. The value of skewness is zero to two decimal places, which suggests, as did the box plot in Figure 10.3, that the distribution is very symmetric. The

```
STEM AND LEAF PLOT OF VARIABLE:
    STUDENT , N =    48

MINIMUM IS:          -1.918
LOWER HINGE IS:      -0.751
MEDIAN IS:           -0.000
UPPER HINGE IS:       0.751
MAXIMUM IS:           1.918

 -1      9955
 -1      3332
 -0    H 99777777
 -0    M 44211111
  0      11111244
  0    H 77777799
  1      2333
  1      5599
```

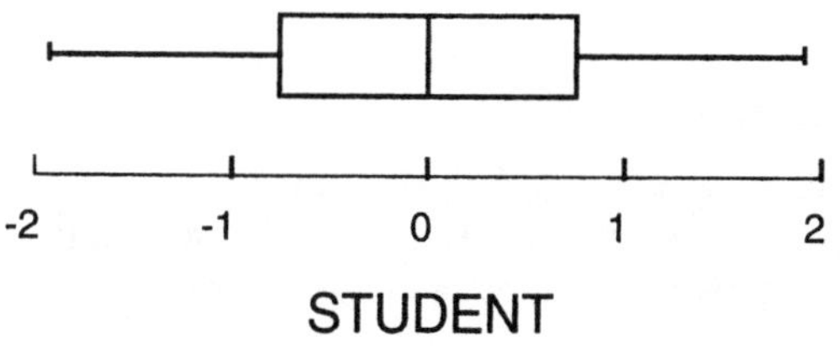

STUDENT

Figure 10.3 Stem and leaf display and box plot of the studentized residuals from the ANOVA on the TEACHING METHOD data set in Table 10.3.

value of kurtosis in Table 10.12 is –.98, suggesting that the distribution is flat-topped, as it appeared in Figure 10.3. The *p*-values for kurtosis with small sample sizes in Snedecor and Cochran (1982, Appendix 20ii) indicate that this value is significant, $p < .05$, one-tailed.

Table 10.12 Skewness and kurtosis for the residuals from the ANOVA on the TEACHING METHOD data set in Table 10.3.

```
TOTAL OBSERVATIONS:       48

                       RESIDUAL

   N OF CASES               48
   SKEWNESS(G1)           -0.00
   KURTOSIS(G2)           -0.98
```

10.6.1.3 Remedies

If the residuals are not normally distributed, the p-values from the F tests and t tests in the regression output may be too small. Fortunately, F's and t's are robust to violations of the normality assumption, so moderate departures from normality can often be ignored if the other assumptions are met. In such a case, the reader should be warned that the p-values are not accurate. If the deviation from normality is substantial, the most common remedies are (1) to transform the dependent variable, or (2) perform the analysis on the medians rather than the means, which is beyond the scope of this book. Transforming dependent variables proceeds in the same fashion as transforming the outcome variable covered in Chapter 7. Most data analysts would consider the normality assumption adequately met in the TEACHING METHOD data set, at least close enough not to warrant the interpretive complexity that would be added by transforming the scores.

10.6.2 Equal Variance

The second assumption underlying ANOVA is that the residuals have the same variance within each cell of the design (see section 8.6.2). You can usually assess this assumption both graphically and with formal tests. However, formal tests such as *Levene's test* are not applicable for extremely small cell sizes, such as the cell sizes of two in the present example. Therefore, in this section I will limit the discussion to graphical assessment. Levene's test is discussed in sections 8.6.2 and 9.6.2.

When the number of cells is large, it becomes difficult to pick out which cells have large and small variances from plots such as Figure 9.1. When the number of data points is small, the problem is further compounded by the fact that the variances themselves will be highly variable, even when they are equal in the population. Thus, rather than compare variances among cells, we are limited to looking for patterns in the variances across cells. This raises the further issue of how the cells are to be ordered, that is, across what variable do you want to look for patterns. Below I show two reasonable ways to order the cells: (1) by their expected values, and (2) by levels of factors in the design.

10.6.2.1 Plots of Residuals Versus Estimated Values

A very common violation of the equal variance assumption occurs when the variance of the data increases with the magnitude of the dependent variable. Fortunately, this type of violation can often be fixed by transformations up or down the ladder (see Chapter 7). Plots of the studentized residuals against the cell means, their expected values, can sometimes uncover such violations. To generate this plot, open the file containing the residuals (saved during the computation of the ANOVA in Table 10.3) and type

```
>GRAPH
>PLOT STUDENT*ESTIMATE
```

The resulting plot for the example data is shown in Figure 10.4. The variance does not appear to change in a systematic manner as the means of the cells increase, so we can conclude that there is no evidence that the equal variance assumption is violated in this fashion.

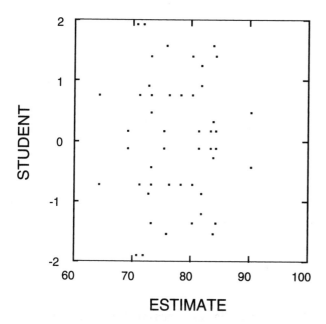

Figure 10.4 Plot of the studentized residuals against the cell means from the ANOVA on the TEACHING METHOD data set.

10.6.2.2 Plots of the Residuals by Factors

Another way to order the cells of the design is to look for patterns among the variances in plots of the residuals against individual factors or combinations of factors. Many such plots may be possible and, if you generate enough of them, you are likely to uncover patterns due to chance. Therefore, you should limit yourself to plots of factors or combinations of factors that you suspect might differ in their variance.

For example, one might suspect that the experimental groups in the TEACHING METHODS data set might have larger variance than the control groups. Because there are only two data points per cell, you cannot profitably assess this suspicion by plotting the residuals across all of the individual cells. Instead, you can group the residuals for the experimental and control groups by simply plotting them against the treatment factor. To do this you must get the residuals and the factor codes into the same file, which you can do by abutting the original data file called, say, TEACH.DAT, with the file containing the residuals, NESTED.RES, with the following series of commands

```
>DATA
>USE TEACH.DAT(TREAT) NESTED.RES(STUDENT)
>SAVE NEWFILE
>RUN
>SYSTAT
>EDIT NEWFILE
```

The variables in parentheses are the only columns that SYSTAT will put into the new file, cleverly called NEWFILE. If you leave off the parentheses, SYSTAT will put all of the variables in both files into the new file. (Note that if you are using a Mac it is usually easier just to cut and paste variables from one file to another.)

You can now create the desired plot by typing the commands

```
>GRAPH
>PLOT STUDENT*TREAT
```

This plot is shown in Figure 10.5. This plot does not show any clear indication that the variances are different in the two treatment groups.

Another plot of the same data that can be easier to diagnose are box plots of the residuals within each level of a factor. For example, to create box plots of the studentized residuals for each of the two treatment conditions, type

```
>BOX STUDENT*TREAT
```

This plot is shown in Figure 10.6, and although the box is a little larger in the experimental condition, the difference is not large enough to cause concern.

Nested Fixed Factors ANOVA 355

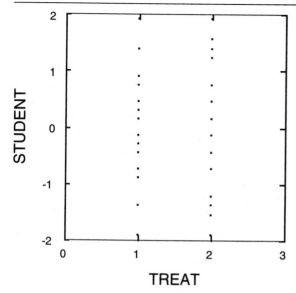

Figure 10.5 Plot of the studentized residuals against the two treatment groups in the TEACHING METHOD data.

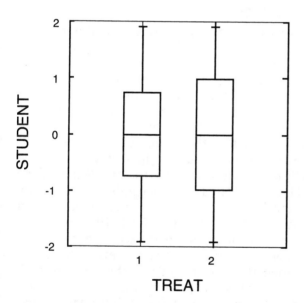

Figure 10.6 Box plots of the studentized residuals against the two treatment groups in the TEACHING METHOD data set.

10.6.2.3 Remedies

If you could see a clear pattern in Figure 10.4, the most common remedy would be to perform a *variance stabilizing transformation* on the dependent variable (Snedecor & Cochran, 1982, pp. 287–292). Transforming variables is covered in Chapter 7. If you find a pattern in plots like those in Figures 10.5 or 10.6, there may not be much that you can do. You could not, for example, transform the experimental group and leave the control group untransformed. Fortunately, F and t tests are quite robust against violations of this assumption, and although the p-values will not be exactly accurate, you should not be overly concerned with the accuracy of your p-values: The estimates of effects and means do not depend on this assumption. When the violation is serious enough to bring even the approximate validity of the significance tests into question, perhaps the variance of the groups should be analyzed as a dependent variable. For example, the treatment may cause pupils to become more variable (less stable?) independently of any effect it might have on average performance. Bartlett's and Levene's tests can be adapted to this type of analysis.

10.6.3 Independence

The third assumption underlying ANOVA is that the errors are uncorrelated with one another. When the normality assumption is met, uncorrelated errors are also *independent*, so this assumption is often called the independence assumption. When errors are correlated, the estimates of means and effects are still unbiased, but the population variance is underestimated. This produces F's and t's that are too large, and confidence intervals that are too small.

As stated in the previous chapter, in ANOVA correlated errors are usually impossible to detect in the data. Whether one should worry about correlated errors depends almost entirely on the experimental design. For example, if the pupils in each treatment condition in the TEACHING METHOD study did not have an opportunity to interact with each other during the course of the experiment, and if they weren't instructed by the same teacher, then their errors are probably uncorrelated. However, it is very common in such studies to teach and test pupils as a group, allowing them to interact during the course of the study. When this is the case, the errors within cells are almost certainly not independent.

If only the pupils *within* a condition are allowed to interact with each other, then the problem is mainly one of degrees of freedom: You don't really have one *df* within each cell. Instead, you should probably treat the mean of the pair of pupils within each condition as the data point, and perform the ANOVA on the basis of this design (in this case, repeated measures on classrooms).

If the pupils are allowed to interact between conditions, that is, the experimental pupils can play with, have lunch with, and otherwise interact with pupils in the control condition, then not only are their errors correlated, but you no longer have a good measure of the effect of the new teaching method. For example, while discussing their experiences on the playground, the experimental group may have boasted about their special treatment, thus making the controls feel left out and perform more poorly on the test as a consequence. There are no transformations or redesigns of the analysis that can recover the effect of the new teaching method. This is an experimental design problem that must be avoided when the data are collected: It is beyond even the versatility of a powerful program like SYSTAT to solve.

10.6.4 Errors Are Unbiased

The fourth assumption underlying ANOVA is that the errors are unbiased. For almost all ANOVAs, this assumption can be considered true by definition (see section 8.6.4).

10.7 Confidence Intervals

SYSTAT does not directly provide confidence intervals in ANOVA, but they can be obtained easily from estimates of means and variances that SYSTAT does provide. So long as the ANOVA has no random factors, the confidence intervals can be computed in the manner described in section 8.7, so that discussion will not be repeated here.

10.8 Graphing the Results

This section shows how to create graphs of means or effects in nested designs, such as those in the TEACHING METHOD data set, with error bars around those means or effects. More information and types of graphs can be found in section 8.8.

10.8.1 Bar Chart with Error Bars

When the independent variable is categorical and does not have a meaningful scale, that is, the ordering of the categories is arbitrary, it is standard to plot the means of the categories using a bar graph with error bars (see section 9.8.1). For example, the means for the two treatment conditions have standard errors equal to 0.952 (from Table 10.3). So to plot the means for the schools, along with errors bars showing ±1 standard error of the mean using the appropriate pooled variance for the variance estimate (section 8.7.1), you must first create a new column containing the standard errors. To do this type

```
>LET SE_TREAT = .952
```

Now you can create the desired plot using the commands

```
>GRAPH
>BAR SCORE * TREAT/ERROR=SE_TREAT
```

The option *ERROR=SE_TREAT* tells SYSTAT to include error bars based on the values found in the column SE_TREAT. The resulting graph is shown in Figure 10.7. If you wanted to use the axis labels *control* and *experimental*, for example, instead of "1" and "2", you could create a character column containing those labels and use it in the plot. You can also use other types of error bars, using separate variance estimates, confidence

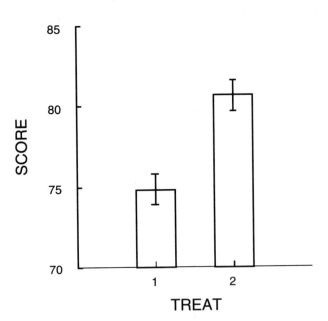

Figure 10.7 Bar graph showing the means of the control and experimental groups in the TEACHING METHOD data set, with error bars of ±1 SE of the mean, where the SEs are based on the pooled variance from the ANOVA.

limits, and so on. See section 3.2.2.5.

In the same manner, you can make a bar chart of the means of the six schools. These means have standard errors equal to 1.649 (from Table 10.3, based on the pooled error estimate). So to plot the means for males and females, along with error bars showing ±1 standard error of the mean, you must first create a new column containing the standard errors by typing

```
>LET SE_SCHOO = 1.649
```

Now you can create the desired plot using the commands

```
>GRAPH
>BAR SCORE * SCHOOL/ERROR=SE_SCHOO
```

This plot is shown in Figure 10.8.

10.8.2 Line Chart with Error Bars

When an independent variable is on a meaningful scale, it is common to graph the means using a line graph. As for bar graphs in the previous section, if you want to include error bars you will need to create a new column in your data set that contains the values for those error bars. For example, suppose

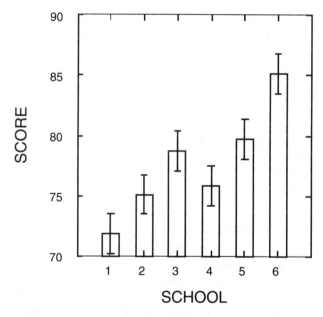

Figure 10.8 Bar graph showing the means of the six schools in the TEACHING METHOD data set, with error bars of ±1 SE of the mean, where the standard errors are based on the pooled variance from the ANOVA.

that the schools are ordered by the SES of their communities and you want plot the means along with error bars based on the pooled error estimate. First you must create a new column containing the values 1.649, as was done in the previous section (taken from Table 10.3) and then plot the means for the groups, along with error bars, using the commands

```
>GRAPH
>CPLOT SCORE * SCHOOL/ERROR=SE_SCHOO LINE
```

The resulting graph is shown in Figure 10.9.

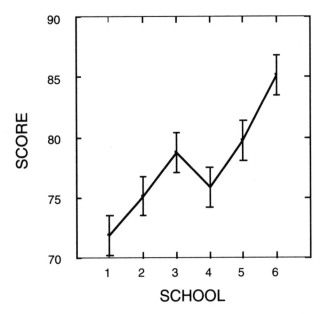

Figure 10.9 Line graph of the means of the six schools in the TEACHING METHOD data set, including standard error bars based on the pooled error variance from the ANOVA.

10.8.3 Two-Way Plots of Means

SYSTAT allows you to plot more than one dependent variable in the same graph. For example, suppose that you want to create separate bars for the treatment and control groups within each school of the TEACHING METHOD data set. First, you must create two new columns in the data set containing only the test scores of the control and experimental groups, respectively. This can be done easily using the commands

```
>IF TREAT = 1 THEN LET CONTROL = SCORE
>IF TREAT = 2 THEN LET EXPER = SCORE
```

Each of these columns will have missing data, but this won't affect the graph. To include the appropriate error bars you need to create yet another column, this time containing the values of the standard error, for example. The standard error for the experimental and control groups within each school based on the pooled error variance is 2.332, as shown in Table 10.3. Thus, the new column can be created by typing

```
>LET SE_2WAY = 2.332
```

With these new columns you can now create the bar graph by typing, for example

```
>GRAPH
>BAR CONTROL EXPER*SCHOOL,
ERROR1=SE_2WAY,SE_2WAY YMIN=70 YMAX=90,
FILL=7,4   YLAB="SCORE"
```

The resulting bar graph is shown in Figure 10.10 (see section 9.8.3 for an example of menu commands). The *error1* option does two things: (1) it tells SYSTAT to show only the upper limit on the error bar, which is useful when the bars in the graph contain some kind of color or pattern, and (2) it specifies the columns containing the error bar values for controls and experimentals, respectively, separated by a comma. In this case the error bar values are contained in the same column, but both still must be specified. The *ymin* and *ymax* options were set by trial and error to make the graph look good. The *fill* option specifies the numbers corresponding to the desired patterns used to fill the bars. Finally, the *ylab* option allows you to specify the label of the Y-axis. Because SYSTAT thinks you have two dependent variables in this plot, it does not automatically know what to call the Y-axis.

You could have grouped the bars in this graph by treatment group rather than by school, but this would require a little more work. First, you must create six new columns containing the scores for each school with commands such as

```
>IF SCHOOL = 1 THEN LET S1 = SCORE
>IF SCHOOL = 2 THEN LET S2 = SCORE
>IF SCHOOL = 3 THEN LET S3 = SCORE
>IF SCHOOL = 4 THEN LET S4 = SCORE
>IF SCHOOL = 5 THEN LET S5 = SCORE
>IF SCHOOL = 6 THEN LET S6 = SCORE
```

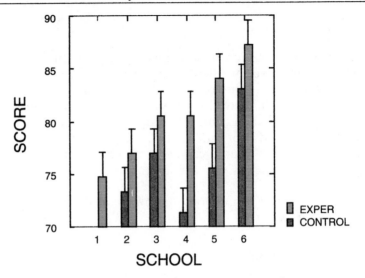

Figure 10.10 Bar graph showing the means of experimental and control groups within each school in the TEACHING METHOD data set.

(See section 2.2.2.) With these new columns you can now create the bar graph by typing

```
>BAR S1 S2 S3 S4 S5 S6 * TREAT/ YMIN=70,
YMAX=90 FILL=2,3,4,5,6,7 YLAB="SCORE"
```

The resulting bar graph is shown in Figure 10.11. This graph makes the overall difference between the experimentals and controls a little easier to see than in Figure 10.10.

If you prefer a line graph over a bar graph, you can generate one using the same columns for the schools that you just created. Simply type

```
>CPLOT S1 S2 S3 S4 S5 S6 * TREAT/YMIN=65,
YMAX=90 YLAB="SCORE" LINE=1,3,5,7,9,11
```

The options are identical to those for the bar chart above, except that the *line* option replaces the *fill* option. The resulting graph is shown in Figure 10.12. This plot does a better job than the bar graph in Figure 10.11 of making explicit the fact that the experimentals out performed the controls in every school.

10.8.4 Plotting Interactions

The two-way plots in the previous section showed the means of the cells corresponding to each school-by-treatment combi-

Nested Fixed Factors ANOVA 363

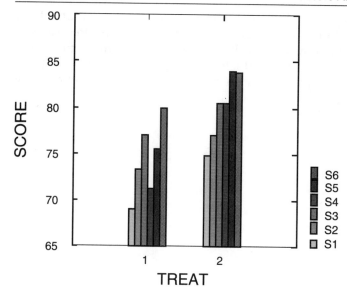

Figure 10.11 Bar graph showing the means of each school within the control (1) and experimental (2) groups in the TEACHING METHOD data set.

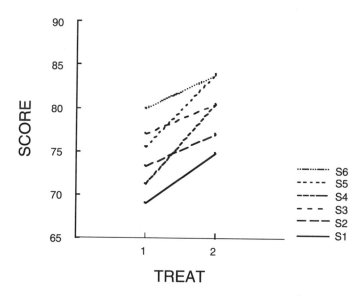

Figure 10.12 Line graph connecting the means of the control (1) and experimental (2) groups for schools in the TEACHING METHOD data set.

nation. Such plots are often mistakenly called "plots of the interaction," but in fact they contain both main effects as well as interaction. If you really want a plot of the interaction itself, you must use the residuals, which can be derived from Table 10.3. For instructions on how to plot real interactions, see section 9.8.3.

11
Random Factors ANOVA: Repeated Measures

This chapter shows how to perform an analysis of variance in SYSTAT with two or more factors, one of which is a random factor. Technically, truly random factors are factors that have an infinite number of levels, only a finite number of which have been randomly sampled in an experiment. In practice, we usually consider a factor random when the number of levels selected for the experiment is small relative to the number of levels in the population, and a different random set of these levels would be selected in future (hypothetical) replications of the experiment. The most common random factor is subjects: We assume that the number of subjects in our experiment is small relative to the number in the relevant population and that we would get different subjects in future replications. When the number of levels of a factor in the study is approximately equal to the number in the population, it is safer to assume that the factor is fixed, and use the SYSTAT procedures discussed in Chapter 9.

In "fixed factor" designs, you can still usually identify a random factor, but this factor is confounded with the error in the smallest cells of the design. For example, in a between-subjects design in which each subject is measured only once, *subject* is still a random factor, but each subject's effect is confounded with error, so the subject effects cannot be estimated apart from error in the analysis. Any individual differences among subjects get lumped together into the mean square within cells.

However, in *repeated measures* designs, each level of the random factor (e.g., each subject) is measured repeatedly. When each level of the random factor is measured more than once, the random effects can be estimated separately from within cell error by averaging across cells. Sometimes, repeated measures designs are referred to as *within subjects* designs, and factors on which subjects are measured repeatedly are called *within subjects factors*.

Means and effects are estimated in the same manner in repeated measures analyses as they are in fixed factor analyses (at least when you have balanced designs—there is little consensus on what to do with unbalanced repeated measures designs). Furthermore, the computations of sums-of-squares and degrees of freedom are identical in designs with random factors and designs with only fixed factors. What differs when you add a random factor is that you have a new source of noise against which to measure possible effects, so the significance tests and confidence intervals use different variance estimates than they would in a fixed-only design. This all leads to the following: You *could* analyze a repeated measures design exactly as though it were a fixed-factor design, using the procedures in Chapters 9 and 10. In this case, the "random" factor, for example subjects, would be coded in the same manner as any other factor and could be crossed or nested with other factors. The means, effects, and ANOVA table that you would get from this analysis would be correct except for some of the the F's, p-values, and standard errors. These you would need to compute yourself, which is very easy if you are familiar with expected mean square procedures.

Fortunately, if you do not know how to choose error terms for F tests, you can get SYSTAT to analyze the design properly as a repeated measures design. This chapter shows how to do this. Many of the procedures are the same as in fixed factor designs; the most striking difference is in the way the data file must be set up. This is the topic of the first section. Once you get the data file correct, the rest of the analysis should follow easily.

11.1 The Example Data Set

11.1.1 The Choice Experiment

Imagine a situation in which you have two buttons on a control board in front of you, with a small light above each button. When the left light comes on, your task is to move your dominant index finger from a central resting position and push the button on the left. When the right light comes on, you do the same but push the button on the right. Simple choice response times, such as how fast you can push the buttons in response to the lights is is a well-known correlate of scores on intelligence tests. Suppose, for some theoretical reason, you wanted to know what difference it makes in response times for people to be cued about the side on which the light will appear, so that they can anticipate their choice. To do this you might add three new "warning" lights to the board: a new light on the left that indicates that the button light will appear on the left, a new light on the right that indicates that the button light will appear on the right, and a center light that indicates a 50/50 chance of the left or right button lights coming on. Subjects are not allowed to move their fingers before the button lights come on, and if they do, no button light comes on for that trial. Furthermore, the delay between when the warning light comes on and when the target light comes on can be varied: Suppose that you used two delays, 0.2 seconds and 0.5 seconds (half a second is a long time in this type of study) You suspect that the longer the warning, the better subjects will be able to anticipate the light. Subject response times are measured in milliseconds (msec).

This may be a little detailed for an example data set, but this is a realistic type of study, and it provides us with a repeated measures design with two crossed within subjects factors: warning location (left, center, right) and warning delay (.2 sec and .5 sec). There are many trials for each of the six combinations, presented in random order for each subject, but it is common to use the averages of trials within cells as the data points. Therefore, you have six numbers for each of your, say, 15 subjects. This chapter shows how to analyze such a data set in SYSTAT.

11.1.2 The Data File

In all of the preceding chapters, each variable in the data set was given a single column in the data file, as shown in Table 11.1. (The response time data in Table 11.1 was generated from the first three data columns with the command LET RT=SUBJECT*10 + LOCATION*100 + DELAY*25 + 200 + 100*zrn.) However, this is *not* the way you set up a data file for a repeated measures design in SYSTAT. Instead, we need to give SYSTAT the data in a format with all of the observations on a given subject in a single row. This means that we need at least six columns in the example data file: one for each of the six

Table 11.1 The first 15 cases in the CHOICE data set.

	SUBJECT	LOCATION	DELAY	RT
1	1	1	1	282
2	1	1	2	228
3	1	2	1	281
4	1	2	2	423
5	1	3	1	344
6	1	3	2	563
7	2	1	1	425
8	2	1	2	275
9	2	2	1	410
10	2	2	2	316
11	2	3	1	373
12	2	3	2	171
13	3	1	1	260
14	3	1	2	369
15	3	2	1	469

repeated measures. Furthermore, these six columns need to be ordered hierarchically by the two within subjects factors.

The appropriate data set is shown in Table 11.2. The first row in Table 11.2 contains all six measurements on subject 1. The first column for subject 1 contains the entry 282, which is that subject's response time for location$_1$delay$_1$ in Table 11.1. The second column for subject 1 contains the entry 228, which is that subject's response time for location$_1$delay$_2$. Reading on across the first row we find the responses times for location$_2$delay$_1$ (281), location$_2$delay$_2$ (423), location$_3$delay$_1$ (344), and location$_3$delay$_2$ (563). Thus, the six response times are hierarchically organized with location varying from 1 to 3 and delay varying from 1 to 2 within each location. It is very important to be clear about the structure of these rows because SYSTAT will need to be told how to divide the rows. Note that the column labels in Table 11.2 code the repeated measures condition that they contain: column "L1D1" contains the data for the condition with location=1 and delay=1; this type of labelling is not necessary, but can prove to be very useful in keeping track of variables.

If you are entering your data by hand, you should enter it in the form of Table 11.2. If you are getting your data from another program, perhaps text output from the program that collected the data, it may come in the form of Table 11.1.

Table 11.2 All 15 subjects, six measures each, in the CHOICE data set.

	L1D1	L1D2	L2D1	L2D2	L3D1	L3D2
1	282	228	281	423	344	563
2	425	275	410	316	373	171
3	260	369	469	346	446	390
4	223	376	266	313	394	249
5	351	290	296	431	208	483
6	451	473	453	490	524	399
7	132	191	346	402	452	376
8	397	239	369	432	470	443
9	523	375	391	410	456	501
10	336	322	374	535	494	504
11	116	171	490	347	428	327
12	350	250	333	478	487	285
13	300	240	431	410	377	636
14	452	338	300	390	500	502
15	471	496	390	496	575	549

Unfortunately, SYSTAT is of no help when changing the data from the format shown in Table 11.1 to the format shown in Table 11.2. If you have only a hundred or so data points it is probably easiest just to retype them into a new data file (which is what I did to get Table 11.2). If you have a large number of data points, you should get help from some other program to change the file format. The easiest way is to write a small program or script, perhaps within a spreadsheet, to do the work, but not everyone has access to an appropriate program or knows how to program in a computer language. Perhaps future versions of SYSTAT will contain some apparatus for performing this necessary, and very labor intensive, operation.

11.2 Looking at the Data

11.2.1 Scatterplots

In repeated measures designs there are no really good ways to look at scatterplots in SYSTAT. About the best you can do is plot the repeated measures against subjects and use this plot to check for observations that are extreme relative to the data set as a whole. To create a scatterplot of the raw data by subject, you can either specify each column individually, L1D1 through L3D2, or you can let SYSTAT do the work for you by using a hyphen to indicate a range of columns. This latter method is used in the following commands:

```
>GRAPH
>PLOT L1D1-L3D2 * SUBJECT/YMIN=0,
YLAB="RESPONSE TIME (MSEC)"
```

"L1D1-L3D2" tells SYSTAT to use the data in columns L1D1 through L3D2 as the repeated measures values (unlike some other programs that would plot a separate graph for each variable). The resulting scatterplot is shown in Figure 11.1. When you find an extreme case in this plot, look at the rest of the data for that subject and see if the extreme case is out of line for that subject. None of the data points in Figure 11.1 stand radically apart from the rest of the data. It is also a good idea to look at each subject, i.e., at each vertical slice, to see whether any data points stand out within a slice. This diagnostic procedure is mainly aimed at ensuring that there are no gross errors in the data set. Each point within a vertical slice of this graph comes from a different condition, so they are not expected to have any particular distribution.

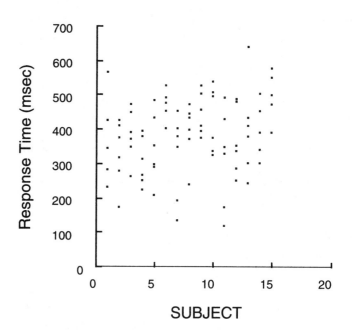

Figure 11.1 Scatterplot of the raw data in the CHOICE data set.

11.2.2 Subject Profiles

A better way to look at repeated measures data is to look at separate plots of each subject's data. SYSTAT offers an easy way to do this using profile icons. Each subject gets a separate polygon whose height above the X-axis is determined by the value of the variables specified in the *icon* command. The variables are equally spaced from left to right across the polygon in the order that they are listed in the icon command, so whether the actual shape of the profiles has any obvious meaning depends on the ordering of the variables. The following command shows profiles using the order of the repeated measures in the data set:

```
>GRAPH
>ORIGIN  0  -20    [Macintosh only]
>ICON  L1D1-L3D2/ROW=3,
TITLE="Subject Profiles" HEIGHT=4 IN,
WIDTH=5 IN PROFILE
```

On a Macintosh, the *origin* command moves the plot into the viewing area of the screen, which is necessary because we changed the height and width of the plot. The *row* option tells SYSTAT how many rows of polygons you would like. The *profile* option at the end of the command is the one that tells SYSTAT to use profile icons. The resulting graph is shown in Figure 11.2. This plot is useful for finding extreme observations, identifying patterns in the profiles, and comparing between subject variability. However, when the variables are not ordered on a quantitative scale, which these are not, the patterns may be hard to interpret. There does appear to be a tendency to increase from left to right in these profiles, which would make sense given the way the variables are defined. The first subject on the left in the second row appears to have somewhat less variability in response times than the rest.

It is often a good idea to reorder the variables and look at new profile plots. For example, the profiles in Figure 11.2 are ordered by location and by delay within location. To look at profiles ordered by delay and location within delay, reorder the variables in the icon command:

```
>ORIGIN  0  -20    [Macintosh only]
>ICON L1D1 L2D1 L3D1 L1D2 L2D2 L3D2/,
TITLE="Subject Profiles" ROW=3 HEIGHT=4 IN,
WIDTH=5 IN PROFILE
```

In this command, the three delay$_1$ conditions come first and the three delay$_2$ conditions come last, which means that they

MACINTOSH

① Choose the menu item **Graph/Icons/Profile** and select the variables L1D1, L1D2, L2D1, L2D2, L3D1, and L3D2 from the variable list.

② Click on **Rows**, type "3" into the box, and click **OK**.

③ Click on **Title**, type "Subject Profiles" into the box, and click **OK**.

④ Click on the **ruler**, type "4" into the **Height** box and "5" into the **Width** box, type "0" into the **X Origin** box and "–20" into the **Y Origin** box, and click **OK**.

⑤ Click **OK**.

MS-DOS

① Choose the menu item **Graph/Icon/Type/Prof**.

② Choose the menu item **Graph/Icon/Variables/** and select the variables L1D1, L1D2, L2D1, L2D2, L3D1, and L3D2 from the variable list.

③ Choose the menu item **Graph/Icon/Options/ Rows** and type "3" into the box.

④ Choose the menu item **Graph/Icon/Options/ Height**, type "4" into the box and hit <ret>, then select "IN".

⑤ Choose the menu item **Graph/Icon/Options/ Width**, type "5" into the box and hit <ret>, then select "IN".

⑥ Choose the menu item **Graph/Icon/Go!**

will be plotted in that order from left to right in the profile. This plot is shown in Figure 11.3.

When one or more subjects have much larger means than the others, their values may dominate, making the other subjects profiles low and flat. To compensate for this you can standardize the profiles within subjects by including the *stand* option after the slash in the *icon* command line, so that each subject's profile is visible. In the present example this was not necessary. Using this option will tend to obscure between subject variability.

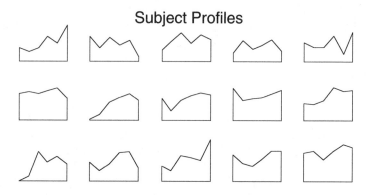

Figure 11.2 Profile polygons for each subject in the CHOICE data set, ordered by location and delay within location.

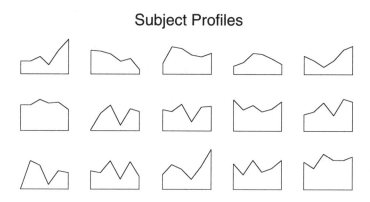

Figure 11.3 Profile polygons for each subject in the CHOICE data set, ordered by delay and location within delay.

11.3 Repeated Measures ANOVA

The *anova* command for repeated measures look quite a bit different from the anova commands in previous chapters. Immediately after the command word, enter the names of all of the repeated measures columns in hierarchical order (see section 11.1). If the columns are in the appropriate order in the data set you can abbreviate this variable list by specifying the first and last repeated measures column, separated by a dash. Thus, "L1D1-L3D2", for example, tells SYSTAT to include all of the columns between L1D1 and L3D2, inclusive. Next comes the slash, followed by option specifications. The most important option is the *repeat* option, which allows you to specify the number of levels of each within subjects factor in top-down hierarchical order. For example, in the CHOICE data set the columns are ordered by location at the highest level and then delay within location. Location has three levels and delay has two levels, so the appropriate setting for the *repeat* option is "3,2". Finally, you may give names to the within subject factors to be used in the ANOVA output. These names follow the *names* option in the same order that their levels are specified in the *repeat* option, and the names must be enclosed in quotes. As with any ANOVA, you can save residuals and choose from standard or extended output.

The following commands generate a repeated measures ANOVA on the CHOICE data set with two within subjects factors and show the results in standard output:

```
>MGLH
>ANOVA L1D1-L3D2 /REPEAT=3,2 NAMES="LOCATION",
"DELAY"
>SAVE "CHOICE.RES"/MODEL
>PRINT MEDIUM
>ESTIMATE
```

The results are shown in Table 11.3. You should always examine the "repeated measures factors and levels" table near the top of the output to ensure that the *repeat* and *names* options divided the within subjects factors appropriately.

The location effect and location*delay interaction both have more than one *df* in their sums-of-squares, and for multi-*df* effects SYSTAT reports standard *F* tests as well as Greenhouse-Geisser and Huynh-Feldt probability adjustments and various multivariate tests (shown at the bottom of the output). These may be useful on some rare occasions, but generally these multi-*df* effects do not address questions of scientific interest. (Note that one *df* effects, such as the delay effect in

MACINTOSH

① Choose the menu item **Stats/MGLH/Fully Factorial (M)ANOVA...**

② Select L1D1, L1D2, L2D1, L2D2, L3D1, and L3D2 from the **Dependent Variable(s)** list.

③ Click on **More...**

④ Turn on **Repeat** and type "3,2" into the box.

⑤ Type ' "location","delay" ' (including the double-quotes) into the **Names** box.

⑥ Turn on **Means & Std. Errors** and **Save Residuals**.

⑦ Click **OK**.

⑧ Type "choice.res" into the **Save residuals as** box and click **OK**.

MS-DOS

① Choose the menu item **Utilities/Output/Results/Medium**.

② Choose the menu item **Statistics/MGLH/MGLM/Model/Variables**, hit <esc>, and type "L1D1-L3D2 = constant" into the box.

③ Choose the menu item **Statistics/MGLH/MGLM/Model/Options/Repeat** and type "3,2" into the box.

④ Choose the menu item **Statistics/MGLH/MGLM/Model/Options/Names** and type ' "location","delay" ' (including the double-quotes) into the box.

⑤ Choose the menu item **Statistics/MGLH/MGLM/Estimate/Save**, hit <esc>, and type the save filename, e.g., "choice.res" into the box.

⑥ Choose the menu item **Statistics/MGLH/MGLM/Estimate/Go!**

Table 11.3, do not require these additional statistics.) Therefore, for most practical purposes these adjustments and multivariate tests can simply be ignored.

The standard output also shows sets of orthogonal polynomial contrasts for each multi-*df* effect. These automatic contrasts can save you considerable effort if polynomial trends are of interest in your data. We will turn to contrasts in the next section.

The ERROR terms in the repeated measures output in Table 11.3 are all interactions between the random factor, subjects, and the effect being tested. For example, the ERROR term for the location effect (MS=7733.930) is really the subject × location interaction. The ERROR term for the linear trend in location (MS=9600.386) is really the subject × linear-trend-in-location interaction.

11.4 Contrasts

As I've said in previous chapters, F tests with more than one degree of freedom in the numerator, like the location and location*delay tests in Table 11.3, are usually of little scientific interest because (1) if they are significant, they only tell us that some differences exist, not which means are different, and (2) if they are not significant, this does not mean that none of the means are different or that there are no significant trends among the means. The following section show how to test specific predictions with contrasts on within subjects factors in SYSTAT.

11.4.1 Planned Comparisons

11.4.1.1 Main Effects

Suppose that your theory predicted that location 1, in which the warning light appears on the left and therefore is first processed by a subject's right hemisphere, would have a faster response time than either of the other two locations, which would be approximately equal. This hypothesis can be expressed using the contrast weights –2, 1, and 1, respectively, for the means of the three locations (see section 8.4.1 for how to derive contrast weights). These contrast weights can be entered into SYSTAT directly to test the desired hypothesis. To test this hypothesis, first compute the repeated measures ANOVA as was done in Table 11.3, and then type the commands

Table 11.3 Repeated measures ANOVA on the CHOICE data set showing standard results.

```
NUMBER OF CASES PROCESSED:        15

DEPENDENT VARIABLE MEANS
     L1D1          L1D2          L2D1          L2D2          L3D1          L3D2
    337.933       308.867       373.267       414.600       435.200       425.200

REPEATED MEASURES FACTORS AND LEVELS
                  DEPENDENT VARIABLES
WITHIN FACTOR    1  2  3  4  5  6
location         1  1  2  2  3  3
delay            1  2  1  2  1  2

RESIDUALS HAVE BEEN SAVED

UNIVARIATE AND MULTIVARIATE REPEATED MEASURES ANALYSIS

WITHIN SUBJECTS
---------------

SOURCE           SS            DF      MS           F         P         G-G       H-F

location         176964.622    2       88482.311    11.441    .2E-03    .6E-03    .3E-03
ERROR            216550.044    28      7733.930

GREENHOUSE-GEISSER EPSILON:           0.8403
HUYNH-FELDT EPSILON       :           0.9420

delay            12.844        1       12.844       0.002     0.966     .         .
ERROR            96473.822     14      6890.987

GREENHOUSE-GEISSER EPSILON:           .
HUYNH-FELDT EPSILON       :           .

location
*delay           19887.022     2       9943.511     1.375     0.269     0.270     0.269
ERROR            202548.311    28      7233.868

GREENHOUSE-GEISSER EPSILON:           0.8890
HUYNH-FELDT EPSILON       :           1.0000
```

Table 11.3 (continued)

```
SINGLE DEGREE OF FREEDOM POLYNOMIAL CONTRASTS
---------------------------------------------
```

POLYNOMIAL TEST OF ORDER   1 (LINEAR)

SOURCE	SS	DF	MS	F	P
location	171093.600	1	171093.600	17.822	.854012E-03
ERROR	134405.400	14	9600.386		
location *delay	1363.267	1	1363.267	0.139	0.715
ERROR	137020.733	14	9787.195		

POLYNOMIAL TEST OF ORDER   2 (QUADRATIC)

SOURCE	SS	DF	MS	F	P
location	5871.022	1	5871.022	1.001	0.334
ERROR	82144.644	14	5867.475		
location *delay	18523.756	1	18523.756	3.958	0.067
ERROR	65527.578	14	4680.541		

MULTIVARIATE REPEATED MEASURES ANALYSIS

TEST OF: location		HYPOTH. DF	ERROR DF	F	P
WILKS' LAMBDA=	0.435	2	13	8.455	0.004
PILLAI TRACE =	0.565	2	13	8.455	0.004
H-L TRACE =	1.301	2	13	8.455	0.004

TEST OF: location *delay		HYPOTH. DF	ERROR DF	F	P
WILKS' LAMBDA=	0.772	2	13	1.916	0.186
PILLAI TRACE =	0.228	2	13	1.916	0.186
H-L TRACE =	0.295	2	13	1.916	0.186

```
>HYP
>WITHIN = "LOCATION"
>CONTRAST
>-2 1 1
>TEST
```

The output from this procedure is shown in Table 11.4. The value of the contrast C is given in the output as "null hypothesis contrast ABC'" (where ABC just refers to matrices that SYSTAT uses in computing the contrast). This value, 354.667, is equal to $\Sigma \lambda_{ij} \overline{X}_{ij}$, where the λ_{ij} are the *six* contrast weights and the $\overline{X}_{ij}$ are the means of the *six* location × delay conditions. Six? Yes: When we specified, for example, a contrast weight of –2 for the mean of location$_1$, SYSTAT used that weight once for each occurrence of a condition involving location$_1$. In this example it used the –2 weight twice—once for the mean of location$_1$delay$_1$ and once for the mean of location$_1$delay$_2$. Then SYSTAT uses a somewhat unusual formula to find the "sum of squares" for the contrast:

$$SS_C = nC^2 = 15 \times 354.667^2 = 1{,}886{,}830.214 \tag{11.1}$$

where n is the number of cases entering into each of the means being tested by the contrast. This value is shown in the output in Table 11.4. This formula is unusual because it does not scale the sum of squares down by $\Sigma \lambda_{ij}^2$ as is commonly done (e.g., see Equation 10.1). Instead, SYSTAT scales the error term for the F test up by $\Sigma \lambda_{ij}^2$, which ultimately accomplishes the same end. The value of the F test would be the same by either method. However, SYSTAT always uses the local error term for the contrast, so if you wanted to use the global error term you might want to scale down the SS$_C$ by $\Sigma \lambda_{ij}^2$ yourself.

SYSTAT also offers a second way to test contrasts in repeated measures designs that allows you to compare any combinations of within subjects means directly. Instead of using the *contrast* command, you give the contrast weights for each repeated measures column to the *cmatrix* command. For example, to test the same hypothesis in Table 11.4 using the *cmatrix* command, type

```
>HYP
>CMATRIX
>-2 -2 1 1 1 1
>TEST
```

MACINTOSH

After you have computed the overall ANOVA:

① Choose the menu item **Stats/MGLH/Test of Effects...**

② Select location from the **Within Subjects** variables list.

③ Click on the **Coefficients** button and type "–2 1 1" into the coefficients box.

④ Click **OK**.

MS-DOS

After you have computed the overall ANOVA:

① Choose the menu item **Statistics/MGLH/MGLM/ Hypothesis/Contrast/ Matrix** and type "–2 1 1" into the numeric array box.

② Choose the menu item **Statistics/MGLH/MGLM/ Hypothesis/Within** and type "location" into the box.

③ Choose the menu item **Statistics/MGLH/MGLM/ Hypothesis/Estimate/Go (Test)!**

MACINTOSH

After you have computed the overall ANOVA:

① Choose the menu item **Stats/MGLH/User Defined Contrasts...**

② Click on **More...**

③ Type "–2 –2 1 1 1 1" into the **C Matrix** box.

④ Click **OK**.

MS-DOS

After you have computed the overall ANOVA:

① Choose the menu item **Statistics/MGLH/MGLM/ Hypothesis/Cmatrix** and type "–2 –2 1 1 1 1" into the box.

② Choose the menu item **Statistics/MGLH/MGLM/ Hypothesis/Go (Test)!**

The weights in this command put –2's on the two location$_1$ columns and +1's on the other four columns. This method has the disadvantage that you have to work out how to pair columns with weights, but it has the advantage of being able to test contrasts on effects other than main effects. We will use it to test contrasts on simple effects and interactions below.

To find the effect size r for the contrast, plug the F and df values from Table 11.4 into the following formula:

$$r = \sqrt{\frac{F}{F + df_{\text{denom}}}} = \sqrt{\frac{14.167}{14.167 + 14}} = 0.71 \quad (11.2)$$

11.4.1.2 Simple Effects

Sometimes you may want to ask a specific question about an effect within a single level of one factor. For example, you might ask whether the contrast computed above, that location$_1$ was faster than the average of the other two locations, is

Table 11.4 Test of the contrast predicting that location 1 would be faster than the other two in the CHOICE data set.

```
HYPOTHESIS.

C MATRIX
       1           2           3           4           5           6
   -2.000      -2.000       1.000       1.000       1.000       1.000

NULL HYPOTHESIS CONTRAST ABC'
              354.667

                      -1
INVERSE CONTRAST A(X'X)   A'
                0.067

TEST OF HYPOTHESIS

        SOURCE         SS          DF         MS              F           P

    HYPOTHESIS  1886826.667         1   1886826.667      14.167       0.002
         ERROR  1864597.333        14    133185.524
```

significant within delay$_2$. You can test this effect, and any contrast among cell means in the example data set, by designating the specific cell means in user-defined contrasts using the *cmatrix* command. This command allows you to specify contrast weights for each repeated measures column. For example, the columns in the CHOICE data set that contain delay$_2$ conditions are columns 2, 4, and 6. These three columns contain location$_1$, location$_2$, and location$_3$, respectively, so they should be paired with the weights –2, 1, and 1, respectively, which correspond to the desired contrast. The other three columns, containing delay$_1$ conditions, should receive weight zero. Therefore, the commands for testing this contrast are (see section 11.4.1.1 for menu descriptions):

```
>HYP
>CMATRIX
>0 -2 0 1 0 1
>TEST
```

The output is shown in Table 11.5. As you can see, this simple effect contrast on delay$_2$ conditions is significant. The effect size for the contrast is

$$r = \sqrt{\frac{F}{F + df_{denom}}} = \sqrt{\frac{13.963}{13.963 + 14}} = 0.71 \quad (11.3)$$

Table 11.5 Test of a contrast on the simple effect of location at delay 2, predicting that the mean of location 1 would be lower than the means of locations 2 and 3 in the CHOICE data set.

HYPOTHESIS.

C MATRIX

1	2	3	4	5	6
0.000	-2.000	0.000	1.000	0.000	1.000

TEST OF HYPOTHESIS

SOURCE	SS	DF	MS	F	P
HYPOTHESIS	739704.067	1	739704.067	13.963	0.002
ERROR	741658.933	14	52975.638		

Thus, the effect size for this simple effect is about the same size as that for the overall contrast (Table 11.4) on the three location means, averaged across delay.

11.4.1.3 Interaction Effects

As with main effects, tests of interaction effects with more than one df in the numerator do not test specific hypotheses. They may tell you that there are some differences among the interaction effects, but will not tell you where. When they are not significant, this does not mean that there are no significant contrasts among the interaction effects. For example, in Table 11.3 the interaction between location and delay was not significant, $p = .269$, but it would be a mistake to conclude from this that the interaction was just noise. Suppose, for example, that you had predicted a priori, on the basis of theoretical considerations, that the longer delay condition would yield faster response times than the shorter delay condition when the warning light appeared in the left or the right locations, but that this difference would be larger and in the opposite direction when the light appeared in the central location. This is a contrast on the interaction between location and delay. For locations 1 and 3, the weights for delay conditions 1 and 2 are +1 and –1, respectively; for location 2 the weights for delay conditions 1 and 2 are –2 and +2, respectively. These weights are shown in the cells in Table 11.6.

Table 11.6 Contrast weights predicting that the longer delay condition would yield faster response times than the shorter delay when the warning light appeared on the left or the right, but that this difference would be larger and in the opposite direction when the light appeared in the center.

	Location 1	Location 2	Location 3
Delay 1	+1	–2	+1
Delay 2	–1	2	–1

SYSTAT will not allow you to compute a contrast on the interaction directly. However, when the cell sizes are all equal and the contrast weights sum to zero in both the rows and the columns, as they do for this contrast, a contrast on the means is identical to a contrast on the interaction. (If you want to compute the contrast on the interaction residuals, you must find the residuals yourself from the means given in Table 11.3

using mean polish procedures [Rosenthal & Rosnow, 1991, pp. 367–378; see also section 11.7.3].) Thus, you can compute the contrast on the interaction by pairing the contrast weights with the appropriate columns in the *cmatrix* command:

```
>HYP
>CMATRIX
>1 -1 -2 2 1 -1
>TEST
```

(See section 11.4.1.1 for menu descriptions.) The contrast output is shown in Table 11.7. The contrast is not significant "two-tailed" (see section 2.2.3.3.1), but because we specifically predicted the direction of this effect a priori, arguably we are justified in dividing the *p*-value in Table 11.7 by two, and we find that this one-tailed value is significant, $p = .036$. The effect size *r* for this interaction contrast can be computed in the usual manner:

$$r = \sqrt{\frac{F}{F + df_{\text{denom}}}} = \sqrt{\frac{3.958}{3.958 + 14}} = 0.53 \quad (11.4)$$

This example illustrates why you should not conlude, on the basis of a nonsignificant multi-*df F* test, such as the test of the interaction in Table 11.3, that there are no significant effects lurking within that multi-*df* effect.

Table 11.7 Test of the contrast on the interaction using the weights in Table 11.6.

HYPOTHESIS.

C MATRIX

1	2	3	4	5	6
1.000	-1.000	-2.000	2.000	1.000	-1.000

TEST OF HYPOTHESIS

SOURCE	SS	DF	MS	F	P
HYPOTHESIS	222285.067	1	222285.067	3.958	0.067
ERROR	786330.933	14	56166.495		

11.4.2 Polynomials

Sometimes the specific hypotheses that you want to test may be sets of orthogonal polynomials. In fact, the hypothesis tested in the previous section (that delay$_2$ would yield faster response times than delay$_1$ in location$_1$ and location$_3$, but this difference would be larger and in the opposite direction in location$_2$) turns out to be the "quadratic × linear" interaction for location-by-delay. SYSTAT computes these polynomial contrasts automatically when it generates the repeated measures ANOVA, so we could have saved ourselves the trouble of testing the contrast had we noticed that it was an interaction of polynomials. This interaction contrast was shown in the ANOVA output in Table 11.3, along with all of the other orthogonal polynomials for within subjects main effects and interactions. Thus, there is no need to test these yourself in repeated measures designs.

11.4.3 Error Terms

11.4.3.1 Local Error Terms

When SYSTAT computes a contrast on within subjects factors, it tests the contrast using the "local" error term, which is the error term specific to that contrast. This is the error term you would use if you were to calculate a contrast score c for each subject using the formula $c = \Sigma \lambda_{ij} X_{ij}$ (where the λ_{ij} are the contrast weights and the X_{ij} are the values in each repeated measures column for each subject) and then computed a one-sample t-test to determine whether the mean of these contrast scores was significantly different from zero. The t that you would get from this procedure would be the square root of the F test in SYSTAT's output.

The advantages of using the local error term are that the F tests for orthogonal contrasts will be independent and will truly be distributed as F given the null hypothesis and usual ANOVA assumptions. The disadvantage is that the df for the error term is limited to the number of subjects minus 1. When the error terms for different orthogonal contrasts are not too different, it may be possible to pool them to increase your degrees of freedom (see the following section).

11.4.3.2 Pooled, or "Global," Error Terms

As discussed in the previous section, SYSTAT tests within-subject contrasts using the *local error term*, which is the error term specific to the contrast being tested. However, this error term is limited in its *df* to the number of subjects minus 1. An alternative is to use the global error term, which pools together the error terms for all orthogonal contrasts on the same effect. This global error term is usually considered appropriate when the error terms being pooled do not differ by more than a ratio of about 2:1 or so. Because this error term is based on more *df*, it can give a better estimate of the true population variance, and will produce a more powerful test. However, the validity of significance tests using pooled error terms depends on the homogeneity of covariances assumption being adequately met (see section 11.6.1.3). However, if you have reason to think that the assumption is adequately met, or if you are in desperate need of error *df* and a lack of power is more of a concern than violation of the assumption, the pooled error term may be your best choice. However, in such a case, you should realize that you are accepting the lesser of two evils.

For example, in Table 11.3 the quadratic location × linear delay contrast was not quite significant "two-tailed," $F_{(1,14)} = 3.958$, $p = .067$, but the 14 *df* in the denominator means that this test had relatively low power. Because the mean square for this error term, 4680.541, differs only by about a factor of two from the mean square for the linear location × linear delay error term, 9787.195, we might pool together the two error terms to test both contrasts. (Note: If you are justified in using the pooled error term to test one of the contrasts, you should use it to test both; you are not justified, for example, in using pooled error terms only when they produce more significant tests and using local error terms otherwise!) To pool these error terms, divide the sum of their sums-of-squares by the sum of their degrees of freedom:

$$MS_{pooled} = \frac{SS_1 + SS_2}{df_1 + df_2} \quad (11.5)$$

$$= \frac{65527.578 + 137020.733}{14 + 14} = 7233.868$$

Note that this is exactly equal to the mean square for the error term of the location × delay interaction in Table 11.3. In balanced designs, just as orthogonal contrasts divide the sum-of-squares for the multi-*df* main effect or interaction from which they are taken, the local error terms for those contrasts divide the sum-of-squares for the *error term* for the multi-*df* main effect or interaction. Thus, the global error term for the contrasts on the interaction in Table 13.3 is simply the usual error term for the interaction itself.

Any set of error terms in which the minimum and maximum mean squares do not differ by more than about a factor of two can be pooled, assuming that homogeneity of covariances is met across the columns contributing to those error terms (actually, this assumption is not likely to be met, so you should only do this when power is the overriding consideration). In Table 11.3, the error terms for location, delay, and location × delay are all close in magnitude (these sums-of-squares already pool over the error terms for the contrasts). So, we might pool them all using a simple extension of Equation 11.5:

$$MS_{pooled} = \frac{SS_1 + SS_2 + SS_3}{df_1 + df_2 + df_3}$$

$$= \frac{216550.044 + 96473.822 + 202548.311}{28 + 28 + 28}$$

$$= 6137.764 \quad (11.6)$$

This pooled error term could now be used to test all of the effects in Table 11.3!

There are pros and cons for doing this, as mentioned above. This pooled error term has substantially more *df* than any of its constituents, which can result in much more powerful tests. And if the terms that are pooled together all estimate the same error variance, then the pooled error term is a better estimate of the real variance in the population. However, when your *df* are above 25 to 30 or so, the gain in power from any extra *df* is unlikely to be worth the likely violation of the homogeneity of covariances assumption that you are committing by pooling error terms. Furthermore, if you use the same pooled error term for more than one *F* test, these *F* tests will not be independent: If the error term underestimates the true error variance, then the *F*'s will all be too large; if the error term overestimates the true error variance, then the *F*'s will all be too small. Fortunately, by pooling over a lot of *df* by including terms that do not differ by more than a factor of about 2, the degree of over- or underestimation will hopefully be minimal. In practice, researchers are often happier living with a number of dependent F tests based on a lot of df than with a number of independent F tests each based on woefully few df. See section 11.6.1.3 for further discussion of error term issues in repeated measures.

11.5 Post-Hoc Tests
11.5.1 Marginal Means

SYSTAT does not support post-hoc tests on within subject factors. You can, however, test pairwise difference among the marginal means of the repeated measures factor using the *cmatrix* command, as described in sections 11.4.1.1 and 11.4.1.2. For example, to test the three pairwise differences among the three location means, you would first compute the overall repeated measures ANOVA and then use the follow commands:

```
>HYPOTHESIS
>CMATRIX
>-1 -1 1 1 0 0
>TEST
>CMATRIX
>-1 -1 0 0 1 1
>TEST
>CMATRIX
>0 0 -1 -1 1 1
>TEST
```

(See section 11.4.1.1 for menu descriptions.) The three outputs from these commands are shown in Table 11.8. Note that the p-values in these tests are not corrected for the number of tests performed. To use a Bonferroni correction, simply multiply the p-values by the number of pairwise tests, in this case three. The resulting p-values for the three comparisons in Table 11.8 are .039, .003, and .156, respectively.

Note that all of these pairwise comparisons in Table 11.8 used local error terms for the contrasts (see section 11.4.3.2). If you wanted to use a global error term, you could do so, finding the appropriate global error term from the ANOVA in Table 11.3 (in this example, the global error term would be the error term for the location effect, MS=7733.930) and then using this global error term to recompute the F tests in Table 11.8.

Table 11.8 Post-hoc pairwise differences among the three location means in the CHOICE data set, using local error terms.

Location₁ Versus Location₂

HYPOTHESIS.

C MATRIX

1	2	3	4	5	6
-1.000	-1.000	1.000	1.000	0.000	0.000

TEST OF HYPOTHESIS

SOURCE	SS	DF	MS	F	P
HYPOTHESIS	298497.067	1	298497.067	8.083	0.013
ERROR	517010.933	14	36929.352		

Location₁ Versus Location₃

HYPOTHESIS.

C MATRIX

1	2	3	4	5	6
-1.000	-1.000	0.000	0.000	1.000	1.000

TEST OF HYPOTHESIS

SOURCE	SS	DF	MS	F	P
HYPOTHESIS	684374.400	1	684374.400	17.822	0.001
ERROR	537621.600	14	38401.543		

Location₂ Versus Location₃

HYPOTHESIS.

C MATRIX

1	2	3	4	5	6
0.000	0.000	-1.000	-1.000	1.000	1.000

TEST OF HYPOTHESIS

SOURCE	SS	DF	MS	F	P
HYPOTHESIS	78916.267	1	78916.267	4.516	0.052
ERROR	244667.733	14	17476.267		

11.5.2 Cell Means

You can also test pairwise differences among cell means defined by combinations of within subjects factors. For example, if you wanted to test all pairwise differences among the six cell means in Table 11.3 using local error terms you would use commands of the form

```
>HYPOTHESIS
>CMATRIX
>-1  1  0  0  0  0
>TEST
>CMATRIX
>-1  0  1  0  0  0
>TEST
>CMATRIX
>-1  0  0  1  0  0
>TEST
>CMATRIX
>-1  0  0  0  1  0
>TEST
...
```

... and so on for 15 tests (see section 11.4.1.1 for menu descriptions). The p-values in these outputs will not be adjusted, so they should all be multiplied by 15. Some of these pairwise differences confound delay and location effects. In fact, the third of these contrasts above confounds the delay effect, the location effect, and the interaction! By conducting many tests you increase your chances of finding spurious effects, and because many of these pairwise differences will be hard to interpret anyway, you should not test pairwise differences indiscriminately.

You can sometimes use global error terms for testing pairwise comparisons, but because some of these pairwise comparisons cut across main effects and interactions these global error terms may not be easy to compute. Furthermore, the tests using global error terms are only valid under the equal covariances assumption (see section 11.6.1.3). To determine which effects are involved in a pairwise comparison (or any contrast, for that matter) simply arrange the contrast weights in a factorial table and perform a mean polish on those weights. Any place you find nonzero residuals you know that the contrast involves that effect. Two examples should illustrate this procedure.

Consider the pairwise comparison between $location_1 delay_1$ and $location_2 delay_1$. This comparison uses contrast weight of

−1 and +1 on the means of those two conditions and weights of zero on all the rest. These weights are shown in the two-way table in the top panel of Table 11.9. To compute the mean polish, find the predicted row and column effects by summing across the rows and columns, and then dividing these sums by the numbers of rows and columns, respectively. For example, the first column in the top panel of Table 11.9 sums to −1, so the predicted column effect is −1/2 = −0.5. Note that any rows or columns that sum to zero will have predicted effects equal to zero. Finally, find the predicted interaction effects by subtracting the corresponding row and column effects from the contrast weights in each cell. For example, in the upper left cell (location$_1$delay$_1$) the weight is −1, the corresponding row effect is zero, and the corresponding column effect is −0.5. Therefore, the predicted interaction effect for this cell is −1 − 0 − (−0.5) = −0.5. This and the rest of the interaction effects are shown in the completed mean polish table in the bottom panel of Table 11.9.

The effects in this mean polish table make explicit the predicted involvement of the different main effects and interactions in this pairwise comparison. The pairwise comparison of location$_1$delay$_1$ and location$_2$delay$_1$ involves both the main effect of location and the location × delay interaction. Thus, the global error term for this comparison must involve components of the error terms for both of these effects. Fortunately, these error terms are very close in magnitude (see Table 11.3), so we would not hesitate in pooling them using Equation 11.5:

$$\text{MS}_{\text{pooled}} = \frac{SS_1 + SS_2}{df_1 + df_2} \tag{11.7}$$

$$= \frac{216550.044 + 202548.311}{28 + 28} = 7483.899$$

and then using this pooled error term to test the pairwise comparison on an F with 1 and 56 df.

As a second example, consider the pairwise comparison between location$_1$delay$_1$ and location$_2$delay$_2$. This comparison uses contrast weights of −1 and +1 on the means of those two conditions and weights of zero on all the rest. These weights are shown in the two-way table in the top panel of Table 11.10. The completed mean polish table for these contrast weights is shown in the bottom panel of Table 11.10. The

Table 11.9 The top panel shows the contrast weights for the pairwise comparison of location$_1$delay$_1$ with location$_2$delay$_1$ in the CHOICE data set. The bottom panel shows the mean polish table for those weights.

Weights

	Location 1	Location 2	Location 3
Delay 1	−1	+1	0
Delay 2	0	0	0

Mean Polish Table

	Location 1	Location 2	Location 3	
Delay 1	−.5	.5	0	0
Delay 2	.5	−.5	0	0
	−.5	.5	0	0

effects in this mean polish table make explicit the predicted involvement of the main effects of location and delay, as well as the location × delay interaction. Thus, the global error term for this comparison must involve components of the error terms for all three of these effects. The error terms for these effects are all close in magnitude, so we could pool them, which was actually done above in Equation 11.6, and then use this pooled error term to test the pairwise comparison on an F with 1 and 84 df.

Table 11.10 The top panel shows the contrast weights for the pairwise comparison of location$_1$delay$_1$ with location$_2$delay$_2$ in the CHOICE data set. The bottom panel shows the mean polish table for those weights.

Weights

	Location 1	Location 2	Location 3
Delay 1	−1	0	0
Delay 2	0	+1	0

Mean Polish Table

	Location 1	Location 2	Location 3	
Delay 1	−.167	−.167	.333	−.333
Delay 2	.167	.167	−.333	.333
	−.5	.5	0	0

11.6 Checking Assumptions

The assumptions underlying significance testing in repeated measures differ depending on whether the test is of a contrast or of a multi-*df* effect. In this section, assumptions underlying multi-*df* effects are discussed first, and then the assumptions underlying tests of contrasts are discussed.

11.6.1 Assumptions Underlying Omnibus (Multi-*df*) Tests

Repeated measures ANOVA makes the usual four assumptions about the distribution of errors within each cell—that they are normally distributed, have constant variance, are independent, and are unbiased—and it also makes a fifth assumption—that the covariances of the repeated measures are homogeneous. The unbiased assumption is met by definition in ANOVA and so is not discussed further. Each of the other assumptions is discussed in this section in the context of tests of multi-degree of freedom effects, such as *F* tests with more than one *df* in the numerator.

11.6.1.1 Normality and Equal Variance

The normality assumption states that, within each cell of the design, the errors have a normal distribution. However, in designs with only one observation per cell, which is the case with most repeated measures designs, the error in each cell is confounded with the effect for the highest order interaction. Thus, the normality assumption cannot be assessed by looking at the residuals from the ANOVA. Likewise, the equal variance assumption cannot be assessed because there is no estimate of the error variance within each cell. Thus, these two assumptions cannot be assessed in this type of design. Fortunately, (1) *F* and *t* tests are quite robust to violations of these assumptions, and (2) we are usually not interested in using estimates of σ^2 alone as the error term for any of the tests, but instead use mean squares for interactions with a random factor as error terms; σ^2 contributes to the these mean squares, but it is often overwhelmed by the interaction component, so the effects of any violations of the assumptions are further reduced. In general, nobody worries about these assumptions in repeated measures designs with one observation per cell.

When you do have multiple observations per cell, you can compute the usual within cell variance and assess the validity of the assumptions by using diagnostic methods on the residuals. These methods are the same in this case as they are in fixed effects designs as discussed in sections 8.6 and 9.6, so those discussions will not be repeated here. However, (1) because the within cell variance usually only serves as the error term for effects involving the random factor, such as subject differences, which we are rarely interested in testing, and (2) because there is often reason to doubt that two residuals within the same cell that come from the same subject are independent, even after removing their mean, it is often best just to average over the observations within each cell and use these averages as data points, returning us to a design with one observation per cell.

11.6.1.2 Independence

It is because multiple observations on a single subject are not independent that we use repeated measures designs. When the subject effects and interactions are removed from the data, we hope that most, if not all, of the dependency among observations will also be removed, and the independence assumption will be met. Generally, you would only worry about this assumption if you had reason to believe that, after the subject effects were removed, the residuals were still correlated. This might be a concern, for example, in a design in which there are two or more observations in each cell for each subject and these observations are collected in close proximity in time. As mentioned above, in such a case it is often best just to average over the observations within each cell and use these averages as data points, turning the design into one with only one observation per cell. Designs with only one observation per cell are less prone to violations of this assumption, but poor experimental designs can still create problems. For example, you would not want to have the observations for each successive level of a factor collected successively in time. This could cause dependencies among the observation that might or might not be absorbed by estimates of effects, and, if they were, they would confound those effects.

Thus, the independence assumption in repeated measures ANOVA, as with between subjects designs, is not an assumption that is usually assessed by looking at the data. Rather, it is assessed by looking at the experimental design itself. A well-designed repeated measures experiment should minimize the impact of dependencies among the observations.

11.6.1.3 Homogeneity of Covariances

The F tests for within subjects factors in repeated measures designs make a fifth assumption that was not relevant for tests of between subjects factors. When the *df* for the numerator of the *F* test is 1, as is the case for factors with only two levels or for contrasts, and a local error term is used, this assumption is always met. Therefore, when you are only reporting 1-*df F* tests, this assumption need not be relevant to you. However, if you are using pooled error terms (see section 11.4.3.2), or you are determined to report *F* tests with more than one *df* in the numerator, read on.

This assumption can be stated in various ways, and it is called by a variety of different names. One is the assumption of *homogeneity of treatment differences*. Stated this way, the assumption is that the variances (across subjects) of the pairwise differences between levels of the within subject factor are all equal in the population. For example, in the CHOICE data set we could compute, for each subject, all of the pairwise differences between levels of the location factor. I will illustrate this in two steps: First, compute the mean for each location condition for each subject using commands such as

```
>LET L1 = (L1D1+L1D2)/2
>LET L2 = (L2D1+L2D2)/2
>LET L3 = (L3D1+L3D2)/2
```

(See section 2.2.1 for menu commands and discussion of math functions.) These new columns of means are shown in the first three columns of Table 11.11. Second, compute all three of the pairwise differences, D_i, between these columns:

```
>LET D1 = L1 - L2
>LET D2 = L1 - L3
>LET D3 = L2 - L3
```

These three columns are shown in columns four, five, and six of Table 11.11. The homogeneity of treatment-difference variances can now be stated in terms of these three columns: Their variances should be equal. To find out if this is the case, have SYSTAT compute the variances of these three columns (see section 3.1.1 for menu descriptions):

```
>STATS
>STATISTICS D1 D2 D3/VARIANCE
```

These variances are shown in Table 11.12. Clearly, the first two are similar but the third is much smaller. As with homogeneity of variances in general, a rough rule for when the variances are different is when the ratio of the largest to the smallest is more than about 2:1. The variances in Table 11.12 are on the borderline of this violation criterion.

Table 11.11 The means for the three location conditions for each subject in the CHOICE data set are shown in the first three columns, and the pairwise differences between these means are shown in the last three columns.

L1	L2	L3	D1	D2	D3
255.0	352.0	453.5	−97.0	−198.5	−101.5
350.0	363.0	272.0	−13.0	78.0	91.0
314.5	407.5	418.0	−93.0	−103.5	−10.5
299.5	289.5	321.5	10.0	−22.0	−32.0
320.5	363.5	345.5	−43.0	−25.0	18.0
462.0	471.5	461.5	−9.5	0.5	10.0
161.5	374.0	414.0	−212.5	−252.5	−40.0
318.0	400.5	456.5	−82.5	−138.5	−56.0
449.0	400.5	478.5	48.5	−29.5	−78.0
329.0	454.5	499.0	−125.5	−170.0	−44.5
143.5	418.5	377.5	−275.0	−234.0	41.0
300.0	405.5	386.0	−105.5	−86.0	19.5
270.0	420.5	506.5	−150.5	−236.5	−86.0
395.0	345.0	501.0	50.0	−106.0	−156.0
483.5	443.0	562.0	40.5	−78.5	−119.0

This assumption can also be stated in terms of the equality of population covariances among the columns. In this form it is called the assumption of *equal covariances*, the *sphericity* assumption, or the *circularity* assumption.

Yet another way of stating this assumption, which is relevant to SYSTAT output, is that the off-diagonals of the error matrices for the population are zero. SYSTAT will output these matrices when you use extended output for repeated measures designs. This output is too long to merit display here, so I just show the relevant parts of this output for the CHOICE data set in Table 11.13. If the sphericity assumption is met, the off-

Table 11.12 Variances of the columns of pairwise differences shown in Table 11.11.

```
TOTAL OBSERVATIONS:     15


                            D1              D2              D3

N OF CASES                  15              15              15
VARIANCE              9232.338        9600.386        4369.067
```

diagonals in the matrices that SYSTAT calls the "error sum of product" matrices should be close to zero. (Note that the sum of the diagonal of this matrix is related to the mean square for error in the univariate test in Table 11.3; this sum is called the *trace* of the matrix.) As you can see, neither of the off-diagonals in Table 11.13 are very close to zero, although the off-diagonal for the location × delay matrix is small relative to the size of the diagonals. How small is small? This is hard to tell by eye. Tests for violations of this assumption exist, such as Mauchly's sphericity test, but you would probably be better off using adjusted p-values, which are described next.

Table 11.13 Repeated measures ANOVA on the CHOICE data showing excerpts from the extended results output.

```
           + TRIALS FACTOR: Location      +


ERROR SUM OF PRODUCT MATRIX    G = CE'EC'
                       1              2
            1     268810.800
            2     -78618.710    164289.289



           + TRIALS INTERACTION OF +
           +      Location         +
           +      Delay            +


ERROR SUM OF PRODUCT MATRIX    G = CE'EC'
                       1              2
            1     137020.733
            2       1848.060     65527.578
```

Rather than worry about whether the assumption is violated or not, another method is to assume that it is violated (it usually is) and adjust the p-values accordingly. SYSTAT gives two of these adjustments, the Greenhouse-Geisser (G-G) adjustment and the Huynh-Feldt (H-F) adjustment. The difference in the p-values between these adjustments and the original p-values can give you an idea of how badly the equality of covariances assumption is violated. In general, the Greenhouse-Geisser adjustment tends to be too conservative and the

Huynh-Feldt is too liberal (Maxwell & Delaney, 1989, p. 479), so the correct p-value is probably somewhere in between. If you are fortunate enough to find that they both lead to the same conclusion, then you are home free. If you are determined to report an F test with more than one df in the numerator, you should always use one of these p-value adjustments rather than the original p-value. F tests are not robust to violations of the sphericity assumption, and it is not unusual for the unadjusted p-values to be two to three times too small, i.e., "too significant" (Maxwell & Delaney, 1989).

In Table 11.3, the p-value for the location effect is .002. However, the unequal variance in Table 11.12 and the large off-diagonal (–78618.710) in Table 11.13 indicate that this p-value is probably too small. The Huynh-Feldt adjusted p in Table 11.3 is .003, which is 50% larger than the original p. The Greenhouse-Geisser adjusted p is larger still, .006, which is three times as large as the original p. Fortunately, all three of these p-values indicate a significant effect.

For the location × delay interaction, the off-diagonal in Table 11.13 is small relative to the diagonal values. Consistent with this, the G-G and H-F adjustments to the p-value for this effect in Table 11.3 do not differ very much from the unadjusted p-value. Thus, it appears that the equal covariances assumption may not be violated (for practical purposes) for this interaction.

As mentioned above, when you test contrasts using a local error term, the equal covariances assumption is always met. However, when you use pooled error terms for the contrast, the validity of pooling the error terms depends on this assumption holding for the columns being used to obtain those error terms. Because this assumption is almost never met, the only time you should use a pooled error term for contrasts on within subjects factors is when you are badly in need of df. As mentioned above, in such a case you should realize that you are accepting the lesser of two evils.

11.6.2 Assumptions Underlying Tests of Contrasts

When you test contrasts on within subjects factors using local error terms, you avoid most of the difficulties surrounding assumptions, because within subjects contrasts essentially redefine error variance as "within cell" error, just like in a between subjects design.

To illustrate, consider the linear trend in location, which SYSTAT computed automatically in Table 11.3. The usual way of thinking about this contrast is as a contrast on the *means* of the three locations using the weights –1, 0, and 1. Another way of thinking about this contrast is as a test of the mean of the contrast scores for each subject. On this view, each subject has three means, one for each location, and these means are multiplied by the contrast weights and summed, to give a contrast score for each subject. For example, the means of the three locations for the first subject in Table 11.2 are (282+228)/2 = 255, (281+423)/2 = 353, and (344+563)/2 = 453.5. Therefore, the linear contrast score for this subject is (–1)(255) + (0)(353) + (1)(453.5) = 198.5. Using SYSTAT's math functions (section 2.2.1) you can find this contrast score for all of the subjects:

```
>LET C = (-1)*(L1D1+L1D2)/2 + 0*(L2D1+L2D2)/2,
+ 1*(L3D1+L3D2)/2
```

These contrast scores are shown in Table 11.14.

Table 11.14 Linear contrast scores for the 15 subjects in Table 11.2.

Subject	C
1	198.5
2	–78.0
3	103.5
4	22.0
5	25.0
6	–0.5
7	252.5
8	138.5
9	29.5
10	170.0
11	234.0
12	86.0
13	236.5
14	106.0
15	78.5

These contrast scores may now serve as data points. The test of the linear contrast is just a test that the mean of the contrast scores is equal to zero, which can be accomplished using a *one-sample t*-test. To compute this test manually, use the statistics command (section 3.1.1) to find the mean, N, and variance of the contrast scores:

>STATS
>STATISTICS C / MEAN N VARIANCE

The output from this command is shown in Table 11.15. The variance of these contrast scores, 9600.386, is exactly the mean square for error that SYSTAT used to test the linear contrast on the location main effect in Table 11.3! The mean of the contrast scores, $\bar{C}$, is 106.8. To continue with the *t*-test

$$t = \frac{\bar{C}}{SE_C} = \frac{\bar{C}}{\sqrt{s_C^2 / N}}$$

$$= \frac{106.8}{\sqrt{9600.386 / 15}} = 4.222 \qquad (11.8)$$

This *t* on 14 degrees of freedom is exactly the square root of the *F* test of the linear contrast in Table 11.3. Thus, as I have tried to demonstrate with this example, an *F* test on a within subjects factor using a local error term is identical to a one-sample *t* test of the mean of the contrast scores.

Table 11.15 The mean, N, and variance for the contrast scores in Table 11.14.

TOTAL OBSERVATIONS:	15
	C
N OF CASES	15
MEAN	106.800
VARIANCE	9600.386

Another way to compute one-sample *t*-tests in SYSTAT is to make use of SYSTAT's *matched-pairs t*-test command (see also section 12.2.3). With this command, SYSTAT assumes that you have two columns of data with values that are paired within each row, for example, two scores per subject. The *t*-test, then, tests the significance of the difference between the two columns. We have one column of data, shown in Table 11.14, but what is the second column? To compute a one-sample *t*-test,

MACINTOSH

① Choose the menu item **Stats/Stats/t-test...**
② Click on the **Paired** button if it is not already selected.
③ Select the two paired columns from the left **Variables** list, e.g., C and ZERO.
④ Click **OK**.

MS-DOS

① Choose the menu item **Statistics/Stats/Ttest/ Dependent/Variables/** and select C and ZERO from the variables list.
② Choose the menu item **Statistics/Stats/Ttest/ Dependent/Go!**

you can first create a new column in SYSTAT that contains all zeros, using a command such as

```
>LET ZERO = 0
```

and then compute a matched pairs *t*-test using the column of data paired with the column of zeros. The commands for the matched-pairs *t*-test are simply

```
>STATS
>TTEST  C  ZERO
```

and the output from this command is as follows:

```
PAIRED SAMPLES T-TEST ON  C VS ZERO WITH 15 CASES

MEAN DIFFERENCE =      106.800
SD DIFFERENCE =         97.982
T =   4.222 DF =    14 PROB = .854012E-03
```

This is exactly what we computed above by hand.

The assumptions underlying such a *t* test, and therefore also the *F* test, are the usual four assumptions: the errors are normally and equally distributed, they are independent, and they are unbiased. The equal distribution assumption is trivially met, because there is only one cell in this design. The independence assumption should be met so long as the data for different subjects is uncorrelated, that is, the subjects did not interact with each other, and so on. The unbiased errors assumption is met by definition.

Therefore, the only assumption that one might wish to examine (and hardly anyone ever does) is the normality assumption, which can be diagnosed in the usual manner. Because the residuals are just the contrast scores minus their mean, you can assess the normality assumption by looking at plots and statistics of the raw contrast scores. Figure 11.4 shows a stem and leaf display and a box plot of the contrast scores. Although the distributions of contrast scores is not perfectly symmetric, this is expected for small numbers of scores. These plots certainly show no evidence for any substantial departure from normality.

You could compute skewness and kurtosis for these contrast scores and generate a normal probability plot, but this would be overkill. For small N these diagnostics would not tell you anything that you could not see in the stem and leaf display or box plot. When the normality assumption appears to be substantially violated, the usual transformations could be used to make the scores look more normal, but these transforma-

```
STEM AND LEAF PLOT OF VARIABLE:   C,   N = 15
MINIMUM IS:            -78.000
LOWER HINGE IS:          27.250
MEDIAN IS:              103.500
UPPER HINGE IS:         184.250
MAXIMUM IS:             252.500

-0     7
-0     0
 0   H 222
 0     78
 1   M 003
 1   H 79
 2     33
 2     5
```

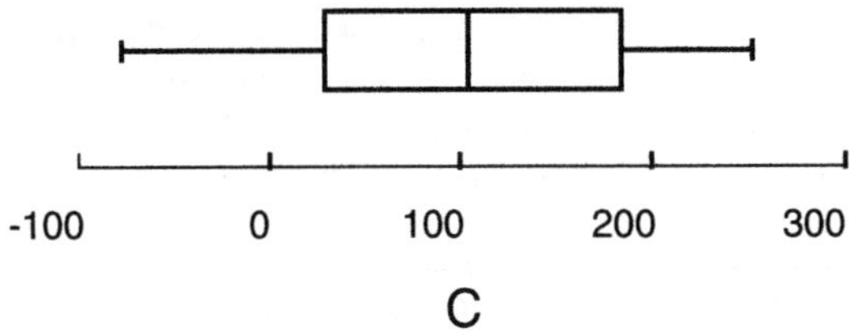

Figure 11.4 Stem and leaf display and box plot of the contrast scores for the linear contrast on location in the CHOICE data set.

tions would translate back into very complicated transformations of the original data (in Table 11.2). Therefore, you might be better off in such a case using a nonparametric test of some sort. Fortunately, the normality assumption will rarely be violated badly enough to justify such measures.

As you can see by comparing sections 11.6.1 and 11.6.2, the assumptions underlying contrasts are far simpler, and far more likely to be met, than the assumptions underlying omnibus F tests.

11.7 Graphing the Results

You will probably want to present a graph of each effect of interest in your data set. Unfortunately, SYSTAT's graphics routines do not know about the hierarchical structure of the repeated measures columns, so you will need to create new columns containing the means of the conditions that you wish to graph. This section shows how to make graphs for the two within subjects factors in the CHOICE data set. More information and types of graphs can be found in sections 3.2.2 and 8.8.

11.7.1 Within Subjects Main Effects

The means for each within subjects condition are shown at the top of the output in Table 11.3. When you only have one within subjects factor, you can simply enter these means into a new column in your data set, and then enter a second column containing condition labels. Then you can create plots of the condition means by simply plotting the first column against the second. When you have more than one within subjects factor, you must create a new column containing condition labels for each within subjects factor. For the CHOICE data set, this means one new column with the location condition labels, one new column with the delay condition labels, and a third column with the condition means from Table 11.3. These three columns are shown in Table 11.16.

Table 11.16 The within subject condition means and their condition labels in the CHOICE data set.

Location	Delay	Mean
1.000	1.000	337.933
1.000	2.000	308.867
2.000	1.000	373.267
2.000	2.000	414.600
3.000	1.000	435.200
3.000	2.000	425.200

With these new columns you can now create bar charts of the means for the two factors by typing

```
>GRAPH
>BAR MEAN * LOCATION / AXES=2
>BAR MEAN * DELAY / AXES=2 YMIN=375 YMAX=385
```

These bar charts are shown in Figure 11.5. (See section 3.2.2 for more information on two-variable plots and section 3.2.4 for more on controlling the look of these plots.)

If you prefer a line graph, you can use these same columns to create such graphs of the means for the two factors by typing

```
>GRAPH
>CPLOT MEAN * LOCATION / AXES=2
>CPLOT MEAN * DELAY / AXES=2 YMIN=375 YMAX=385
```

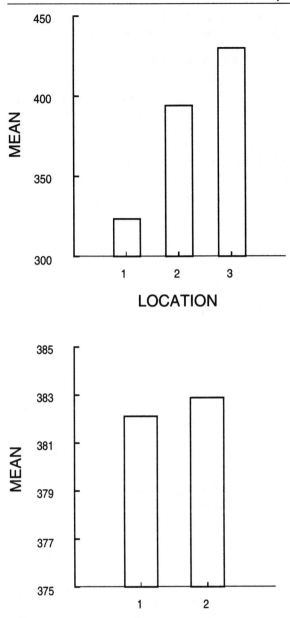

Figure 11.5 Bar graphs showing the means of the location and delay conditions in the CHOICE data set.

These line graphs are shown in Figure 11.6. See section 3.2.4 for more information on controlling the look of these graphs.

11.7.2 Two-Way Plots of Means

SYSTAT allows you to plot more than one dependent variable in the same graph. For example, suppose that you want to create separate bars for the two delay conditions within each location condition in a bar graph. First, you must create two new columns in the data set containing only the data for delay$_1$ and delay$_2$, respectively. This can be done easily using the commands

```
>IF DELAY = 1 THEN LET DELAY1 = MEAN
>IF DELAY = 2 THEN LET DELAY2 = MEAN
```

(See section 2.2.2.) These new columns are shown in Table 11.17. Each of these columns will have missing data, but this won't affect the graph.

Table 11.17 The within subject condition means and their condition labels in the CHOICE data set.

Location	Delay	Mean	Delay1	Delay2
1.000	1.000	337.933	337.933	.
1.000	2.000	308.867	.	308.867
2.000	1.000	373.267	373.267	.
2.000	2.000	414.600	.	414.600
3.000	1.000	435.200	435.200	.
3.000	2.000	425.200	.	425.200

With these new columns you can now create the desired bar graph by typing, for example

```
>GRAPH
>BAR   DELAY1 DELAY2 * LOCATION/ YLAB="MEAN RT",
FILL=4,7
```

The resulting bar graph is shown in Figure 11.7. The *ylab* option specifies the desired label for the Y-axis, and the *fill* option specifies the choice of patterns used to fill the bars. (See section 3.2.3 for more information on three-variable plots and section 3.2.4 for more on controlling the look of these plots.)

If you prefer a line graph over a bar graph, you can generate one using the same columns for delay$_1$ and delay$_2$ that you just created. Simply type

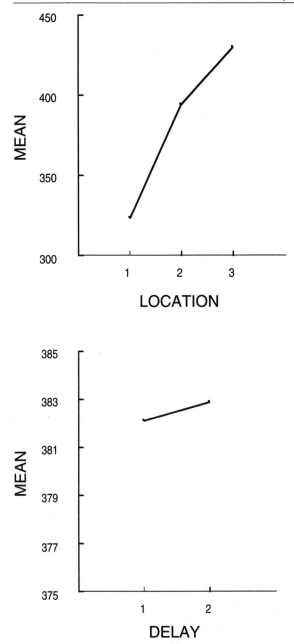

Figure 11.6 Line graphs showing the means of the location and delay conditions in the CHOICE data set.

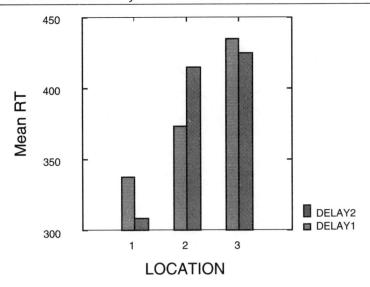

Figure 11.7 Bar graph showing the means of delay conditions within each location condition in the CHOICE data set.

```
>GRAPH
>CPLOT  DELAY1 DELAY2 * LOCATION/,
YLAB="MEAN RT" LINE=1,6
```

The options are identical to those for the bar chart above, except that the *line* option replaces the *fill* option. The resulting graph is shown in Figure 11.8.

11.7.3 Plotting Interactions

The two-way plots in the previous section showed the means of the cells corresponding to each location/delay combination. Such plots are often mistakenly called "plots of the interaction," but, in fact, they contain both main effects as well as interaction (Rosenthal & Rosnow, 1991, pp. 365–367). If you really want a plot of the interaction itself, you must use the interaction residuals, which SYSTAT does not provide for within subjects factors. Instead, you will have to start with the condition means and perform the mean polish decomposition yourself (which is easy using a spreadsheet). For example, to plot the two-way interaction between location and delay in the CHOICE data set, enter the six condition means from the top of the output in Table 11.3 into a two-way table like the one shown in the top panel of Table 11.18. Then compute the row means, column means, and grand mean for this table. Finally,

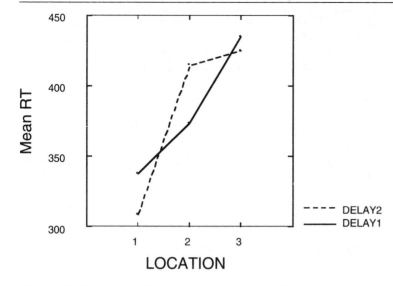

Figure 11.8 Line graph showing the means of delay conditions within each location condition in the CHOICE data set.

to find the interaction residuals, subtract from each cell its row mean, its column mean, and then add back to it the grand mean. The resulting interaction residuals, or "effects," are shown in the lower panel in Table 11.18.

Next, in a SYSTAT data worksheet, make one column with the location labels, one column with the interaction residuals corresponding to delay$_1$, and one column with the interaction residuals corresponding to delay$_2$. These columns are shown in Table 11.19. With these columns you can now plot the interaction with the commands

```
>GRAPH
>CPLOT DELAY1 DELAY2 * LOCATION/YLAB=,
"RESIDUAL EFFECT" AXES=2 LINE=1,6
```

This plot is shown in Figure 11.9. Plotting the interaction residuals can help prevent common misinterpretations of the interaction, which often arise from looking at plots of the means, such as Figures 11.7 or 11.8, instead of plots of the residual interaction effects. Often, you will find that the pure interaction itself is not of much interest, and that what you are really interested in are differences among means, as in Figure 11.8. But if you really want to interpret the interaction, Figure 11.9 is the graph you want.

Table 11.18 The top panel shows the means for the six conditions in the CHOICE data set, along with their row means, column means, and grand mean. The bottom panel shows the interaction residuals for those means.

Means

	Location 1	Location 2	Location 3	Row Means
Delay 1	337.933	373.267	435.200	382.133
Delay 2	308.867	414.600	425.200	382.889
Col. Means	323.400	393.934	430.200	*382.511*

Mean Polish Table

	Location 1	Location 2	Location 3
Delay 1	14.911	−20.289	5.378
Delay 2	−14.911	20.289	−5.378

Table 11.19 A data file showing the columns needed to plot the two-way interaction for the within subjects factors in the CHOICE data set.

LOCATION	DELAY1	DELAY2
1	1.000	14.911	−14.911
2	2.000	−20.289	20.289
3	3.000	5.378	−5.378

11.8 Unequal Cell Sizes

In repeated measures ANOVA, unequal cell size and unbalanced data problems can be either extremely simple or extremely difficult, depending on whether you have unequal cell sizes or completely missing cells. In a repeated measures design with multiple observations per cell, unequal cell sizes pose difficulties for deciding just how to weight the marginal means. As I recommended above, one approach, which is a type of unweighted means approach, is to average over the observations within each cell and perform all of your analyses using those means as data points. Some of those means will be based on larger n's, and thus be better estimated than others, but this is rarely worth worrying about—those cell means are still your best estimates of the population cell means. You can now carry on with the analysis with equal cell sizes of one.

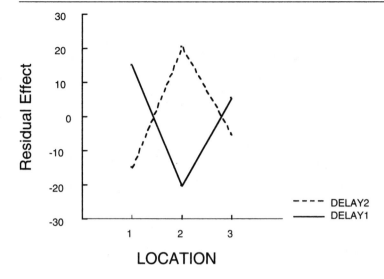

Figure 11.9 Plot of the two-way interaction residuals for the within subjects factors in the CHOICE data set.

If, however, after averaging across observations within cells you have some cell sizes of zero, your problem is much more difficult. When you have missing cells, you lack some of the cell means necessary for computing contrast scores, and so on. The right thing to do here would be to use a sophisticated multiple-imputation method to fill in the missing cells based on the rest of the data (Little & Rubin, 1987; Rubin, 1987). Unfortunately, these methods are not yet available in any commercial computer packages, including SYSTAT. Until such methods are available, people will probably continue the notorious practice of filling in missing cells with the subject's condition means and reducing the *df* accordingly. This practice is notorious because if the cells are missing in a systematic way (i.e., more in some conditions than others) it can make effects appear where they do not exist.

Moral: don't have missing cells.

12

Nesting in Repeated Measures

Chapter 10 discussed designs with nested factors, and Chapter 11 discussed designs with repeated measures. Quite often, you will run into repeated measures designs that also have some type of nesting, and it is the purpose of this chapter to help you analyze such designs in SYSTAT. I divide these designs into two types: those that have nesting between subjects and those that have nesting within subjects. The former requires nothing that hasn't been covered in previous chapters, so I will mainly just illustrate an analysis of an example data set and refer you to previous chapters for more discussion. The latter type requires a few additional considerations on testing contrasts on within subject factors, and these are presented in this chapter. If you have a design with nesting both between and within subjects, you will want to read both sections of this chapter.

12.1 Nesting Between Subjects

12.1.1 An Example Data Set

In section 11.1 I presented a hypothetical experiment in which each of 15 subjects was tested in a choice response time task under six conditions. Now suppose that the subjects were divided into three age categories: young (1), middle aged (2), and elderly (3). In this experiment, subjects are nested in their age group. This data set is shown in Table 12.1. (See section 11.1 on entering data for repeated measures ANOVA.)

Table 12.1 Fifteen subjects, six measures each, with five subjects per age group in the CHOICE data set.

	Age	L1D1	L1D2	L2D1	L2D2	L3D1	L3D2
1	1	282	228	281	423	344	563
2	1	425	275	410	316	373	171
3	1	260	369	469	346	446	390
4	1	223	376	266	313	394	249
5	1	351	290	296	431	208	483
6	2	451	473	453	490	524	399
7	2	132	191	346	402	452	376
8	2	397	239	369	432	470	443
9	2	523	375	391	410	456	501
10	2	336	322	374	535	494	504
11	3	116	171	490	347	428	327
12	3	350	250	333	478	487	285
13	3	300	240	431	410	377	636
14	3	452	338	300	390	500	502
15	3	471	496	390	496	575	549

12.1.2 ANOVA with Nested Between Subjects Effects

This type of design can be thought of as consisting of two separate parts: a between subjects part (for example, age groups) and a within subjects part (for example, location and delay). The between subjects part uses a single data point for each subject, the subject's mean across the six repeated measures, and then goes on to carry out a purely between subjects ANOVA, just like the ones discussed in Chapters 8 and 9. The within subject part ignores the between subjects effects entirely and computes the within subjects effects only. Therefore, you could perform a repeated measures ANOVA with between subjects nesting in two parts, one part a fully between subject analysis using the procedures from Chapters 8 and 9, and one part a fully within subjects analysis using the procedures in Chapter 11. This analysis would leave out only the interactions among between and within subjects factors, which are discussed below. Fortunately, SYSTAT makes it quite easy for you to compute both parts at once in the same output. The only difference between these commands and the usual repeated measures *anova* commands is that we have included age as a categorical variable in the analysis. (If you have two between subjects factors, one of which is nested in the other you can use the *model* command; see Chapter 10.) Otherwise, the commands are identical to those in Chapter 11 (see section 11.3):

```
>MGLH
>CATEGORY AGE
>ANOVA L1D1-L3D2 / REPEAT = 3,2  NAMES =,
"LOCATION","DELAY"
>PRINT MEDIUM
>ESTIMATE
```

The results are shown in Table 12.2 with *medium* output (means and standard errors). The first thing you should always examine is the "repeated measures factors and levels" table near the top of the output to ensure that the *repeat* and *names* options divided the within subjects factors appropriately.

The different parts of this output are discussed in the following sections.

12.1.2.1 Between Subjects

As mentioned above, the between subjects part of the analysis essentially uses a single data point for each subject, the subject's mean across the six repeated measures, and then goes on to carry out a between subjects ANOVA on those data points, just like in the analyses in Chapters 8 and 9. To bring this point home, Table 12.3 shows the means across the repeated measures for the 15 subjects in Table 12.1. You can perform a one-way ANOVA on these data with the usual commands (see section 8.3):

```
>MGLH
>CATEGORY AGE
>ANOVA S_MEAN
>PRINT SHORT
>ESTIMATE
```

This ANOVA output is shown in Table 12.4. If you compare this output with the between subjects part of the output in Table 12.2, you find that the F test and p-value are exactly the same in both tables. The only difference is that the sums of squares and mean squares in Table 12.2 are six times as large as those in Table 12.4, because the latter averaged over the six repeated measures and therefore used only one-sixth the n of the former. Note that this is irrelevant for the degrees of freedom in the two tables.

Thus, the analysis of the between subjects part of the data proceeds in exactly the same manner as the data set covered in Chapter 8. You can examine the between subjects data using

MACINTOSH

① Choose the menu item **Stats/MGLH/Fully Factorial (M)ANOVA...**
② Select L1D1, L1D2, L2D1, L2D2, L3D1, and L3D2 from the **Dependent Variable(s)** list.
③ Select AGE from the **Factor(s)** list.
④ Click on **More...**
⑤ Turn on **Repeat** and type "3,2" into the box.
⑥ Type ' "location","delay" ' (including the double-quotes) into the **Names** box.
⑦ Turn on **Means & Std. Errors** and **Save Residuals**.
⑧ Click **OK**.
⑨ Type "choice.res" into the **Save residuals as** box and click **OK**.

MS-DOS

① Choose the menu item **Utilities/Output/Results/Medium**.
② Choose **Statistics/MGLH/MGLM/Category/Variables/** and select AGE from the variable list.
③ Choose **Statistics/MGLH/MGLM/Model/Variables**, hit **<esc>**, and type "L1D1-L3D2 = constant" into the box.
④ Choose **Statistics/MGLH/MGLM/Model/Options/Repeat** and type "3,2" into the box.
⑤ Choose **Statistics/MGLH/MGLM/Model/Options/Names** and type ' "location","delay" ' (including the double-quotes) into the box.
⑥ Choose **Statistics/MGLH/MGLM/Estimate/Save**, hit **<esc>**, and type the save filename, e.g., "choice.res" into the box.
⑦ Choose **Statistics/MGLH/MGLM/Estimate/Go!**

Table 12.2 Repeated measures ANOVA on the CHOICE data set with age as a between subjects factor.

```
LEVELS ENCOUNTERED DURING PROCESSING ARE:
AGE
        1.000        2.000        3.000

NUMBER OF CASES PROCESSED:     15
```

DEPENDENT VARIABLE MEANS

L1D1	L1D2	L2D1	L2D2	L3D1	L3D1
337.933	308.867	373.267	414.600	435.200	425.200

REPEATED MEASURES FACTORS AND LEVELS
DEPENDENT VARIABLES

WITHIN FACTOR	1	2	3	4	5	6
Location	1	1	2	2	3	3
Delay	1	2	1	2	1	2

LEAST SQUARES MEANS.

AGE = 1.000   N OF CASES = 5.000

	L1D1	L1D2	L2D1	L2D2	L3D1	L3D2
LS. MEAN	308.200	307.600	344.400	365.800	353.000	371.200
SE	57.122	46.259	32.121	26.314	31.090	58.754

AGE = 2.000   N OF CASES = 5.000

	L1D1	L1D2	L2D1	L2D2	L3D1	L3D2
LS. MEAN	367.800	320.000	386.600	453.800	479.200	444.600
SE	57.122	46.259	32.121	26.314	31.090	58.754

AGE = 3.000   N OF CASES = 5.000

	L1D1	L1D2	L2D1	L2D2	L3D1	L3D2
LS. MEAN	337.800	299.000	388.800	424.200	473.400	459.800
SE	57.122	46.259	32.121	26.314	31.090	58.754

UNIVARIATE AND MULTIVARIATE REPEATED MEASURES ANALYSIS

BETWEEN SUBJECTS

SOURCE	SS	DF	MS	F	P
AGE	76933.356	2	38466.678	2.201	0.153
ERROR	209746.467	12	17478.872		

Table 12.2 (*continued*)

```
WITHIN SUBJECTS
---------------


SOURCE              SS           DF      MS            F          P         G-G     H-F

Location         176964.622      2     88482.311     10.981     .4E-03    0.001   .4E-03
Location*AGE      23155.111      4      5788.778      0.718     0.588     0.566   0.588
ERROR            193394.933     24      8058.122

GREENHOUSE-GEISSER EPSILON:         0.8317
HUYNH-FELDT EPSILON        :        1.0000

Delay                12.844      1        12.844      0.002     0.968       .       .
Delay*AGE          1688.022      2       844.011      0.107     0.900       .       .
ERROR             94785.800     12      7898.817

GREENHOUSE-GEISSER EPSILON:           .
HUYNH-FELDT EPSILON        :          .

Location
*Delay            19887.022      2      9943.511      1.225     0.311     0.308   0.311
Location
*Delay*AGE         7739.511      4      1934.878      0.238     0.914     0.891   0.914
ERROR            194808.800     24      8117.033

GREENHOUSE-GEISSER EPSILON:         0.8584
HUYNH-FELDT EPSILON        :        1.0000

SINGLE DEGREE OF FREEDOM POLYNOMIAL CONTRASTS
---------------------------------------------

POLYNOMIAL TEST OF ORDER    1 (LINEAR)

SOURCE              SS           DF      MS              F           P

Location        171093.600       1    171093.600       18.434      0.001
Location*AGE     23030.800       2     11515.400        1.241      0.324
ERROR           111374.600      12      9281.217

Location
*Delay            1363.267       1      1363.267        0.119      0.736
Location
*Delay*AGE          90.133       2        45.067        0.004      0.996
ERROR           136930.600      12     11410.883
```

Table 12.2 (*continued*)

POLYNOMIAL TEST OF ORDER   2 (QUADRATIC)

SOURCE	SS	DF	MS	F	P
Location	5871.022	1	5871.022	0.859	0.372
Location*AGE	124.311	2	62.156	0.009	0.991
ERROR	82020.333	12	6835.028		
Location *Delay	18523.756	1	18523.756	3.841	0.074
Location *Delay*AGE	7649.378	2	3824.689	0.793	0.475
ERROR	57878.200	12	4823.183		

MULTIVARIATE REPEATED MEASURES ANALYSIS

TEST OF: Location		HYPOTH. DF	ERROR DF	F	P
WILKS' LAMBDA=	0.382	2	11	8.917	0.005
PILLAI TRACE =	0.618	2	11	8.917	0.005
H-L TRACE =	1.621	2	11	8.917	0.005

TEST OF: Location*AGE		HYPOTH. DF	ERROR DF	F	P
WILKS' LAMBDA=	0.785	4	22	0.706	0.596
PILLAI TRACE =	0.215	4	24	0.721	0.586
H-L TRACE =	0.273	4	20	0.683	0.612
THETA =	0.214 S = 2, M =-0.5, N = 4.5 PROB =				0.522

TEST OF: Location *Delay		HYPOTH. DF	ERROR DF	F	P
WILKS' LAMBDA=	0.751	2	11	1.826	0.207
PILLAI TRACE =	0.249	2	11	1.826	0.207
H-L TRACE =	0.332	2	11	1.826	0.207

TEST OF: Location *Delay*AGE		HYPOTH. DF	ERROR DF	F	P
WILKS' LAMBDA=	0.883	4	22	0.354	0.839
PILLAI TRACE =	0.117	4	24	0.374	0.825
H-L TRACE =	0.133	4	20	0.332	0.853
THETA =	0.117 S = 2, M =-0.5, N = 4.5 PROB =				0.734

the means in Table 12.3 just as you would for any one-way ANOVA (see section 8.3). You can test contrasts, compute post-hoc tests, find confidence intervals, and assess the validity of the assumptions underlying the between subjects F tests using the same methods as described in Chapters 8 and 9. The between subjects part of the output contains nothing that we haven't seen in earlier chapters.

Table 12.3 The age group and mean for each subject in the CHOICE data set.

	Age	Mean
1	1.000	353.500
2	1.000	328.333
3	1.000	380.000
4	1.000	303.500
5	1.000	343.167
6	2.000	465.000
7	2.000	316.500
8	2.000	391.667
9	2.000	442.667
10	2.000	427.500
11	3.000	313.167
12	3.000	363.833
13	3.000	399.000
14	3.000	413.667
15	3.000	496.167

Table 12.4 One-way ANOVA on AGE, the between subjects part of CHOICE data set.

```
LEVELS ENCOUNTERED DURING PROCESSING ARE:
AGE
           1.000          2.000          3.000

DEP VAR: S_MEAN     N: 15  MULTIPLE R: 0.518  SQUARED MULTIPLE R: 0.268

                       ANALYSIS OF VARIANCE

SOURCE        SUM-OF-SQUARES   DF   MEAN-SQUARE    F-RATIO       P

AGE                12822.226    2      6411.113      2.201   0.153

ERROR              34957.744   12      2913.145
```

12.1.2.2 Within Subjects

Most of the within subjects part of the analysis is identical to what the output would look like for a purely within subjects design, ignoring the between subjects factors. The differences are that (1) the error terms from the purely within subject design are now divided into pieces, and (2) there are now interactions between within subjects factors and between subjects factors. As an example of the first difference, in Table 11.3, which did not have a between subjects factor, the error term for the location main effect was the location*subjects interaction, with sum-of-squares equal to 216,550.044 and *df* equal to 28. In Table 12.2 this term has now been divided into two pieces: (1) a location* age interaction with sum-of-squares equal to 23,155.111 and *df* equal to 4, and (2) a location* subjects(age) interaction with sum-of-squares (denoted "ERROR") equal to 193,394.933 and *df* equal to 24.)The notation "subjects(age)" is to be read "subjects nested in age.") Note that both the sums-of-squares and the degrees of freedom add up to the values for the location*subjects interaction in Table 11.3: 23,155.111 + 193,394.933 = 216,550.044, and 4 + 24 = 28. By removing the location*age interaction from the error term for location in the between subjects analysis, you hope to obtain a more precise estimate of the error variance and a more accurate *p*-value.

In a similar manner, the delay*subjects interaction in Table 11.3 (the error term for delay) is divided into delay*age and delay*subjects(age)—the latter is the error term for delay*age in Table 12.2. Finally, the location*delay*subjects interaction in Table 11.3 (the error term for location*delay) is divided into location*delay*age and location*delay* subjects(age)—the latter is the error term for location*delay*age in Table 12.2. Thus, all of the error terms in this new repeated measures ANOVA in Table 12.2 have fewer *df* than the corresponding error terms in the analysis with no between subjects factors in Table 11.3. The hope is that this loss of *df* will be more than made up for by the increase in precision that is gained by removing the interactions with age from those error terms.

The sums-of-squares for the within subjects main effects and interactions are unchanged by the addition of between subjects factors. For example, the location effect in Table 11.3 had mean square equal to 176,964.622 and *df* equal to 2, which is exactly what it has in Table 12.2. The only change is in the error term for this effect, as just described.

As discussed in Chapter 11 (section 11.3), the multivariate statistics and adjustments to *p*-values in tables such as Table 12.2 can be ignored if you do not want to assess multi-*df* effects and instead are interested in contrasts. I turn to contrasts next.

12.1.3 Contrasts Involving Nested Between Subjects Effects

F tests with more than one degree of freedom in the numerator, like the location and location*age tests in Table 12.2, are usually of little scientific interest. In their stead, you should compute contrasts to test specific hypotheses about your data. Contrasts that only involve within subjects factors are computed using the same procedures in SYSTAT that were described in section 11.4, so that discussion will not be repeated here. However, contrasts involving between subjects factors are not handled in the same way, so I will discuss them here.

12.1.3.1 Between Subjects Effects

Suppose that your theory predicted that there would be a linear increase in response times across the three age groups. You would think that you could test this contrast in the usual manner, using the weights –1, 0, and 1 and the SYSTAT procedures described in sections 8.4.1 and 8.4.2, for example. However, if you try this you will find that SYSTAT computes six contrasts, one contrast within each of the repeated measures columns! If you want to test the hypothesis in SYSTAT, you can do so by taking advantage of a feature of repeated measures discussed in section 11.3: The between subjects part of the analysis is identical to a between subjects ANOVA on the mean of the repeated measures columns. Therefore, to test this hypothesis, first create a new column of subject means using a command such as

```
>LET S_MEANS = AVG(L1D1-L3D2)
```

which is what I did to get the column of means shown in Table 12.4. Note that the hyphen, when used within the AVG() function, specifies a range of columns to average across, not subtraction. Then you can compute the between subjects ANOVA in the usual manner, as described in section 12.1.2.1 (see also section 8.3). The output for this ANOVA was shown in Table 12.4. Now you can ask SYSTAT to compute the desired contrast in the usual manner (see sections 8.4.1 and 8.4.2):

```
>HYPOTHESIS
>EFFECT = AGE
>CONTRAST
>-1 0 1
>TEST
```

This contrast is shown in the top panel of Table 12.5. The linear trend is not significant. In a similar manner, you can test the quadratic trend in age using the contrast weights –1, 2, and –1:

```
>HYPOTHESIS
>EFFECT = AGE
>CONTRAST
>-1 2 -1
>TEST
```

The quadratic contrast is shown in the bottom panel of Table 12.5. This contrast is also not statistically significant. Do the sums-of-squares add up the way they should? The SS for the linear contrast is 7691.378 and the SS for the quadratic contrast is 5130.848, so the sum is 12,822.226. This is the sum-of-squares for the age effect in Table 12.4. And if you recall that the number of data points going into each mean in this analysis is six, you can multiply this sum-of-squares by 6 to get the sum-of-squares for the age effect in the repeated measures output in Table 12.2: 76,933.356.

Thus, by redefining the dependent variable as the mean of the six repeated measures, we have simplified the contrast procedures to the same ones we saw earlier in Chapter 8, section 4. If we had two or more between subjects factors, we could test contrasts on their interactions just as we did in Chapter 9, section 4. By creating one new column of means in the data set we have avoided having to test these contrasts by hand.

12.1.3.2 Interactions of Between and Within Subjects Factors

The age effect in Table 12.2 is a between subject effect, so contrasts can be tested on the age effect in the manner described in the previous section. The location effect is a within subject effect, so contrasts can be tested on location using SYSTAT's usual repeated measures contrast procedures (see section 11.4). What about an effect like the location × age interaction? This interaction involves both a between and a within subjects factor, so neither of the foregoing contrast

Table 12.5 Linear contrast on the age effect in the CHOICE data set.

Linear Trend

```
TEST FOR EFFECT CALLED:     AGE

A MATRIX
                         1            2            3
                     0.000       -2.000       -1.000

TEST OF HYPOTHESIS

      SOURCE          SS        DF         MS           F          P

  HYPOTHESIS     7691.378        1    7691.378       2.640      0.130
       ERROR    34957.744       12    2913.145
```

Quadratic Trend

```
TEST FOR EFFECT CALLED:     AGE

A MATRIX
                         1            2            3
                     0.000        0.000        3.000

TEST OF HYPOTHESIS

      SOURCE          SS        DF         MS           F          P

  HYPOTHESIS     5130.848        1    5130.848       1.761      0.209
       ERROR    34957.744       12    2913.145
```

procedures will work. The following trick enables you to get SYSTAT to test contrasts on such an interaction using the proper local error term.

Suppose you had a very specific hypothesis that the linear trend in location would increase from the youngest to the middle age group, and then with the oldest age group drop back down to a level approximately one-third of the way towards the youngest group. In numbers, we can represent this as an increase from one to four, and then a drop back to three. The mean of these numbers is 8/3=2.667, so we can find the appropriate contrast weights by subtracting 2.667 from each of these numbers: –1.667, 1.333, and .333. Although it makes no difference for the analysis, I would now multiply these weights by 3 to get rid of the decimal paces, so the weights for the age conditions are –5, +4, and +1, as shown in the "left margin" (second column) of Table 12.6. The linear trend in location can be represented with the usual weights –1, 0, and +1, as shown in the "top margin" (first row) of Table 12.6. To get the contrast weights for the interaction contrast, then, you simply multiply the weights in the margins pairwise to get the weights in the cells. The resulting contrast weights are shown in the cells of Table 12.6.

Table 12.6 The contrast weights for the hypothesis that the linear trend in location will increase from the youngest to the middle age group, and then with the oldest age group drop back down to a level approximately one-third of the way towards the youngest group.

	λ_{ij}	Location 1	Location 2	Location 3
		–1	0	+1
Age 1	–5	5	0	–5
Age 2	+4	–4	0	4
Age 3	+1	–1	0	1

You could proceed to find the means for each condition by hand and compute the contrast with usual contrast formulas, but there is an easier way in SYSTAT that will also give you the proper local error term for the contrast. First, compute a contrast score for each subject using the weights for the within subject factor. In this case, the within subject factor is location, so you should multiply the three location means by the linear weights, –1, 0, and 1, respectively. (You could do this in one step using the weights –1, –1, 0, 0, 1, 1, respectively, across the six repeated measures columns, but for illustrative purposes I will take two steps.) To find the three location means for each subject type

```
>LET L1 = (L1D1+L1D2)/2
>LET L2 = (L2D1+L2D2)/2
>LET L3 = (L3D1+L3D2)/2
```

These three columns are shown in Table 12.7. Now you can compute the contrast score (CS) for each subject by multiplying each of these columns by the corresponding linear weights:

```
>LET CS = (-1)*L1 + (0)*L2 + (1)*L3
```

Table 12.7 Location condition means and linear contrast scores (CS) for the CHOICE data set.

	Age	L1	L2	L3	CS
1	1	255.0	352.0	453.5	198.5
2	1	350.0	363.0	272.0	−78.0
3	1	314.5	407.5	418.0	103.5
4	1	299.5	289.5	321.5	22.0
5	1	320.5	363.5	345.5	25.0
6	2	462.0	471.5	461.5	−0.5
7	2	161.5	374.0	414.0	252.5
8	2	318.0	400.5	456.5	138.5
9	2	449.0	400.5	478.5	29.5
10	2	329.0	454.5	499.0	170.0
11	3	143.5	418.5	377.5	234.0
12	3	300.0	405.5	386.0	86.0
13	3	270.0	420.5	506.5	236.5
14	3	395.0	345.0	501.0	106.0
15	3	483.5	443.0	562.0	78.5

This column is shown in the last column of Table 12.7.

Now you are in position to compute the desired interaction contrast. First, compute a one-way ANOVA on these contrast scores with age as the main effect. The appropriate commands are

```
>CATEGORY AGE
>ANOVA CS
>PRINT SHORT
>ESTIMATE
```

(See section 8.3.) Finally, compute a contrast on the age effect in this analysis using the weights −5, 4, and 1:

```
>HYPOTHESIS
>EFFECT = AGE
>CONTRAST
>-5, 4, 1
>TEST
```

(See section 8.4.1.) The output for this contrast is shown in Table 12.8. This contrast is not significant, $F(1,12) = 1.564$, $p = .235$.

Table 12.8 Test of the hypothesis corresponding to the contrast weights in Table 12.6.

```
TEST FOR EFFECT CALLED:     AGE


A MATRIX

                      1              2              3

                  0.000         -6.000          3.000

TEST OF HYPOTHESIS

        SOURCE         SS        DF         MS            F          P

    HYPOTHESIS   14516.743        1   14516.743        1.564      0.235
         ERROR  111374.600       12    9281.217
```

12.2 Nesting Within Subjects

12.2.1 An Example Data Set

In section 11.1 I presented a hypothetical experiment in which each of 15 subjects was tested in a choice response time task under six conditions. In the analyses so far, we have assumed that the location and delay factors, both within subjects factors, were fully crossed. However, suppose that different delay times were used within each location condition, so that the delay effect is *nested* in the location factor, not crossed with it. The layout of the data in the data file is the same as that shown in Table 11.1 (or Table 12.1 without the AGE column).

12.2.2 ANOVA with Nested Within Subjects Effects

Although SYSTAT makes no provisions for analysis of this type of data, the usual procedures in SYSTAT can be used to get all of the numbers we need. This section goes to some lengths to show how this can be done, with the result that you obtain a lot of multi-*df* significance tests. *If you are only interested in contrasts, you may skip this section entirely.*

The first step is to run the ANOVA as though delay were fully crossed with location, which is exactly what we did in Chapter 11. The relevant parts of the ANOVA output from Table 11.3 are reproduced here in Table 12.9.

Table 12.9 Excerpts from the repeated measures ANOVA on the CHOICE data set with fully crossed within subjects factors (from Table 11.3).

```
WITHIN SUBJECTS
---------------


SOURCE            SS           DF       MS           F         P        G-G       H-F

location       176964.622      2    88482.311      11.441    .2E-03    .6E-03    .3E-03
ERROR          216550.044     28     7733.930

delay              12.844      1       12.844       0.002    0.966       .         .
ERROR           96473.822     14     6890.987

location
*delay          19887.022      2     9943.511       1.375    0.269     0.270     0.269
ERROR          202548.311     28     7233.868


SINGLE DEGREE OF FREEDOM POLYNOMIAL CONTRASTS
---------------------------------------------

POLYNOMIAL TEST OF ORDER    1 (LINEAR)

SOURCE            SS           DF       MS           F         P

location       171093.600      1    171093.600     17.822   .854012E-03
ERROR          134405.400     14      9600.386

location
*delay           1363.267      1      1363.267      0.139    0.715
ERROR          137020.733     14      9787.195


POLYNOMIAL TEST OF ORDER    2 (QUADRATIC)

SOURCE            SS           DF       MS           F         P

location         5871.022      1      5871.022      1.001    0.334
ERROR           82144.644     14      5867.475

location
*delay          18523.756      1     18523.756      3.958    0.067
ERROR           65527.578     14      4680.541
```

Table 12.10 ANOVA table for the repeated measures design in which delay is nested in location. The last column shows the terms in the unnested analysis shown in Table 12.9 that were used to reconstruct this table. Names of error terms are shown underneath in brackets.

Source	Sum-of-Squares	df	Origin in Table 12.9
Subjects	?	14	*not shown*
Location	176964.622	2	`location`
ERROR *[loc. × subjects]*	216550.044	28	`location ERROR`
ERROR *[delay(loc.) × subjects]*	299022.133	42	`delay ERROR + location*delay ERROR`

When you analyze a design with nested factors as fully crossed, you will obtain a number of lines in the output that do not belong. This is because you cannot estimate interactions between one factor and another in which it is nested, yet the fully crossed analysis gives lines for those interactions. For example, if delay is nested in location, you cannot estimate the location*delay interaction; therefore, the location*delay interaction in Table 12.9 should not be there. Similarly, the location*delay*subject interaction, which is shown as the error term for location*delay in Table 12.9, should not be there. Where do these undesired sums-of-squares and *df* belong?

The answers are shown in Table 12.10. We know that the delay(location) effect should have 3 *df* in this design (1 *df* within each level of location). But the delay effect in Table 12.9 has only 1 *df*; the other two were pulled out by SYSTAT and called the location*delay interaction. Thus, to reconstruct the delay(location) sums-of-squares we simply add together the delay and location*delay sums-of-squares from Table 12.9. The *df* add in a similar manner. Furthermore, we know that the delay(location) × subject interaction should have 42 *df* in this design; in Table 12.9 these *df* are split into two parts and called the ERROR for the delay effect with 14 *df* and the ERROR for the location*delay interaction with 28 *df*. Thus, to reconstruct the delay(location)*subject interaction sums-of-squares, we simply add together these two error sums-of-squares from Table 12.9, and add the *df* in a similar manner. The remaining main effect, location, is unchanged by the nesting. Table 12.10 summarizes the reconstruction of the appropriate terms for the nested design.

To summarize, we can't ask SYSTAT to nest within subject

factors, but we can ask it to cross within subject factors. Therefore, if we pretend that the within subjects factors are crossed, we can get SYSTAT to give us an ANOVA like the one shown in Table 12.9. This ANOVA has terms that do not belong in the nested design, but we can get the proper terms by adding the terms that don't belong to the terms that do not have the required *df*. The general rule is to pool together the nested effect(s) with the spurious interaction between that effect and the effect it is nested under. So, in the example, delay gets pooled with location*delay, and delay*subjects gets pooled with location*delay*subjects. The remaining terms are unaffected by the nesting.

12.2.3 Contrasts Involving Nested Within Subjects Effects

If you are not interested in seeing all of the multi-*df* significance tests for the within subjects effects, there is a simple alternative to the previous section. Instead of running the ANOVA as a fully crossed design, treat the *a* levels of the nested factor across the *b* levels of the nesting factor as though they were all levels of single factor with $a \times b$ levels. For example, if you only have the two within subject factors, and one with two levels is nested in the other with three levels, let SYSTAT run the ANOVA as though you had just one within subject factor with six levels. Then you can use the usual contrast procedures to pull out any of the effects in which you are interested.

In the CHOICE data set we are assuming that the delay factor with two levels is nested within the location factor with three levels, so the recommendation is to run the ANOVA as a repeated measures analysis with a single within subjects factor with six levels. This is very easy to do with the following commands (see section 11.3):

```
>MGLH
>ANOVA L1D1-L3D2  /REPEAT=6   NAMES="Condition"
>PRINT SHORT
>ESTIMATE
```

This output is shown in Table 12.11. I chose the generic name "condition" to represent the single within subjects factor consisting of the six delay-within-location conditions. Unless one of the polynomial contrasts in this table happens to correspond to a hypothesis of interest, which is unlikely, none of the significance tests in this table are of interest. Now, however, we are in a position to test contrasts that are of interest.

Table 12.11 Repeated measures ANOVA on the CHOICE data set treating the six repeated measures columns as six levels of a single within subjects factor called "Condition."

```
DEPENDENT VARIABLE MEANS
    L1D1        L1D2        L2D1        L2D2        L3D1        L3D2
   337.933     308.867     373.267     414.600     435.200     425.200
```

UNIVARIATE AND MULTIVARIATE REPEATED MEASURES ANALYSIS

WITHIN SUBJECTS

SOURCE	SS	DF	MS	F	P	G-G	H-F
Condition	196864.489	5	39372.898	5.346	.3E-03	0.002	.4E-03
ERROR	515572.178	70	7365.317				

```
GREENHOUSE-GEISSER EPSILON:      0.6911
HUYNH-FELDT EPSILON       :      0.9451
```

SINGLE DEGREE OF FREEDOM POLYNOMIAL CONTRASTS

POLYNOMIAL TEST OF ORDER   1 (LINEAR)

SOURCE	SS	DF	MS	F	P
Condition	157259.524	1	157259.524	14.247	0.002
ERROR	154528.133	14	11037.724		

POLYNOMIAL TEST OF ORDER   2 (QUADRATIC)

SOURCE	SS	DF	MS	F	P
Condition	1139.051	1	1139.051	0.119	0.735
ERROR	134069.330	14	9576.381		

Table 12.11 (continued)

POLYNOMIAL TEST OF ORDER 3 (CUBIC)

SOURCE	SS	DF	MS	F	P
Condition	31348.148	1	31348.148	6.359	0.024
ERROR	69013.607	14	4929.543		

POLYNOMIAL TEST OF ORDER 4

SOURCE	SS	DF	MS	F	P
Condition	6095.238	1	6095.238	1.003	0.334
ERROR	85096.048	14	6078.289		

POLYNOMIAL TEST OF ORDER 5

SOURCE	SS	DF	MS	F	P
Condition	1022.528	1	1022.528	0.196	0.664
ERROR	72865.059	14	5204.647		

MULTIVARIATE REPEATED MEASURES ANALYSIS

TEST OF: Condition		HYPOTH. DF	ERROR DF	F	P
WILKS' LAMBDA=	0.306	5	10	4.534	0.020
PILLAI TRACE =	0.694	5	10	4.534	0.020
H-L TRACE =	2.267	5	10	4.534	0.020

For example, to test the linear trend in location, we could pair the contrast weight –1 with the two location$_1$ conditions, the weight 0 with the two location$_2$ conditions, and the weight +1 with the two location$_3$ conditions. However, to illustrate how the contrast sums-of-squares add up to the values in the overall ANOVA output, it is necessary to use normalized contrast weights (where the positive weights sum to 1 and the negative weights sum to –1). Recall that multiplying or dividing a set of contrast weights by a constant has no effect on the size of the F test of the contrast, because the error term is also multiplied or divided by the same constant. To normalize a set of contrast weights you simply divide each of the weights by a particular constant, the square root of the sum of the squared contrast weights: $\sqrt{\Sigma\lambda^2}$. For example, if you wanted to normalize the contrast weights –1, –1, 0, 0, 1, and 1, you would divide each of the weights by

$$\sqrt{\Sigma\lambda^2} = \sqrt{(-1)^2 + (-1)^2 + 0^2 + 0^2 + 1^2 + 1^2} = 2 \qquad (12.1)$$

This results in the weights –.5, –.5, 0, 0, .5, and .5, respectively.

We can test the linear trend in location using these weights with the following *cmatrix* command (e.g., see section 11.4.1.2):

```
>HYPOTHESIS
>CMATRIX
>-.5 -.5 0 0 .5 .5
>TEST
```

This output is shown in the upper panel of Table 12.12. This contrast uses the local error term and is highly significant.

To test the quadratic trend in location we could use the contrast weights –1, 2, and –1, paired with the appropriate location conditions, but, again, for illustrative purposes we will use normalized weights. The constant by which we should divide the above weights is

$$\sqrt{\Sigma\lambda^2} = \sqrt{(-1)^2 + (-1)^2 + 2^2 + 2^2 + (-1)^2 + (-1)^2} = 3.4641 \qquad (12.2)$$

Thus, the contrast weights for the six conditions are –.288675, –.288675, .57735, .57735, –.288675, and –.288675 (I'm using a lot of decimal places to help illustrate a point; in practice, three or so should suffice). Plugging these into the *cmatrix* command:

```
>HYPOTHESIS
>CMATRIX
>-.288675 -.288675 .57735 .57735 -.288675,
-.288675
>TEST
```

This output is shown in the lower panel of Table 12.12. This contrast also uses the local error term but is not significant.

The sums-of-squares for these two contrasts are the same as those for the polynomial contrasts in Table 12.9 (and Table 11.3). Similarly, the sums-of-squares for error are nearly identical in the two tables. (Any differences are due to rounding error.) This illustrates that the factor in which another is nested is not affected by whether you properly regard the nested factor as nested, or treat it as crossed. However, this does matter for the nested factor itself, as we see next.

Suppose that you wanted to test the simple effects of delay, nested within each of the three location conditions; that is, you want to compare $delay_1$ with $delay_2$ within $location_1$ and so on. This can be accomplished readily using the *cmatrix* command. Again, using the weights -1 and $+1$ would give the correct F tests, but to see how the sums-of-squares add up we must normalize the weights. To do this we divide each of the

Table 12.12 Linear and quadratic contrasts on the location conditions in the CHOICE data set.

Linear Trend

C MATRIX

1	2	3	4	5	6
-0.500	-0.500	0.000	0.000	0.500	0.500

TEST OF HYPOTHESIS

SOURCE	SS	DF	MS	F	P
HYPOTHESIS	171093.600	1	171093.600	17.822	.854012E-03
ERROR	134405.400	14	9600.386		

Quadratic Trend

C MATRIX

1	2	3	4	5	5
-0.289	-0.289	0.577	0.577	-0.289	-0.289

TEST OF HYPOTHESIS

SOURCE	SS	DF	MS	F	P
HYPOTHESIS	5871.017	1	5871.017	1.001	0.334
ERROR	82144.568	14	5867.469		

weights by the square root of 2, which gives the weights −.7071, .7071, and four zeros. Thus, the following commands test the three simple effects of delay within each level of location:

```
>HYPOTHESIS
>CMATRIX
>-.7071 .7071 0 0 0 0
>TEST
>HYPOTHESIS
>CMATRIX
>0 0 -.7071 .7071 0 0
>TEST
>HYPOTHESIS
>CMATRIX
>0 0 0 0 -.7071 .7071
>TEST
```

(See section 11.4.1.2.) The results of these three contrasts are shown in the top, middle, and bottom panels of Table 12.13, respectively. All three of these contrasts use the appropriate local error term, and none is significant.

If we sum the sums-of-squares for these three simple effects we get 19,899.878, which is within rounding error of the sum-of-squares for delay-nested-in-location in Table 12.10. Pooling the three error terms for these contrasts gives 299,022.318, which is the error term for delay(location) in Table 12.10. If we now divide the former over its $df(3)$ by the latter over its $df(42)$, we find $F(3,42) = .931$, which is the test of the main effect of delay-nested-in-location. Note that if you are not interested in this multi-df effect, you would not need to use the normalized contrasts weights in the foregoing analysis.

As an aside, note that each of the simple effect contrasts of delay nested in location could have been tested using SYSTAT's *matched-pairs t*-test command (see also section 11.6.2). With this command, SYSTAT assumes that you have two columns of data with values that are paired within each row, for example, two scores per subject. The *t*-test, then, tests the significance of the difference between the two columns. In the present example, the two columns are just the two delay columns within a given level of location. To compute a matched pairs *t*-test between any two of these columns type

```
>STATS
>TTEST    COLUMN1 COLUMN2
```

Table 12.13 Tests of the simple effects of delay at the three levels of location using normalized contrast weights.

Delay₁ versus Delay₂ within Location₁

C MATRIX

1	2	3	4	5	6
-0.707	0.707	0.000	0.000	0.000	0.000

TEST OF HYPOTHESIS

SOURCE	SS	DF	MS	F	P
HYPOTHESIS	6336.537	1	6336.537	1.334	0.267
ERROR	66514.508	14	4751.036		

Delay₁ versus Delay₂ within Location₂

C MATRIX

1	2	3	4	5	6
0.000	0.000	-0.707	0.707	0.000	0.000

TEST OF HYPOTHESIS

SOURCE	SS	DF	MS	F	P
HYPOTHESIS	12813.341	1	12813.341	2.659	0.125
ERROR	67471.708	14	4819.408		

Delay₁ versus Delay₂ within Location₃

C MATRIX

1	2	3	4	5	6
0.000	0.000	0.000	0.000	-0.707	0.707

TEST OF HYPOTHESIS

SOURCE	SS	DF	MS	F	P
HYPOTHESIS	750.000	1	750.000	0.064	0.805
ERROR	165036.102	14	11788.293		

MACINTOSH

① Choose the menu item **Stats/Stats/t-test...**

② Click on the **Paired** button if it is not already selected.

③ Select the two paired columns from the left **Variables** list, e.g., L1D1 and L1D2.

④ Click **OK**.

MS-DOS

① Choose the menu item **Statistics/Stats/Ttest/ Dependent/Variables/** and select L1D1 and L1D2 from the variables list.

② Choose the menu item **Statistics/Stats/Ttest/ Dependent/Go!**

For example, to compute the matched pairs t-test between $delay_1$ and $delay_2$ within $location_1$, type

```
>STATS
>TTEST L1D1 L1D2
```

The output from this command is shown in Table 12.14. Note that the square of the t in this table, $1.155^2 = 1.334$, is exactly the same as the F test for this effect in Table 12.13. All contrasts using local error terms can be computed using t-tests in a manner similar to this.

12.3 Unequal Cell Sizes

The same considerations regarding unequal cell sizes and unbalanced designs that were mentioned at the end of Chapter 11 (section 11.8) are relevant here. Missing cells are a big problem for which you probably do not have the software to solve. Unequal numbers of observations within cells can be handled by averaging over the observations within cells and treating those means as raw data points.

In addition, there is one type of imbalance that is quite common in the designs raised in this chapter: unequal numbers of subjects nested in levels of some other factor, such as age. For example, what if you had three subjects in the youngest age group, six subjects in the middle age group, and five subjects in the oldest age group? It turns out that this is not a difficult problem if you are willing to give up tests of multi-df effects. By working with the means for each subject you can turn the age effect into a one-way ANOVA, as was done in Table 12.4, and analyze it using the methods in Chapter 8. Fortunately, one-way ANOVAs are not adversely affected by unequal cell sizes. If you have two between subject effects, use a two-way unweighted means analysis with the subject means as data points.

The approach to analyzing within subject effects is not affected at all by unequal numbers of subjects within nests; simply define contrast scores in the usual manner and test

Table 12.14 Matched-pairs t-test between $delay_1$ and $delay_2$ nested in $location_1$.

```
PAIRED SAMPLES T-TEST ON      L1D1     VS     L1D2    WITH     15 CASES

MEAN DIFFERENCE =       29.067
SD DIFFERENCE   =       97.479
T =         1.155 DF =  14 PROB =        0.267
```

them with local error terms. For interactions of between and within subject factors you can also use contrast scores, once again turning the analysis into a test of contrasts in a one-way ANOVA (as done in section 12.1.3.2) or in an unweighted means ANOVA (section 9.9).

In general, if you remember that one-way ANOVAs and contrasts using local error terms can be analyzed in usual ways even with unbalanced designs and unequal cell sizes, you can get yourself out of most difficulties by redefining your problem as a contrast and/or a one-way ANOVA.

13
Contingency Tables

When your dependent variable is categorical and those categories are not on any meaningful scale, the analyses in the foregoing chapters will often be suboptimal. For example, if you ask a group of people a yes or no question, you *could* code "yes" as 1 and the "no" category as 0 and perform an ANOVA using this 0/1 variable as the dependent variable, but you would know for certain that the normality assumption would be violated. In many cases this violation will not change the conclusions that you would draw from the results, but a better type of analysis is available in SYSTAT: *table analysis*. In this type of analysis, rather than using individual data points as the basis of the analysis, you instead use *counts* or *percentages* of data points within categories and ask whether these counts or percentages differ across categories. So, for example, you might add up the number of yes's and no's, and test to see whether these numbers differ significantly from chance. SYSTAT makes this type of analysis very easy.

SYSTAT offers two formats for data files that can be used for table analysis: one observation per case and counts. These two methods are identical from an analysis standpoint, but one or the other may be more convenient from the data entry standpoint. I begin this chapter by discussing these two formats and then turn to the analyses.

13.1 The Data File

13.1.1 One Observation per Case

If you have collected the data for a table analysis, it probably came in one observation at a time, and you eventually need to summarize these observations within categories. For example, suppose that you are a telephone pollster and you have randomly called a number of people in a given city and asked them a series of categorical questions: What is their political party affiliation, what is their gender, and how do they feel about a new bond levy facing voters in the fall. When you called each person, you wrote down their responses to each question, and now you want to enter them into a table analysis in SYSTAT. One way to do this is to open a new data editor window, create a column corresponding to each of the questions, and then enter each subject's responses into a separate row of the file. The first few examples of such a data file are shown in Table 13.1.

Table 13.1 The first five cases of the POLLSTER data set, with one case per observation.

	PARTY$	GENDER$	RESPONSE$
1	Democrat	male	yes
2	Republican	female	no
3	Republican	female	yes
4	Libertarian	male	no
5	Democrat	female	yes

13.1.2 Counts

Suppose instead of writing down a new row of responses for each person that you interview, you instead draw 12 boxes, one for each combination of party, gender, and response, and when you call a person you simply put a new tick mark in the appropriate box. After the data is collected you then count up the number of ticks in each of the 12 boxes. SYSTAT can perform table analyses directly on these counts, rather than entering the entire raw data. The advantage is that you will often have less typing to do and SYSTAT will work with the file a little faster. You must decide whether this savings outweighs the task of tallying the categories by hand.

To enter counts into SYSTAT you will need one column for each of the questions (variables) that you asked, just as in the previous section, and one additional column to contain the counts within each category. Table 13.2 shows such a data file for the data from the POLLSTER data set, the first five cases of which were shown in Table 13.1. There are 123 observations in this data set, so to enter this data using the one-observation-per-case method in the previous section would have required typing 123 rows of data.

There is one extra operation that you must perform when you enter a data file of this type into a table analysis: You must tell SYSTAT to weight the cases by the value in the column of counts. This can be done very easily using the weight command:

```
>WEIGHT COUNT
```

That's it. You are ready to proceed with the table analysis.

MACINTOSH
① Choose the menu item **Data/Weight...**
② Double-click on the **COUNT** variable to enter it into the box.
③ Click **OK**.

MS-DOS
① Choose the menu item **Data/Weight/Select/** and select COUNT from the variables list.

Table 13.2 The POLLSTER data set, using counts.

	PARTY$	GENDER$	RESPONSE$	COUNT
1	Democrat	male	yes	18
3	Democrat	male	no	9
2	Democrat	female	yes	20
4	Democrat	female	no	8
5	Republican	male	yes	7
6	Republican	male	no	13
7	Republican	female	yes	9
8	Republican	female	no	18
9	Libertarian	male	yes	2
10	Libertarian	male	no	8
11	Libertarian	female	yes	1
12	Libertarian	female	no	10

13.2 One-Way Tables

13.2.1 Generating Tables

The simplest tables of all are one-way tables, which have just one grouping variable. Suppose, for example, that you are interested in determining whether the number of yes responses is different than the number of no responses in the POLLSTER data set. This can be done using a one-way table.

13.2.1.1 Table Format

You have two ways of displaying the data summaries in SYSTAT—table format and list format. Within table format you also have options: You can display the category frequencies or the percentages of the cases accounted for by each category. For example, to show tables with both the raw frequencies and percentages, type the following commands:

MACINTOSH

① Choose the menu item **Statistics/Tables/Tabulate...**
② Double-click on the grouping variable you want, e.g., RESPONSE$, in the left selection list.
③ Turn on **Frequency**, **Percent**, and **Missing**.
④ Click **OK**.

MS-DOS

① Choose the menu item **Stats/Tables/Tabulate/One-way/Variables/** and select RESPONSE$ from the variables list.
② Choose **Stats/Tables/Tabulate/One-way/Options/Miss**.
③ Choose **Stats/Tables/Tabulate/One-way/Options/Frequency**.
④ Choose **Stats/Tables/Tabulate/One-way/Options/Percent**.
⑤ Choose **Stats/Tables/Tabulate/Go!**

MACINTOSH

① Choose the menu item **Stats/Tables/Tabulate...**
② Double-click on the grouping variable you want, e.g., RESPONSE$, in the left selection list.
③ Turn on **List** and **Missing**.
④ Click **OK**.

MS-DOS

① Choose the menu item **Stats/Tables/Tabulate/One-way/Variables/** and select RESPONSE$ from the variables list.
② Choose **Stats/Tables/Tabulate/One-way/Options/Miss**.
③ Choose **Stats/Tables/Tabulate/One-way/Options/List**.
④ Choose **Stats/Tables/Tabulate/Go!**

```
>TABLES  RESPONSE$/FREQ PERCENT MISS
```

The option *freq* tells SYSTAT to generate a table of frequencies, and *percent* tells it to generate a table of percentages. The *miss* option excludes missing cells from the analysis; otherwise, "missing" would be treated as a category, just like "yes" and "no". The resulting tables are shown in Table 13.3.

Table 13.3 Frequencies and percentages of yes's and no's in the POLLSTER data set.

TABLE OF VALUES FOR RESPONSE$

FREQUENCIES

	no	yes	TOTAL
	66	57	123

TABLE OF VALUES FOR RESPONSE$

PERCENTS OF TOTAL OF THIS (SUB)TABLE

	no	yes	TOTAL	N
	53.66	46.34	100.00	123.00

13.2.1.2 List Format

Especially when the number of categories is large, the list format can be very handy. You generate a list of frequencies and percentages by using the *list* option after the slash in the *tables* command:

```
>TABLES  RESPONSE$/LIST MISS
```

The resulting list is shown in Table 13.4. This list contains the frequency information under the column *count*, the percentage information under the column *pct*, and cumulative counts and percentages that may be of interest when the categories are ordered in a meaningful way.

Table 13.4 Frequencies and percentages of yes's and no's in the POLLSTER data set.

COUNT	CUM COUNT	PCT	CUM PCT	RESPONSE$
66	66	53.7	53.7	no
57	123	46.3	100.0	yes

13.2.2 Significance Testing and Confidence Intervals

SYSTAT does not compute significance tests for one-way tables, but it will compute confidence intervals. If your table has only two categories, you can test the significance of the difference in counts using SYSTAT's binomial distribution functions. If your table has more than two categories, you can compute a chi-square test by hand, but this becomes more and more labor intensive as the number of categories increases. When you have more than two categories, you probably should be testing contrasts anyway.

13.2.2.1 Binomial Probabilities

The frequencies in Table 13.3 show that only 57 of 123 people voted for the bond levy. It would take 50%+1 votes for the bond levy to pass. Is the number of yes's significantly below 50%? To compute the probability that this result would occur if the true population was split 50/50 on the bond issue, type

```
>LET P = 2*NCF(57,123,.5)
```

(See section 2.2.3.5a for more information of computing binomial probabilities.) This command fills a new column P with the value 0.47, which is the two-tailed probability of finding 57 or fewer yes's due to chance in this study. Clearly, 57 does not differ significantly from chance, so we can predict only with very little confidence what the outcome of the vote will be.

13.2.2.2 Chi-Square Tests

When the number of categories is greater than two, you could either test the significance of differences among the cell frequencies using probabilities from a multinomial distribution (see your favorite probability textbook for these formulas), or you can use a chi-square test to test differences between the observed and expected frequencies in the table. I will illustrate the chi-square using the two categories from the previous sections, but the formula works for any number of categories. The expected frequency of each category in a one-way table is just the number of data points divided by the number of categories; for the data in Table 13.3 this is 123/2= 61.5. The *df* for the chi-square for one-way tables is just the number of categories minus 1. The formula, with the values from the response variable in the POLLSTER data set, is

$$\chi^2_{(1)} = \frac{\Sigma(\text{observed frequency}_i - \text{expected frequency}_i)^2}{\text{expected frequency}}$$

$$= \frac{(57-61.5)^2 + (66-61.5)^2}{61.5} = 0.659 \quad (13.1)$$

The two-tailed probability corresponding to this chi-square is .417 (see section 2.2.3.4.1), which is in agreement with the binomial probability found in the previous section.

13.2.2.3 Confidence Intervals

SYSTAT will provide confidence intervals around the observed cell percentages if you use the *confi* option after the slash in the tables command. This works with both frequency and percentage tables, but the confidence intervals are easier to relate to the percentage tables. To illustrate, the following command generates a percentage table for the response variable in the POLLSTER data set (same as in Table 13.3) and also provides a table with the 95% confidence interval around the observed cell percentages:

▶ `>TABLES RESPONSE$/CONFI=.95 PERCENT MISS`

The output is shown in Table 13.5. The population value of the percentages of yes's and no's would be contained in 95% of intervals generated in such a manner. Clearly, we cannot make a very good prediction about the outcome of the bond levy: The 95% confidence interval contains outcomes in which the yes's win, as well as outcomes in which the no's get

---

MACINTOSH

① Choose the menu item **Stats/Tables/Tabulate...**
② Double-click on the grouping variable you want, e.g., RESPONSE$, in the left selection list.
③ Turn on **Percent** and **Missing** and type ".95" into the **Confi** box.
④ Click **OK**.

MS-DOS

① Choose the menu item **Stats/Tables/Tabulate/One-way/Variables/** and select RESPONSE$ from the variables list.
② Choose the menu item **Stats/Tables/Tabulate/One-way/Options/Miss.**
③ Choose the menu item **Stats/Tables/Tabulate/One-way/Options/Percent.**
④ Choose the menu item **Stats/Tables/Tabulate/One-way/Options/Confi** and type ".95" into the box.
⑤ Choose the menu item **Stats/Tables/Tabulate/Go!**

Table 13.5 Percentages of yes's and no's in the POLLSTER data set, along with 95% confidence intervals around those percentages.

```
TABLE OF VALUES FOR RESPONSE$

PERCENTS OF TOTAL OF THIS (SUB)TABLE
```

	no	yes	TOTAL	N
	53.66	46.34	100.00	123.00

```
TABLE OF VALUES FOR RESPONSE$

95 PERCENT APPROXIMATE CONFIDENCE INTERVALS SCALED AS CELL PERCENTS
```

	no	yes
	63.66	56.58
	42.92	35.85

nearly two-thirds of the vote. Consequently, it also contains 50%, which is in line with the significance tests above.

13.3 Two-Way Tables

13.3.1 Tables, Significance Tests, and Effect Sizes

13.3.1.1 Table Format

When you ask SYSTAT to generate a two-way table of frequencies of percentages, you can automatically compute significance tests and a large array of effect-size estimates by asking for extended, or *long*, results output. These tables and statistics can be computed as easily as the one-way tables in the previous section. For example, suppose that you want to show the breakdown of yes and no responses in the POLLSTER data set, according to the political party affiliation of the respondent. The following commands generate the appropriate frequency and percentage tables, along with significance tests and effect-size estimates:

MACINTOSH

① Choose the menu item **Edit/Preferences...** and select **Extended** from the **Statistics** drop down list.

② Choose **Stats/Tables/Tabulate...**

③ Double-click on the row variable you want, e.g., RESPONSE$, in the left selection list and the column variable you want, e.g., PARTY$, in the right selection list.

④ Turn on **Column pct, Frequency, Percent, Row pct,** and **Missing.**

⑤ Click **OK.**

MS-DOS

① Choose the menu item **Utilities/Output/Results/Long.**

② Choose **Stats/Tables/Tabulate/Multi-way/Variables/** and select RESPONSE$ and PARTY$ from the variables list; they should be separated by an asterisk (*).

③ Choose **Stats/Tables/Tabulate/Multi-way/Options/Miss.**

④ Choose **Stats/Tables/Tabulate/Multi-way/Options/Frequency.**

⑤ Choose **Stats/Tables/Tabulate/Multi-way/Options/Percent.**

⑥ Choose **Stats/Tables/Tabulate/Multi-way/Options/Rowpct.**

⑦ Choose **Stats/Tables/Tabulate/Multi-way/Options/Colpct.**

⑧ Choose **Stats/Tables/Tabulate/Go!**

▶ >PRINT=LONG
>TABLES RESPONSE$ * PARTY$/FREQ PERCENT COL, ROWPCT MISS

The options are the same as those for one-way tables (except you cannot use the *confi* option), but now you also have the options of computing the percentage of the row totals (*rowpct*) or column totals (*col*) that each cell accounts for. The two-dimensional tables generated by SYSTAT are shown in Table 13.6.

Each of these tables provides a slightly different view of the data. Consider, for example, the upper-left cell in each table. The frequency table tells us that 17 of the respondents were Democrats who responded no to the bond levy question. The total percentage table tells us that these 17 nay-saying Democrats constitute 13.82% of the entire sample. The row percent table tells us that 25.76% of all no responses came from Democrats. And finally, the column percent table tells us that 30.91% of Democrats answered no.

The Pearson chi-square statistic with two degrees of freedom indicates that the observed frequencies deviate from their expected frequencies more than would be expected due to chance. However, as with any multi-degree of freedom test, this chi-square does not tell us which of the cell frequencies account for this effect. To answer more specific questions about these data, we will want to use contrast analysis procedures. See section 13.3.2.

13.3.1.2 List Format

List format provides the same information that is contained in the cells of the frequency and total percentage tables, but does not provide the marginals, the significance tests, or the effect-size estimates. You can generate output in the list format by specifying the list option after the slash in the tables command:

```
>TABLES RESPONSE$ * PARTY$/LIST MISS
```

The output from this command is shown in Table 13.7.

13.3.2 Contrasts in Proportions

When one of your categorical variables has more than two categories, such as the political party variable in the POLLSTER data set, the significance tests generated by SYSTAT will be omnibus tests, with more than one *df* in the effect being tested. This means that when the test is significant you cannot tell which differences among observed frequencies are responsible, and if the test is not significant it does not mean that there are no significant differences among the cells. As always, when faced with this situation you should proceed to test specific hypotheses about the data with contrasts.

Suppose, for example, that we predicted that the Democrats would most strongly favor the bond levy, the Republicans would be second, and the Libertarians would be least supportive, about as far from the Republicans as the Republicans are from the Democrats. The overall chi-square in Table 13.6 tells us that some of the observed frequencies in the table differ from their expected frequencies more than we would expect due to chance, but it tells us nothing about the significance of this particular hypothesis. To test this hypothesis, we can use the contrast weights –1, 0, and +1 in a contrast on the proportions of Libertarians, Republicans, and Democrats, respectively, favoring the bond levy. The formula for a contrast on proportions (Rosenthal & Rosnow, 1991, pp. 538–539) is

$$Z = \frac{\Sigma \lambda_i P_i}{\sqrt{\Sigma S_i^2 \lambda_i^2}} \qquad (13.2)$$

where the λ_i are the contrast weights corresponding to each cell i, the P_i are the proportions for each cell i, and S_i^2 is given by the equation

$$S_i^2 = \frac{P_i(1 - P_i)}{N_i} \qquad (13.3)$$

which is the variance of a proportion.

MACINTOSH

① Choose the menu item **Stats/Tables/Tabulate...**
② Double-click on the row variable you want, e.g., RESPONSE$, in the left selection list and the column variable you want, e.g., PARTY$, in the right selection list.
③ Turn on **List** and **Missing**.
④ Click **OK**.

MS-DOS

① Choose the menu item **Stats/Tables/Tabulate/Multi-way/Variables/** and select RESPONSE$ and PARTY$ from the variables list; they should be separated by an asterisk (*).
② Choose the menu item **Stats/Tables/Tabulate/Multi-way/Options/Miss.**
③ Choose the menu item **Stats/Tables/Tabulate/Multi-way/Options/List.**
④ Choose the menu item **Stats/Tables/Tabulate/Go!**

Table 13.6 Frequencies and percentages of yes and no responses in the POLLSTER data set, cross-tabulated by political party affiliation.

```
TABLE OF RESPONSE$     (ROWS) BY      PARTY$     (COLUMNS)

FREQUENCIES

          Democrat  Libertar  Republic    TOTAL
          -----------------------------------------
no           17        18        31        66

yes          38         3        16        57
          -----------------------------------------
TOTAL        55        21        47       123


TABLE OF RESPONSE$     (ROWS) BY      PARTY$     (COLUMNS)

PERCENTS OF TOTAL OF THIS (SUB)TABLE

          Democrat  Libertar  Republic    TOTAL       N
          -------------------------------------------------
no          13.82     14.63     25.20     53.66     66.00

yes         30.89      2.44     13.01     46.34     57.00
          -------------------------------------------------
TOTAL       44.72     17.07     38.21    100.00
   N           55        21        47       123


TABLE OF RESPONSE$     (ROWS) BY      PARTY$     (COLUMNS)

ROW PERCENTS

          Democrat  Libertar  Republic    TOTAL       N
          -------------------------------------------------
no          25.76     27.27     46.97    100.00     66.00

yes         66.67      5.26     28.07    100.00     57.00
          -------------------------------------------------
TOTAL       44.72     17.07     38.21    100.00
   N           55        21        47       123
```

Table 13.6 *(continued)*

```
TABLE OF RESPONSE$     (ROWS) BY    PARTY$    (COLUMNS)

COLUMN PERCENTS
```

	Democrat	Libertar	Republic	TOTAL	N
no	30.91	85.71	65.96	53.66	66.00
yes	69.09	14.29	34.04	46.34	57.00
TOTAL	100.00	100.00	100.00	100.00	
N	55	21	47	123	

TEST STATISTIC	VALUE	DF	PROB
PEARSON CHI-SQUARE	22.984	2	.102103E-04
LIKELIHOOD RATIO CHI-SQUARE	24.326	2	.522133E-05

COEFFICIENT	VALUE	ASYMPTOTIC STD ERROR
PHI	0.432	
CRAMER V	0.432	
CONTINGENCY	0.397	
GOODMAN-KRUSKAL GAMMA	-0.524	0.119
KENDALL TAU-B	-0.319	0.082
STUART TAU-C	-0.355	0.091
SPEARMAN RHO	-0.335	0.085
SOMERS D (COLUMN DEPENDENT)	-0.357	0.092
LAMBDA (COLUMN DEPENDENT)	0.206	0.091
UNCERTAINTY (COLUMN DEPENDENT)	0.096	0.036

Table 13.7 List format showing the frequencies and percentages of yes and no responses in the POLLSTER data set, broken down by political party affiliation.

COUNT	CUM COUNT	PCT	CUM PCT	PARTY$	RESPONSE$
17	17	13.8	13.8	Democrat	no
18	35	14.6	28.5	Libertarian	no
31	66	25.2	53.7	Republican	no
38	104	30.9	84.6	Democrat	yes
3	107	2.4	87.0	Libertarian	yes
16	123	13.0	100.0	Republican	yes

From which table should we take the percentages (P_i)? The way the hypothesis was stated, "Democrats would most strongly favor the bond levy ...," indicates that we are interested in the proportion of respondents *within* each party that responded yes: These percentages are given in the *column percents* table in Table 13.6. Because yes and no responses add up to 100% in this table, it really doesn't matter which we use, but because the question was stated in terms of favoring the bond levy, we will use yes's. Thus, of the 21 Libertarians, only 14.29% responded yes, so the variance is

$$S_L^2 = \frac{P_L(1-P_L)}{N_L} = \frac{.1429(1-.1429)}{21} = 0.00583$$

Of the 47 Republicans, 34.04% responded yes, so the variance is

$$S_R^2 = \frac{P_R(1-P_R)}{N_R} = \frac{.3404(1-.3404)}{47} = 0.00478$$

Finally, of the 55 Democrats, 69.09% responded yes, so their variance is

$$S_D^2 = \frac{P_D(1-P_D)}{N_D} = \frac{.6909(1-.6909)}{55} = 0.00388$$

We are now in a position to compute the Z test of the contrast from Equation 13.2:

$$Z = \frac{(-1)(.1429) + (0)(.3404) + (1)(.6909)}{\sqrt{(.00583)(-1)^2 + (.00478)(0)^2 + (.00388)(1)^2}} = 5.56$$

This contrast is highly significant. The effect size ϕ (which is Pearson's r on binary data) corresponding to this contrast can be found from the formula

$$\phi = \sqrt{\frac{Z^2}{N}} = \sqrt{\frac{5.56^2}{123}} = .50 \qquad (13.4)$$

There is one loose end to tie up. This contrast shows that there is strong evidence for an increase in yes responses from Libertarians to Democrats, but because the Republican cell was multiplied by a contrast weight of zero, the contrast would be equally large no matter what the proportion of Republicans responding yes turned out to be. Therefore, in order to conclude that our contrast is supported qualitatively by the data, we must show that there is no evidence of significant deviation from the trend being tested. We can do this by testing the significance of the nonlinear trend in the proportions, which, in this case, is just the quadratic. This quadratic trend can be tested with the weights −1, 2, and +1, for the proportions of Libertarians, Republicans, and Democrats responding yes. This is essentially a test of the extent to which the Republicans fall off the line between the Libertarians and Democrats. The variances for the three groups were found above, so all we have to do is plug the new contrast weights into Equation 13.2:

$$Z = \frac{(-1)(.1429) + (2)(.3404) + (-1)(.6909)}{\sqrt{(.00583)(-1^2) + (.00478)(2^2) + (.00388)(-1^2)}} = -0.90$$

This contrast is not significant, $p = .368$ (see section 2.2.3.1.1 for finding exact probabilities from Z). The effect size ϕ corresponding to this contrast is

$$\phi = \sqrt{\frac{Z^2}{N}} = \sqrt{\frac{-0.90^2}{123}} = .007$$

Therefore, we can conclude that there is no evidence for a deviation from the linear trend in yes responses from Libertarians to Republicans to Democrats.

13.4 Three-Way Tables
13.4.1 Generating Tables

SYSTAT will generate two-way slices of multidimensional tables, but it does not provide significance tests or effect-size estimates for multiway tables. This is too bad because these statistics are too labor intensive to compute by hand. Fortunately, omnibus tests on multiway tables are usually of less use than tests of contrasts in proportions on multiway interactions. Generating the tables is discussed in this section; contrasts on multiway tables are discussed in section 13.4.2.

13.4.1.1 Table Format

Analysis of three-way tables, and tables of higher dimension, follow the same basic procedures as the analysis of two-way tables discussed in the previous section. However, SYSTAT does not provide significance tests or effect-size estimate for three-way and higher tables. In this section I will briefly present an analysis of a three-way table to complete the analysis of the POLLSTER data set in Table 13.2. This data set has two grouping variables, the political affiliation and gender of the respondents, and one dependent variable: the person's response to a question about their favorability towards a bond levy.

SYSTAT will give multiway frequency tables, but it does not automatically give all of the multiway percentage tables. Instead, it generates a series of two-way percentage tables for row and column variables within a third variable used as a stratifying variable. Thus, the percentages are always conditioned on the stratifying variable. The *total percentage* tables are conditioned on the stratifying variable only. For example, if you used gender as a stratifying variable, the numbers in the total percentage tables would be of the form, "the percentage of females who ..." The *row percent* tables condition on the stratifying variable and the row variable. If response is the row variable, these numbers would interpreted as "the percentage of no-voting females who ..." Finally, the *column percent* tables condition on the stratifying variable and the column variable. If political party is the column variable, these numbers would be of the from "the percentage of Democratic females who ..." What you *do not* get in any of these tables are percentages of the form "the percentage of Democratic no-voters who are male (or female) is ..." If you want SYSTAT to compute such percentages, simply generate another table without gender as a row or column variable.

Suppose that you do want to use gender as a stratifying variable. To produce two-way RESPONSE$-by-PARTY$ tables within each level of gender, type

```
>TABLES  GENDER$ * RESPONSE$ * PARTY$/FREQ,
PERCENT COL ROWPCT MISS
```

The options are the same as those for two-way tables. The tables generated by this command are shown in Table 13.8. Again, each of these tables provides a slightly different view of the data. Consider the upper-left cell in each table: The frequency tables tell us that eight female Democrats and nine male Democrats responded no to the bond levy question; the total percentage tables tell us that these eight women and nine men constitute 12.12% and 15.79%, respectively, of their genders; the row percent tables tell us that 22.22% of no-voting women and 30% of no-voting men were Democrats; and finally, the column percent tables show that 28.57% of female Democrats said no, while 33.33% of male Democrats said no.

HINT: OUT OF MEMORY?

If your table has a large number of dimensions, SYSTAT may run out of room in memory to perform its calculations. If this happens, try running the analysis on one fewer dimension until you get it to run, and then you can perform the analysis repeatedly, including a different set of variables each time. You can also analyze subsets of variables within levels of the remaining variables by sorting the data set on a subset of your variables and then using the **By groups...** command to perform the table analysis within each level of those variables. Neither of these methods will give you the highest order interactions, but who wants to interpret a six-way interaction anyhow?

13.4.1.2 List Format

List format provides the same information that is contained in the cells of the frequency and total percentage tables, but does not provide the marginals, the significance tests, or the effect-size estimates. You can generate output in the list format by specifying the list option after the slash in the tables command:

MACINTOSH
① Choose the menu item **Stats/Tables/Tabulate...**
② Double-click on the stratifying variable you want, e.g., GENDER$, in the left selection list; the row variable you want, e.g., RESPONSE$, in the right selection list; and the column variable you want, e.g., PARTY$, also in the right selection list.
③ Turn on **Column pct**, **Frequency**, **Percent**, **Row pct**, and **Missing**.
④ Click **OK**.

MS-DOS
① Choose the menu item **Stats/Tables/Tabulate/Multi-way/Variables/** and select GENDER$, RESPONSE$, and PARTY$ from the variables list; they should all be separated by asterisks (*).
② Choose the menu item **Stats/Tables/Tabulate/Multi-way/Options/Miss.**
③ Choose the menu item **Stats/Tables/Tabulate/Multi-way/Options/Frequency.**
④ Choose the menu item **Stats/Tables/Tabulate/Multi-way/Options/Percent.**
⑤ Choose the menu item **Stats/Tables/Tabulate/Multi-way/Options/Rowpct.**
⑥ Choose the menu item **Stats/Tables/Tabulate/Multi-way/Options/Colpct.**
⑦ Choose the menu item **Stats/Tables/Tabulate/Go!**

MACINTOSH

① Choose the menu item **Stats/Tables/Tabulate...**
② Double-click on the stratifying variable you want, e.g., GENDER$, in the left selection list; the row variable you want, e.g., RESPONSE$, in the right selection list; and the column variable you want, e.g., PARTY$, also in the right selection list.
③ Turn on **List** and **Missing**.
④ Click **OK**.

MS-DOS

① Choose the menu item **Stats/Tables/Tabulate/ Multi-way/Variables/** and select GENDER$, RESPONSE$, and PARTY$ from the variables list; they should all be separated by asterisks (*).
② Choose the menu item **Stats/Tables/Tabulate/ Multi-way/Options/Miss**.
③ Choose the menu item **Stats/Tables/Tabulate/ Multi-way/Options/List**.
④ Choose the menu item **Stats/Tables/Tabulate/ Go!**

▶ >TABLES GENDER$ * RESPONSE * PARTY$/LIST MISS

The output from this command is shown in Table 13.9.

13.4.2 Contrasts

Contrasts can be computed on multiway tables in much the same manner that they are computed on two-way tables. For example, suppose that we predicted that the Democrat-Republican difference in favorability towards the bond levy would be greater for females than for males. The proportions of each party favoring the bond levy, broken down by gender, are shown in the second rows of the two *column percent* tables at the end of Table 13.9. These percentages of yes responses are conditioned on both party affiliation and gender. The contrast states that the female/Democrat-female/Republican difference will be greater than the male/Democrat-male/Republican difference, with Libertarians ignored. Thus, the contrast weights for this hypothesis are +1, –1, +1, –1, 0, and 0 for the proportions of yes responses for the female Democrats, female Republicans, male Democrats, male Republicans, female Libertarians, and male Libertarians, respectively. These weights are shown paired with the appropriate percentages from Table 13.9 in Table 13.10.

In the actual computations, any proportions with zero weights can be ignored. The variances for each category with nonzero weights are computed in the following four equations based on Equation 13.3. Of the 28 female Democrats, 71.43% responded yes, so the variance is

$$S_{FD}^2 = \frac{.7143(1 - .7143)}{28} = 0.00729$$

Of the 27 female Republicans, 33.33% responded yes, so the variance is

$$S_{FR}^2 = \frac{.3333(1 - .3333)}{27} = 0.00823$$

Of the 27 male Democrats, 66.67% responded yes, so the variance is

$$S_{MD}^2 = \frac{.6667(1 - .6667)}{27} = 0.00823$$

Table 13.8 Frequencies and percentages of yes and no responses in the POLLSTER data set, cross-tabulated by gender and political party affiliation.

```
TABLE OF RESPONSE$      (ROWS) BY      PARTY$     (COLUMNS)
   FOR THE FOLLOWING VALUES:
          GENDER$      = female
```

FREQUENCIES

	Democrat	Libertar	Republic	TOTAL
no	8	10	18	36
yes	20	1	9	30
TOTAL	28	11	27	66

```
TABLE OF RESPONSE$      (ROWS) BY      PARTY$     (COLUMNS)
   FOR THE FOLLOWING VALUES:
          GENDER$      = male
```

FREQUENCIES

	Democrat	Libertar	Republic	TOTAL
no	9	8	13	30
yes	18	2	7	27
TOTAL	27	10	20	57

```
TABLE OF RESPONSE$      (ROWS) BY      PARTY$     (COLUMNS)
   FOR THE FOLLOWING VALUES:
          GENDER$      = female
```

PERCENTS OF TOTAL OF THIS (SUB)TABLE

	Democrat	Libertar	Republic	TOTAL	N
no	12.12	15.15	27.27	54.55	36.00
yes	30.30	1.52	13.64	45.45	30.00
TOTAL	42.42	16.67	40.91	100.00	
N	28	11	27		66

Table 13.8 *(continued)*

```
TABLE OF RESPONSE$      (ROWS)  BY      PARTY$     (COLUMNS)
   FOR THE FOLLOWING VALUES:
      GENDER$     = male
```

PERCENTS OF TOTAL OF THIS (SUB)TABLE

	Democrat	Libertar	Republic	TOTAL	N
no	15.79	14.04	22.81	52.63	30.00
yes	31.58	3.51	12.28	47.37	27.00
TOTAL	47.37	17.54	35.09	100.00	
N	27	10	20	57	

```
TABLE OF RESPONSE$      (ROWS)  BY      PARTY$     (COLUMNS)
   FOR THE FOLLOWING VALUES:
      GENDER$     = female
```

ROW PERCENTS

	Democrat	Libertar	Republic	TOTAL	N
no	22.22	27.78	50.00	100.00	36.00
yes	66.67	3.33	30.00	100.00	30.00
TOTAL	42.42	16.67	40.91	100.00	
N	28	11	27	66	

```
TABLE OF RESPONSE$      (ROWS)  BY      PARTY$     (COLUMNS)
   FOR THE FOLLOWING VALUES:
      GENDER$     = male
```

ROW PERCENTS

	Democrat	Libertar	Republic	TOTAL	N
no	30.00	26.67	43.33	100.00	30.00
yes	66.67	7.41	25.93	100.00	27.00
TOTAL	47.37	17.54	35.09	100.00	
N	27	10	20	57	

Table 13.8 (continued)

TABLE OF RESPONSE$ (ROWS) BY PARTY$ (COLUMNS)
FOR THE FOLLOWING VALUES:
   GENDER$ = female

COLUMN PERCENTS

	Democrat	Libertar	Republic	TOTAL	N
no	28.57	90.91	66.67	54.55	36.00
yes	71.43	9.09	33.33	45.45	30.00
TOTAL	100.00	100.00	100.00	100.00	
N	28	11	27	66	

TABLE OF RESPONSE$ (ROWS) BY PARTY$ (COLUMNS)
FOR THE FOLLOWING VALUES:
   GENDER$ = male

COLUMN PERCENTS

	Democrat	Libertar	Republic	TOTAL	N
no	33.33	80.00	65.00	52.63	30.00
yes	66.67	20.00	35.00	47.37	27.00
TOTAL	100.00	100.00	100.00	100.00	
N	27	10	20	57	

Finally, of the 20 male Republicans, 65% responded yes, so the variance is

$$S^2_{MR} = \frac{.65\,(1-.65)}{20} = 0.01137$$

Again, leaving out categories with zero weights, we now are ready to compute the Z test of the contrast using Equation 13.2:

Table 13.9 List format showing the frequencies and percentages of yes and no responses in the POLLSTER data, broken down by political party affiliation and gender.

COUNT	CUM COUNT	PCT	CUM PCT	PARTY$	RESPONSE$	GENDER$
8	8	6.5	6.5	Democrat	no	female
10	18	8.1	14.6	Libertarian	no	female
18	36	14.6	29.3	Republican	no	female
20	56	16.3	45.5	Democrat	yes	female
1	57	0.8	46.3	Libertarian	yes	female
9	66	7.3	53.7	Republican	yes	female
9	75	7.3	61.0	Democrat	no	male
8	83	6.5	67.5	Libertarian	no	male
13	96	10.6	78.0	Republican	no	male
18	114	14.6	92.7	Democrat	yes	male
2	116	1.6	94.3	Libertarian	yes	male
7	123	5.7	100.0	Republican	yes	male

$$Z = \frac{(1)(.7143) + (-1)(.3333) + (-1)(.6667) + (1)(.3500)}{\sqrt{(.00729)(1^2) + (.00823)(-1^2) + (.00823)(-1^2) + (.01137)(1^2)}}$$

$$= 0.343$$

This contrast is nowhere near significance. The effect-size ϕ corresponding to this contrast, from Equation 13.4, is

$$\phi = \sqrt{\frac{0.343^2}{102}} = .034$$

Note that the N used in this effect-size calculation is 102, not 123. This is because the contrast did not involve the 21 Libertarians. If the contrast had predicted that the Libertarians would be half-way between the Democrats and Republicans within each gender, the Libertarians would have still received weights of zero, and the chi-square would have come out exactly the same. However, in that case two additional considerations would apply: (1) the effect-size estimate would be based on an N of 123, not 102, and therefore would be smaller, and (2) because the zero weights cause the Libertarians to have zero impact on the contrast, a second contrast would have to

Table 13.10 Percentages of yes responses in the POLLSTER data set, cross-tabulated by gender and political party affiliation. Contrast weights and the numbers in each category are also shown.

		Democrat	Libertarian	Republican
Females	%	71.43	9.09	33.33
	λ_{ij}	+1	0	−1
	N_{ij}	28	11	27
Males	%	66.67	20.00	35.00
	λ_{ij}	−1	0	+1
	N_{ij}	27	10	20

be performed to show that the Libertarians did not, in fact, fall significantly off the line between the Democrats and Republicans. See section 13.3.2 for an example of this latter type of contrast.

13.5 Graphing the Results

More often than not, data is often presented in table form, such as the tables shown earlier in the chapter. However, if you want to present line graphs or bar charts of frequencies or proportions, this can be accomplished easily in SYSTAT using the procedures described in sections 3.2.1 and 3.2.2. Those discussions will not be repeated here, but I will illustrate a few of the possible graphs using the POLLSTER data.

13.5.1 Graphing Single Variables

You can plot the frequencies for single variable, by simply typing the graphics name followed by the variable name. (If you are using a data file with counts, don't forget to weight the cases by those counts; see section 13.1.2) For example, the following commands create a bar graph, a line graph, and a pie chart for the political party variable in the POLLSTER data set:

```
>GRAPH
>BAR PARTY$/WIDTH=5IN
>CPLOT PARTY$/LINE WIDTH=5IN
>PIE PARTY$
```

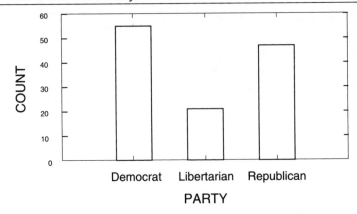

Figure 13.1 Bar graph of the political party variable in the POLLSTER data set.

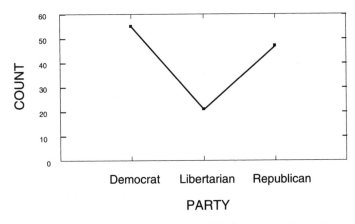

Figure 13.2 Line graph of the political party variable in the POLLSTER data set.

The width option was set to make the X-axis labels fit properly on the plot. These plots are shown in Figures 13.1, 13.2, and 13.3, respectively.

13.5.2 Two-Way Plots

13.5.2.1 Bar Charts and Line Graphs

If you want to put more than one variable in a bar chart or line graph, you must first create a separate dependent variable for each level of one of the variables. This is easiest to do for the variable with the fewest number of levels. For example, if you

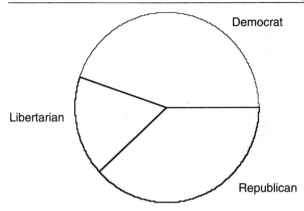

Figure 13.3 Pie chart of the political party variable in the POLLSTER data set.

want to show the frequencies of yes and no responses for each political party in the POLLSTER data set, because the response variable has the fewest levels, first make a new column of counts for yes's and no's with the commands

```
>IF RESPONSE$="YES" THEN LET YES=COUNT
>IF RESPONSE$="NO" THEN LET NO=COUNT
```

(zdee section 2.2.2.) Then you can create the desired graph with the following command (see section 3.2.3.1):

```
>GRAPH
>BAR YES NO * PARTY$/YLAB="Frequency" FILL=1,7,
LEGEND= 50 40 WIDTH=5 IN
```

The resulting bar chart is shown in Figure 13.4. The *ylab* option specified the label for the Y-axis. The *fill* option specified the patterns used to fill the bars and the legend, and the *legend* option specified the location of the legend.

If you prefer a line graph, you can create one using the same procedure as for a bar chart. Using the separate columns of counts for yes's and no's as dependent variables in the *cplot* command (see section 3.2.3.2), type

```
>GRAPH
>CPLOT YES NO * PARTY$/YLAB="Frequency",
LINE=1,6 LEGEND=50 40 WIDTH=5 IN
```

The resulting line graph is shown in Figure 13.5. The options are the same as those for the bar chart above, except that the *fill* option is replaced by the *line* option, which specifies the size of the dashes used for the line in the plot.

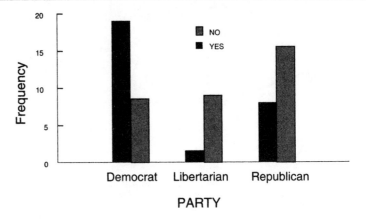

Figure 13.4 Bar chart of the yes and no responses for each political party in the POLLSTER data set.

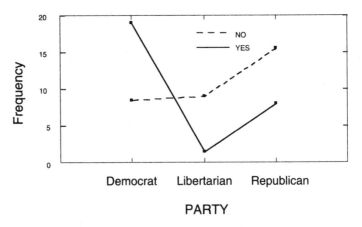

Figure 13.5 Bar chart of the yes and no responses for each political party in the POLLSTER data set.

Note that both the bar chart in Figure 13.4 and the line graph in Figure 13.5 show frequencies of yes's and no's. If you preferred to show percents, you would need to create a column in the data set containing the appropriate percents, which you can obtain from Table 13.6. You probably would not want to plot both the percentage yes and the percentage no because these bars or lines would contain perfectly redundant information.

13.5.2.2 Pie Charts

You cannot put two variables directly into a pie chart, but you can effect virtually the same thing by creating a separate pie chart for each level of one of the variables. For example, if you wanted to create pie charts of party affiliation for males and females separately, you could use the use the *select* command to limit the information in each plot to one gender or the other. The following commands create the two pie charts shown in Figure 13.6:

```
>WEIGHT=COUNT
>SELECT GENDER$= "male"
>PIE PARTY$/ TITLE="Males" HEIGHT=2 IN,
WIDTH=2 IN
>SELECT GENDER$= "female"
>PIE PARTY$/ TITLE="Females" HEIGHT=2 IN,
WIDTH=2 IN
```

13.5.3 Three-Way Percentage Plots

There are many ways to put three variables in a bar chart or line graph, but these graphs will often begin to look so complicated that the data tables would actually be easier to read than the graphs. However, it is often the case that your response variable is binary, such as yes/no, and when this is the case this two-category response variable can be collapsed into a single percentage, for example, *percent yes*. This essentially enables you to present three variables, the response variable plus two grouping variables, in a very readable two-way graph. As with any two-way plots, you must first create a separate dependent variable for each level of one of the grouping variables, preferably the one with the fewest number of levels. For example, if you want to show percent yes for each political party-by-gender combination in the POLLSTER data set, because gender has the fewest levels of the two grouping variables, first make a new column of percentages for males and females. Such a data file is shown in Table 13.11. This table has three columns: party, percent yes for females, and percent yes for males.

Given these columns, you can create a bar chart with the commands (section 3.2.3.1)

```
>GRAPH
>BAR MALE FEMALE * PARTY$/YMIN=0 YMAX=100,
YLAB="PERCENT YES" FILL=1,7 LEGEND=50 40,
WIDTH=5 IN
```

MACINTOSH

① If you haven't already, choose the menu item **Data/Weight...** , double-click on the **COUNT** variable to enter it into the box, and click **OK**.

② Choose the menu item **Data/Select Cases...**, select the variable GENDER$, type ' = "male" ' (including the double-quotes) into the box, and click **OK**.

③ Choose the menu item **Graph/Pie/Pie** and select the variable PARTY$ from the left variable list.

④ Click on the **ruler icon**, type "2" into the **Height** and **Width** boxes, and click **OK**.

⑤ Click **OK**.

MS-DOS

① If you haven't already, choose the menu item **Data/Weight/Select/** and select COUNT from the variables list.

② Choose the menu item **Data/Select/Select/**, select GENDER$ from the variables list, select "=" from the relation list, and type "male", *not including the double-quotes*.

③ Choose the menu item **Graph/Pie/Variables/ Primary/** and select PARTY$ from the variable list.

④ Choose the menu item **Graph/Pie/Options/ Height/**, type "2", hit <ret>, and select "IN" from the units list.

⑤ Choose the menu item **Graph/Pie/Options/ Width/**, type "2", hit <ret>, and select "IN" from the units list.

⑥ Choose the menu item **Graph/Pie/Go!**

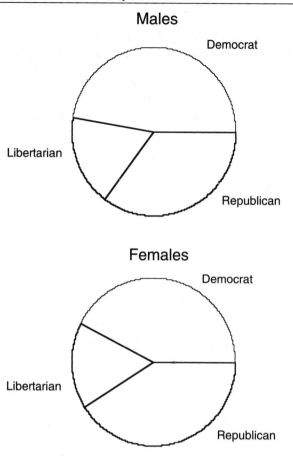

Figure 13.6 Pie charts of the frequencies of each political party for males and females separately in the POLLSTER data set.

The resulting bar chart is shown in Figure 13.7. Given a political affiliation and gender, the bars in this graph show the percentages of yes responses. Of course, the percentages of no responses are just the differences between the tops of the bars and 1.0. Therefore, this graph contains information from three variables: party, gender, and response.

Table 13.11 Data file for creating a "three-way" graph of percent yes against political party and gender.

	PARTY$	FEMALE	MALE
1	Democrat	71.43	.
2	Democrat	.	66.67
3	Libertarian	9.09	.
4	Libertarian	.	20.00
5	Republican	33.33	.
6	Republican	.	35.00

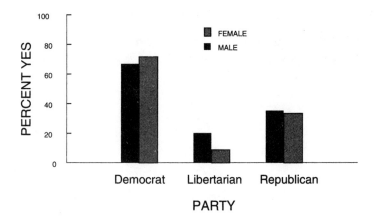

Figure 13.7 Bar chart of percent yes for each political party and gender combination in the POLLSTER data set.

If you prefer a line graph, you can create one using the same procedure as for the bar chart (see section 3.2.3.2). Using the separate columns of percent yes for males and females as dependent variables in the *cplot* command, you can make the line graph by typing

```
>GRAPH
>CPLOT MALE FEMALE * PARTY$/YMIN=0 YMAX=100,
YLAB="PERCENT YES" LINE=1,6 LEGEND=50 40,
WIDTH=5 IN
```

The resulting line graph is shown in Figure 13.8. The options are the same as those for the bar chart above, except that the *fill* option is replaced by the *line* option, which specifies the size of the dashes used for the line in the plot.

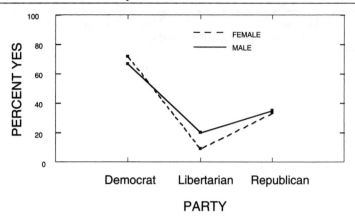

Figure 13.8 Line graph of percent yes for males and females across political parties in the POLLSTER data set.

References

Belsley, D.A., Kuh, E., & Welsch, R.E. (1980). *Regression diagnostics: Identifying influential data and sources of collinearity.* New York: Wiley.

Box, G.E.P., & Cox, D.R. (1964). An analysis of transformations (with discussion). *Journal of the Royal Statistical Society, Ser. B, 26,* 211–246.

Box, G.E.P., & Tidwell, P.W. (1962). Transformations of the independent variables. *Technometrics, 4,* 531–550.

Carroll, R.J., & Ruppert, D. (1988). *Transformation and weighting in regression.* New York: Chapman and Hall.

Conover, W. J., Johnson, M.E., & Johnson, M.M. (1981). A comparative study of tests for homogeneity of variances, with applications to the outer continental shelf bidding data. *Technometrics, 23(4),* 351–361.

Cook, R.D., & Weisberg, S. (1982). *Residuals and influence in regression.* New York: Chapman and Hall.

Daniel, C. (1959). The use of half-normal plots in interpreting factorial two level experiments. *Technometrics, 1,* 311–341.

Daniel, C. (1960). Locating outliers in factorial experiments. *Technometrics, 2,* 149–156.

Daniel, C. (1976). *Applications of statistics to industrial experiments.* New York: Wiley.

Daniel, C., & Wood, F.S. (1980). *Fitting equations to data,* 2nd edition. New York: Wiley.

Estes, W.K. (1991). *Statistical models in behavioral research.* Hillsdale, NJ: Erlbaum.

Frigge, M., Hoaglin, D.C., & Iglewicz, B. (1989). Some implementations of the box plot. *American Statistician*, February, 50–54.

Hoaglin, D.C., Iglewicz, B., & Tukey, J.W. (1986). Performance of some resistant rules for outlier labeling. *Journal of the American Statistical Association, 81*, 991–999.

Kleinbaum, D.G., Kupper, L.L., & Muller, K.E. (1988). *Applied regression analysis and other multivariable methods.* Boston: PWS-Kent.

Little, R.J.A., & Rubin, D.B. (1987). *Statistical analysis with missing data.* New York: Wiley.

Mallows, C.L. (1973). Some comments on C_p. *Technometrics, 15*, 661–676.

Maxwell, S.E., & Delaney, H.D. (1990). *Designing experiments and analyzing data: A model comparison perspective.* Belmont, CA: Wadsworth.

Mosteller, F., & Tukey, J.W. (1977). *Data analysis and regression: A second course in statistics.* Reading, MA: Addison-Wesley.

Rosenthal, R. (1991). *Meta-analytic procedures for social research* (revised edition). Newbury Park, CA: Sage.

Rosenthal, R., & Rosnow, R.L. (1991). *Essentials of behavioral research: Methods and data analysis* (2nd edition). New York: McGraw-Hill.

Rosenthal, R., & Rosnow, R.L. (1985). *Contrast analysis: Focused comparisons in the analysis of variance.* Cambridge: Cambridge University Press.

Rubin, D.B. (1987). *Multiple imputation for nonresponse in surveys.* New York: Wiley.

Snedecor, G.W., & Cochran, W.G. (1980). *Statistical methods* (7th edition). Ames, Iowa: The Iowa State University Press.

Weisberg, S. (1985). *Applied linear regression* (2nd edition). New York: Wiley.

Index

Added variable plots   201–203, 214–216, 223–224, 226
Analysis of variance
   fixed factors
      one-way   238–242
      factorial   283–386
      with nesting   330–337
   repeated measures
      between subjects factors   410–417
      no between subjects factors   373–374
      nesting within subjects   422–425
Arc-tangent hyperbolic function   105
ASCII text files, opening   8–13
Assumptions — *see separate entries*: Normality, Equal variance, Independence, Unbiased errors, Homogeneity of covariances
Autocorrelation
   in ANOVA   241
   in regression   149–150, 151, 199

Bar graphs — *see* Graphs
Bartlett's test   238–239, 264–265
Between subjects analyses in repeated measures   411–415
Binomial probabilities — *see* Probabilities, from binomial distribution
Bonferroni adjustment
   in post-hoc tests   253–255, 301, 347–348
Border plots   77
Box plots
   generating for one variable   63–64, 121–122, 168–169
   generating for two variables   82–83
   contrast scores   398–399
   residuals in ANOVA   259–260, 303–304, 350, 351, 354–355
   residuals in regression   140, 188–190, 192–193
   raw data in ANOVA   238, 263, 283, 305, 329–330
By groups, analyses   59

Categorical variables
   in one-way ANOVA   236
   recoding two variables as one   280–281
Centering a variable   20–22, 185, 225, 227
Character variables
   creating   4–5
   recoding as numeric   236
Chi-square
   distribution, critical values of — *see* Critical values, chi-square
   distribution, probabilities from — *see* Probabilities, from chi-square
   effect-size estimates from   101–103
   tests, in contingency tables   439–440
Circularity assumption — *see* Homogeneity of covariances
Cmatrix command   377–378, 379, 381, 385, 387, 428–430, 431
Coefficient of variation   56–58
Collinearity   165, 177, 179–188, 227–228
   defined   179–180
   diagnostics
      condition indices   177, 182–188, 227–228
      correlation   180–181
      tolerance   181–182
Columns, in data worksheet   4–5
Commands
   hot and cold   7
   using   xvi, xvii, 1–2
Commas, as delimiters in text files   10
Conditional transformations   23–25
Condition indices   182–188
Confidence bands on regression line   158–159
Confidence intervals
   on cells in contingency tables   440–441
   on means   84–85, 268–270, 312
   on regression coefficients
      fitted values   157–158
      intercept   153–154, 204–205
      slope   154–155, 205–206
      prediction   155–157
      regression line   158–159
Contingency tables
   contrasts   443, 446–447, 450, 453–455
   entering data   435–437
   one-way tables   437–441
   two-way tables   441–443
   three-way tables   448–455
   weighting cases   437
Contrasts   242–251, 286–299, 337–345, 374–385
   finding weights   243–244
   normalizing weights   427–428

    on cell means — *see* also Post–hoc tests, cell means
        within subjects factors   387–389
    on interaction effects
        factorial designs   290–294
        nested designs   338–343
        repeated measures
            between subjects   417–418
            within subjects   380–381
            of between and within subjects factors   418–421
    on main effects
        factorial designs   286–289
        nested designs   338
        repeated measures
            between subjects effects   417–418
            within subjects effects   374, 377–378
            nested within subject effects   425–432
        one-way designs   242–251
    on proportions in contingency tables   443, 446–447, 450, 453–455
    on simple effects
        factorial designs   290
        repeated measures, within subjects   378–380
    polynomials
        factorial designs   294–296
        nested designs   343–345
        repeated measures, within subjects   382
        one-way designs   246–248
    scores, for subjects   382, 396–399
    unplanned — *see* Post–hoc tests
Cook's D   137–138, 191–192
Correlated errors — *see* Independence assumption
Correlation
    as effect-size estimate   102–103, 244–245, 447
    Fisher's Z transformation   104–105, 105–107, 108
    in detecting collinearity   180–181
Counts, analyses of — *see* Contingency tables
Critical values, finding from probability
    binomial distribution
        one-tailed   51
        two-tailed, symmetric   52
        two-tailed, asymmetric   52–53
    chi-square distribution
        directional   45–46
        non-directional   44–45
    F distribution
        directional   40
        non-directional   39–40
    normal distribution (Z)
        one-tailed   28–29
        two-tailed   30
    t distribution
        one-tailed   33–34
        two-tailed   34–35

Crossed factors — *see* entries under Analysis of variance using "factorial"
Cross–tabulation — *see* Contingency tables

Data editor window 2
Decimal places, controlling display 3
Delimiters, in text files 8–11
Demeaning data — *see* Centering
Dialog box, expression too long 23
Dollar sign, for character variable labels 4
Double precision, saving files in 6
Dunnett's test 258
Durbin-Watson statistic 149, 199, 241

Effect-size estimates
    from chi-square tests 101–103
    from *t* tests 101–103
    for contrasts 244–245, 247–248, 447, 454
    with contrast weights of zero 247–248
Eigenvalues 177, 182
Entering data 2–6, 235–236, 279–281, 325–327, 367–369, 435–437
    in command window 3
    contingency tables 435–437
    fully-factorial ANOVA 235–236, 279–281
    nested factors ANOVA 325–327
    repeated measures ANOVA 367–369
    variables — *see* entries for Character variables and Numeric variables
Equal covariances assumption — *see* Homogeneity of covariances
Equal variance assumption
    checking in ANOVA 263–266, 305–310, 352–355, 390–391
    checking in regression 144–147, 194–197
    Levene's test 265–266, 307–310
    remedies for violations in ANOVA 267, 310–311, 356
    remedies for violations in regression 147, 197–198, 211–214
    score test 146–147, 196–197
    underlying contrasts 398
Error bars — *see* specific graph type under Graphs
Error sum of product matrices 393–394
Error terms
    local 248, 296–298, 345, 382, 419–420, 428, 430, 433
    pooled, or "global" 248, 250, 296, 345, 382–385, 388–389, 416, 424
    separate 251, 298–299, 345
Estimates of effects, in SYSTAT output 241, 284
Exponential notation
    in data display 3
    in data worksheet 5

Fences, inner and outer   63–64
Finding cases in the data worksheet   73–74, 125, 169–170, 189–190
Fisher's LSD test   252–253
Fisher's *Z* transformation   104–105, 105–107, 108
Format, of numeric displays   3
Functions, mathematical — *see* Transforming data

Geometric mean   218
Global error term — *see* Error terms, pooled or "global"
Graphs — *see* also separate entries under Added variable plots, Box plots, Histograms, Normal probability plots, Scatterplots, Stem and leaf displays
   display options   94–100
      axis values   95–96
      filling bars   96
      legend   96–97
      line types   97
      position   95
      size   95
      sorting categories   97–98
      symbols   98
      title   99
      transposing axes   99–100
   graphing frequencies
      bar graphs
         one variable   455–456
         two variables   456–457, 458
      line graphs
         one variable   455–456
         two variables   457–458
      pie charts   459, 460
   graphing interaction effects   316–318, 362–364, 404–406, 407
   graphing means
      bar graphs
         one variable   83–85, 271–272, 273, 357–359, 400–401
         two variables   89, 90, 312–313, 315–316, 360–362, 363, 402, 404
      line graphs
         one variable   85–89, 272–275, 314–315, 359–360, 400, 403
         two variables   89–92, 316, 317, 362, 363, 402, 404, 405
   graphing percentages
      bar graphs   459–461
      line graphs   461–462
   graphing raw data in ANOVA   237–238, 282–283, 328–330, 369–372
   graphing raw data in regression   120–128, 168–169, 170–175
   graphing residuals — *see* Residuals, graphing
   identifying individual cases   73–74, 125, 169–170, 188–190
Greenhouse-Geisser adjustment   373–374, 394–395

Hinges   65
Histograms
    one variable   60–62, 120–121, 168
    two variables — *see* Kernal density plots
Homogeneity of treatment differences   392
Homogeneity of covariances assumption   391–395
Homoscedasticity — *see* Equal variance assumption
Huynh-Feldt adjustment   373–374, 394–395

If-then commands   23–25
Ill-conditioning — *see* Collinearity
Importing
    files from other programs   13
    text files   8–13
Independence assumption
    in ANOVA   267, 311, 356–357, 391
    in regression
        checking   147–152, 198–200
        remedies   151–152, 200
    underlying contrasts   398
Influence plots   76–77, 126–127
Interquartile range   63

Kernal density plots   80–82, 127–129, 171–172
Kruskal-Wallis test   18
Kurtosis
    computing   56–59
    and normality assumption in ANOVA   260–261, 303–304, 350–351
    and normality assumption in regression   140–141, 193
    in normal probability plots   69–71

Least squares means   241
Levene's test   265–266, 307–310
Leverage   125, 135–137, 190–191
Linearity assumption — *see* Unbiased errors assumption
Local error term — *see* Error terms, local
Logarithmic transformations   22–23
LS MEAN — *see* Least squares means

Mallow's C_p 232–234
Mann-Whitney test   18
Matched-pairs *t*-test   397–398, 430, 432
Mathematical functions — *see* Transforming data
Mauchly's sphericity test   394
Mean, computing   56–58
Mean polish table   284–286, 387–389, 404–406

Mean-square error   241
Median, computing   55, 56–58, 63, 64–65
Menus, on and off in DOS   1
Meta-analysis   101–115
    combining effect-sizes   107–111
    combining probabilities   111–112
    comparing studies   113–115
    displaying results of studies   105–107
    finding effect-size estimates   101–103
    finding probabilities for individual studies   104
Missing cells   406–407, 432–433, 438
Missing data   5
Multicollinearity — *see* collinearity
Multiple-imputation of missing data   407
Multiple-R^2 131–132, 232

Normality assumption
    checking in ANOVA   258–262, 301–304, 350–351, 390–391
    checking in regression   139–142, 192–194
    remedies for violations in ANOVA   262–263, 305, 352
    remedies for violations in regression   142–143, 208–211
    underlying contrasts   398–399
Normal probability plots
    generating   67–68
    interpreting   66, 68–73
    and normality assumption in ANOVA   261–262
    and normality assumption in regression   141–142, 193–194, 210–211, 223, 225, 227
Numeric variables
    creating   5

Opening
    new data worksheet   2–3
    SYSTAT files   8
    text files   8
Operators — *see* Transforming data
Orthogonal polynomials — *see* Contrasts, polynomials
Outliers   63, 64, 65, 122–123, 125

Partial regression leverage plots — *see* Added variable plots
Pearson product-moment correlation — *see* Correlation
Percentages, analyses of — *see* Contingency tables
Phi (ϕ), as effect-size estimate   102, 447, 454
Plots — *see* Graphs and also separate entries under Added variable plots, Box plots, Histograms, Normal probability plots, Scatterplots, Stem and leaf displays
Pooled error term — *see* Error terms, pooled

Post-hoc tests   251–258, 299–301, 345–348, 385–389, 415
   cell means
      Bonferroni adjusted   301, 347–348
      within subjects factors   387–389
   marginal means
      between subjects factors   415
      Bonferroni adjusted   253–255, 347
      Dunnett's test   258
      Fisher's LSD test   252–253
      Scheffé test   257
      Tukey-Kramer HSD test   256, 300
      within subjects factors   385–386
Probabilities
   finding critical values from — see Critical Values
   from binomial distribution
      for contingency tables   439
      exactly x successes   50–51
      one-tailed   46–48
      two-tailed, symmetric   48–49
      two-tailed, asymmetric   49–50
   from chi-square distribution
      directional   42–44, 104
      non-directional   41–42
   from F distribution
      directional   37–38
      non-directional   35–37
   from normal distribution (Z)
      one-tailed   26–28
      two-tailed   28
   from t distribution
      one-tailed   30–32, 104
      two-tailed   32–33
Probability plots — see Normal probability plots
Profile plots   371–372
P-values, finding — see Probabilities

Quotes
   single and double   5
   in Microsoft Excel files   13
   in text files   10

Random factors   365–366
Rankit plots — see Normal probability plots
Rank-ordering data   17–18
Recoding variables
   character to numeric   236
   two categorical variables as one   280–281
Regression
   assumption diagnostics   138–153, 192–204
   case diagnostics   132–138, 188–192
   collinearity — see Collinearity

    error term   130, 131, 176
    multiple   176–179
    polynomial   222, 229–234
    preliminary diagnostics   118–128, 166–175
    simple   130–132
    transforming variables — *see* Transforming data
Repeated measures — *see* Analysis of variance, Repeated measures
Residuals
    assumptions regarding — *see separate entries*: Normality, Equal variance, Independence, Unbiased errors
    graphing in ANOVA   259–260, 261–262, 263–264, 303, 304, 307, 350–351, 352–355
    graphing in regression   133–135, 140–142, 144–146, 149–151, 188–189, 192–196, 211, 212–213
    saving in ANOVA   239, 241, 284
    saving in regression   130, 133, 176, 188
    studentized   133–135, 137, 144–146, 188–190, 192–193, 212, 225–227
Rows, in data worksheet   5–6

Satterthwaite's adjustment — *see* Error terms, separate
Saving
    files   6–7
    residuals and diagnostics
        in ANOVA   239, 241, 284
        in regression   130, 133, 176, 188
    results   7–8
Scatterplots   75–80
    added variable plots   201–203, 214, 216
    border plots   77
    of categorical variables   79–80
    fitting ellipses   78–79, 127–128
    fitting lines   79, 127–128
    influence plots   76–77, 126–127
    residuals in ANOVA   263–264, 303–304, 305, 307, 352–355
    residuals in regression   144–146, 150–153, 154, 195–196, 200, 201, 212–213, 225–227
    raw data in ANOVA   237, 282, 328–329, 369–370
    raw data in regression   125–128, 170–171, 173–175
    scatterplot matrices (SPLOM)   173–175
Scheffé test   257
Scientific notation
    in analysis output   3
    in data worksheet   5
Score test for non–constant variance   146–147, 196–197
SE — *see* Standard errors
Selecting cases in the data worksheet   74–75
Separate error terms — *see* Error terms, separate
Single precision, saving files in   6

Skewness
  computing   56–58
  and normality assumption in ANOVA   260–261, 303–304, 350–351
  and normality assumption in regression   140–141, 193
  in normal probability plots   68–69
Sorting data   15–17
Spaces, as delimiters in text files   10
Sphericity assumption — see Homogeneity of covariances
Standard deviation, computing   56–58
Standard error — see also Confidence intervals, on means
  of estimate   132
  of means   56–58
Standardized regression coefficients   131–132
Standardizing data   18
Statistics, summary
  and one-way ANOVA   238–239
Stem and leaf displays
  generating   64–66, 122–123, 169–170
  contrast scores   398–399
  residuals in ANOVA   259–260, 303–304, 350, 351
  residuals in regression   140, 188–189, 192–193
Stouffer method of combining probabilities   111–112

Table analyses — see Contingency tables
Tabs, as delimiters in text files   10
Text files, opening   8–13
Tolerance   132, 181–182
Trace, of error matrices   394
Transforming data
  centering   20–22, 185, 225, 227
  columns of data   20–25
  conditional transformations   23–25
  ladder of transformations   208–210, 211, 213, 214–215, 222–223
  logical operators   24
  mathematical operators   20
  mathematical functions   22–23
  outcome variable, in regression   208–222
    finding transformations with Box-Cox–Weisberg method   216–222
    linearizing transformations   214–216
    normalizing transformations   208–211
    variance stabilizing transformations   211–214
  predictor variables, in regression   222–234
  relational operators   24
Transposing rows and columns   18–19

t-tests
    effect-size estimates from   102–103
    one-sample ("matched-pairs", "paired")   397–398, 430, 432
    two-sample ("independent", "unpaired")   254
Tukey-Kramer HSD test   256

Unbalanced designs — *see* Unequal n's
Unbiased errors (linearity) assumption
    in ANOVA   267–268, 312, 357
    in regression
        checking   152, 200–203
        remedies   153, 204, 214–216
    underlying contrasts   398
Uncorrelated errors — *see* Independence assumption
Unequal n's
    approximate SS residual   245
    contrasts   321–323
    factorial ANOVA   318–321
    post-hoc tests   257
    repeated measures ANOVA   406–407, 432–433
Unplanned comparisons — *see* Post-hoc tests
Unweighted means ANOVA   318–323, 406, 433

Variance
    computing   56–58
    finding within cell in ANOVA   305–306
    unequal — *see* Equal variance assumption

Weighting cases in contingency tables   437
Whitespace, as delimiters in text files   10–11
Windows, Microsoft   xvii–xix
Within subjects designs — *see* Repeated measures

z-scoring data   18